LANDSCAPES
OF EXPOSURE

Knowledge and Illness in Modern Environments

EDITED BY

*Gregg Mitman, Michelle Murphy,
and Christopher Sellers*

OSIRIS | 19

A Research Journal Devoted to the
History of Science and Its Cultural Influences

Osiris

Series editor, 2002–2012

KATHRYN OLESKO, *Georgetown University*

Volumes 17 to 27 in this series are designed to dissolve boundaries between history and the history of science. They cast science in the framework of larger issues prominent in the historical discipline but infrequently treated in the history of science, such as the development of civil society, urbanization, and the evolution of international affairs. They aim to open up new categories of analysis, to stimulate fresh areas of investigation, and to explore novel ways of synthesizing major historical problems that demand consideration of the role science has played in them. They are written not only for historians of science, but also for historians and other scholars who wish to integrate issues concerning science into courses on broader themes, as well as for readers interested in viewing science from a general historical perspective. Special attention is paid to the international dimensions of each volume's topic.

Cover Illustration:

George D. Clayton prepares a measuring device on a hill overlooking zinc plant in Donora, Pa. (1949). Courtesy of the United States Public Health Service. (National Library of Medicine Order Number A020873).

OSIRIS   2004   SECOND SERIES   VOLUME 19

Introduction:
A Cloud over History

By Gregg Mitman, Michelle Murphy, and Christopher Sellers[*]

W HILE THOUSANDS of government delegates, corporate leaders, and NGO representatives made final travel plans in August 2002 for the United Nations' World Summit on Sustainable Development, in Johannesburg, South Africa, a cocktail unlike any to be served at the diplomatic dinners there was mixing above the southern Indian Ocean. One week before the Rio+10 conference the United Nations Environment Program (UNEP) released findings of the Indian Ocean Experiment (INDOEX), a study begun seven years earlier by a team of 200 European, Indian, and U.S. scientists. Originally funded to carry questions about global climate change from computer model to field, the team had recently found its attention riveted by a two-mile thick cloud of pollution blanketing southern Asia. The consequences of this "Asian haze" for climate, agriculture, and health were just becoming known. "More research is needed," Klaus Toepfer, UNEP executive director, told the press, "but initial findings clearly indicate that this growing cocktail of soot, particles, aerosols and other pollutants [is] becoming a major environmental hazard for Asia. There are also global implications, not least because a pollution parcel like this, which stretches three kilometers high, can travel half way around the globe in a week."[1]

Like a host of predecessors, from carcinogens to acid rain to global warming, the so-called Asian Brown Cloud entered human history through scientific reports. Though its elements had been merging for years, perhaps decades, to form a growing mist, spreading farther across continent and ocean and affecting more and more lives, the Brown Cloud only found a name and international notoriety through the agency

[*] Gregg Mitman: Department of Medical History and Bioethics, University of Wisconsin–Madison, 1300 University Avenue, Madison, WI 53706; gmitman@med.wisc.edu. Michelle Murphy: History Department and Institute for Women Studies and Gender Studies, 100 St. George St., Rm. 2074, Toronto, Ontario, Canada M5S 3G3; michelle.murphy@utoronto.ca. Christopher Sellers: History Department, Stony Brook University, Stony Brook, NY 11794; csellers@notes.cc.sunysb.edu.

The papers gathered in this volume were first presented at the workshop "Environment and Health in Global Perspective," held in Madison, Wisconsin, in April 2002. We would like to thank the National Science Foundation (Grant No. SES-0114570), the University of Wisconsin's Anonymous Fund, and the University of Wisconsin's Department of Medical History and Bioethics, Department of History of Science, and Program in Science and Technology Studies for financial and administrative support. A special thanks also to Victoria Elenes for her assistance with the Web site and organizational details and to Fred Gibbs, Erika Milam, and Carson Burrington for their helpful roles in transcribing comments. We are indebted to Kathryn Olesko for her initial encouragement in pursuing this volume and for her incredible guidance and wisdom along the way.

[1] UNEP, "Regional and Global Impacts of Vast Pollution Cloud Detailed in New Scientific Study," press release, 13 Aug. 2002, http://www.unic.org.in/News/2002/pr/pr86Aug2002.html (quote); UNEP and Center for Clouds, Chemistry, and Climate, The Asian Brown Cloud: Climate and Other Environmental Impacts (Pathumthani, Thailand, 2002), http://www.rrcap.unep.org.

of a privileged few, mainly scientists. The reports of mothers who had brought gasping children into Bombay hospitals whenever a haze had hung over the city, long before the UNEP issued its report, had not been enough. Once the privileged few had noted this dun nimbus, however, it became, with lightning speed, both a real, coherent thing and a serious problem in environmental health in many more people's eyes. By exploring the implications of recent phenomena such as the Asian Brown Cloud, this *Osiris* volume announces a fundamental revision in the way we understand the history of environment and health. Environment and health have long been seen as having separate histories. Given the children's problems in Bombay, for instance, an earlier generation of historians might have pigeonholed the Asian haze as an episode in the history of public health. That rubric, however, hardly does justice to a "discovery" spurred by investigation of global climate and chemistry rather than disease. Signaling a dramatic reversal of the momentum in early-twentieth-century public health "from the environment to the individual," the Brown Cloud suggests how late-twentieth-century developments have outrun long-accepted narratives of public health as well as medical history.[2] The Asian haze also brings into sharp relief persistent assumptions in environmental history and the history of ecology that human health questions are somehow less ecological than those addressing the nonhuman world. Thus the Brown Cloud exemplifies how the once-separate histories of health and of environment have become intertwined in our own time. It points us toward histories that encompass both.[3]

Building upon earlier as well as more recent precedents in medical, science, and environmental history, this edited volume represents a collective effort to fathom more fully the past that has so thoroughly shaped modern environments. If the Brown Cloud is any guide, the long rise and evolution of industrial production over the past three centuries and the concomitant transformations of both cities and farms form a larger historical backdrop for this modern tale. At issue in this volume is modern matter, akin to that of the Brown Cloud, substances neither named nor manufactured until the nineteenth and twentieth centuries. The economic systems spewing out particles and gases and stirring up germs and parasites examined within these pages are those of the past 150 years, in both the industrialized West and other parts of the world. The preponderance of toxic, over infectious, agents of illness in these papers reflects a long-term "epidemiological transition" that reduced the toll of contagious disease, especially in North America and Europe. Though the research here looks at many parts of the globe, the majority of the volume's scholars work in the United States, lending a local shape to our historiographic debates and frames. Nevertheless, we hope this collection will stimulate venturesome longer-term and larger-scale theses, including those that move beyond the seat of a particular Western version of modernity.

---

[2] Though the quote is from Paul Starr, *The Social Transformation of American Medicine: The Rise of a Sovereign Profession and the Making of a Vast Industry* (New York, 1982), 181; it also summarizes a large body of literature, beginning with Hibbert Hill, *The New Public Health* (New York, 1916).

[3] These divergent histories have been less prominent among environmental and medical historians writing on colonial settings. See, e.g., Richard Grove, *Green Imperialism: Colonial Expansion, Tropical Island Edens, and the Origins of Environmentalism, 1600–1860* (New York, 1995); Mark Harrison, "'The Tender Frame of Man': Disease, Climate, and Racial Difference in India and the West Indies," *Bulletin of the History of Medicine* 70 (1996): 68–93; Warwick Anderson, "Immunities of Empire: Race, Disease, and the New Tropical Medicine, 1900–1920," ibid., 94–118; and Michael A. Osborne, "European Visions: Science, the Tropics, and the War on Nature," in *Nature et environnement,* ed. Christophe Bonneuil and Y. Chatelin (Paris, 1996), 21–32.

To aid understanding of the historical "conditions of possibility" for a phenomenon such as the Brown Cloud, we feel the history of environment and health must take "interdisciplinary" as its watchword. In medieval times, Europeans envisioned extensive correspondence between their bodies and the cosmos; in more recent times, the rise of specialist disciplines has tended to sever this connection. Thus today's historians of environment and health face a long tradition of chopping "health" and "environment" into distinct and separate realms of knowledge and practice. In the workshop out of which this volume grew, historians of various stripes as well as anthropologists and geographers came together to explore, discuss, and debate a wide variety of topics and approaches to a history of health and environment. Choices of topic, emphasis, and method revealed productive tensions, not just across the fields of science, medical, and environmental history but also across different veins of scientific expertise: engineering versus medicine, ecology versus public health, bacteriology versus toxicology. We believe a fuller appreciation and understanding of the history of environment and health will come as historians from neighboring fields listen and learn, from one another and from those in other disciplines. Anthropologists and sociologists have shown how interviews and ethnography can highlight the multisited practices and plights of workers, activists, laypersons, and experts. Geographers have offered methodologies that connect landscapes and natural processes with human-made national boundaries and economic systems. In this volume, the inclusion of anthropologists and geographers reveals a field characterized by an emergent traffic among scholarly methodologies, including scientific ones, rather than a legacy of work in a single discipline.

Interest in the history of environment and health has numerous roots outside public health history. Many scholars who recently have moved into this terrain have been motivated and inspired by social movements to combat the inequitable distribution of, and exposure to, environmental hazards. Environmental hazard has emerged as an important wing of worldwide political struggles for social justice and equality. In the United States, for example, the environmental justice movement, which became a major political force in the 1980s and 1990s, shifted the agenda of many environmental groups from land preservation and a generalized "pollution" to confrontation with problems of urban and industrial wastelands and a reckoning with the geography of race and poverty.[4]

---

[4] On environmental justice and racism in an American context, see Robert Bullard, *Dumping on Dixie: Race, Class, and Environmental Quality* (Boulder, Colo., 1990); Giovanna Di Chiro, "Nature as Community: The Convergence of Environment and Social Justice," in *Uncommon Ground: Rethinking the Human Place in Nature,* ed. William Cronon (New York, 1996), 298–321; Michael Egan, "Subaltern Environmentalism in the United States: A Historiographic Review," *Environment and History* 8 (2002): 21–41; Daniel Faber, ed., *The Struggle for Ecological Democracy: Environmental Justice Movements in the United States* (New York, 1998); Robert Gottlieb, *Forcing the Spring: The Transformation of the Environmental Movement* (Washington, D.C., 1993); Dolores Greenberg, "Reconstructing Race and Protest: Environmental Justice in New York City," *Environmental History* 5 (2000): 223–50; George Lipsitz, "The Possessive Investment in Whiteness," *American Quarterly* 47 (1995): 369–466; Eileen Maura McGurty, "From NIMBY to Civil Rights: The Origins of the Environmental Justice Movement," *Environmental History* 3 (1997): 301–23; and Laura Pulido, *Environmentalism and Economic Justice: Two Chicano Struggles in the Southwest* (Tucson, Ariz., 1996). On environmental justice in other national or global contexts, see, e.g., David A. McDonald, ed., *Environmental Justice in South Africa* (Athens, Ohio, 2002); Nancy Peluso and Michael Watts, eds., *Violent Environments* (Ithaca, N.Y., 2001); Richard Peet and Michael Watts, eds., *Liberation Ecologies: Environment, Development, Social Movements* (London, 1996); Ramachandra Guha, *Environmentalism: A Global History* (New York, 2000); Kim Fortun, *Advocacy After Bhopal: Environmentalism, Disaster, New Global Orders* (Chicago, 2001); and Anna Tsing and Paul Greenough, eds., *Nature in the Global South: Environmental Projects in South and Southeast Asia* (Durham, N. C., 2003).

Interest in environment and health has also been inspired by scholarship arising over the past two decades at some especially fertile intersections between environmental and medical history. Historians of urban sanitation such as Martin Melosi and Joel Tarr have brought out the continuing importance of infrastructure, engineering, and an environmental focus to modern public health. Historians of occupational health have woven together workplace and medical histories in a variety of ways, from Arthur McEvoy's proposal for an "ecology of the workplace" to Chris Sellers's argument about the workplace origins of modern environmental health science.[5] Environmental historians responding to the challenge of environmental justice, too, have turned increasingly to the health dimensions of topics such as pollution and industrial wastes, dimensions downplayed by earlier historians of conservation and environmentalism.[6]

Thus contemporary environmental problems, new social movements, and past historical scholarship have inspired an emerging body of research on the subject of environment and health. Not just historians of differing agendas and interests but also anthropologists, sociologists, and geographers have turned to explore the nexus of place, health, and political economy in diverse sites and times around the world. We are struck by the overlap and interchange that has already occurred, which we hope this collection will continue to nourish. Environmental history, for example, through its penchant for broad historical narratives that give agency to both nature and humans, has offered a compelling yet underutilized model to historians of science interested in connecting their field's preoccupation with local sites of knowledge production to narratives that reach across larger spatial and temporal scales.[7] Similarly, the

[5] Martin Melosi, *The Sanitary City: Urban Infrastructure in America from Colonial Times to the Present* (Baltimore, 2000); Joel Tarr, *The Search for the Ultimate Sink* (Akron, Ohio, 1996); Arthur McEvoy, "Working Environments: An Ecological Approach to Industrial Health History," *Technology and Culture* 36 (suppl.) (1995): 145–73; and Christopher Sellers, *Hazards of the Job: From Industrial Disease to Environmental Health Science* (Chapel Hill, N. C., 1997). See also Jacqueline Corn, *Response to Occupational Health Hazards: A Historical Perspective* (New York, 1992); Claudia Clark, *Radium Girls: Women and Industrial Health Reform, 1910–1935* (Chapel Hill, N. C., 1997); Alan Derickson, *Black Lung: Anatomy of a Public Health Disaster* (Ithaca, N.Y., 1998); Gabrielle Hecht, "Rupture Talk in the Nuclear Age: Conjugating Colonial Power in Africa," *Social Studies of Science* 32 (2002): 691–727; Jock McCulloch, *Asbestos Blues: Labour, Capital, Physicians, and the State in South Africa* (Bloomington, Ind., 2002); Michelle Murphy, "Toxicity in the Details: The History of the Women's Office Worker Movement and Occupational Health in the Late Capitalist Office," *Labor History* 41 (2000): 189–213; Randall M. Packard, *White Plague, Black Labor: Tuberculosis and the Political Economy of Health and Disease in South Africa* (Berkeley, Calif., 1989); and David Rosner and Gerald Markowitz, *Deadly Dust: Silicosis and the Politics of Occupational Disease in Twentieth-Century America* (Princeton, N.J., 1991).

[6] Andrew Hurley, *Environmental Inequalities: Class, Race, and Industrial Pollution in Gary, Indiana, 1945–1980* (Chapel Hill, N. C., 1995); Craig Colten and Peter N. Skinner, *The Road to Love Canal* (Austin, Texas, 1996); David Stradling, *Smokestacks and Progressives: Environmentalists, Engineers, and Air Quality in America, 1881–1951* (Baltimore, 1999); Scott Dewey, *Don't Breathe the Air: Air Pollution and U.S. Environmental Politics, 1945–1970* (College Station, Texas, 2000); Stephen Moseley, *The Chimney of the World: A History of Smoke Pollution in Victorian and Edwardian Manchester* (Cambridge, 2001); and Valerie Kutz, *Tainted Desert: Environmental and Social Ruin in the American West* (New York, 1998). For a more thorough historiographic review, see Jeffrey Stine and Joel Tarr, "At the Intersection of Histories: Technology and the Environment," *Technol. Cult.* 39 (1998): 601–41.

[7] Overviews of environmental history include Donald Worster, Alfred Crosby, Richard White et al., "A Roundtable: Environmental History," *Journal of American History* 76 (1990): 1087–147; Alfred Crosby, "The Past and Present of Environmental History," *American Historical Review* 100 (1995): 1177–89; Theodore Steinberg, "Down to Earth: Nature, Agency, and Power in History," *Amer. Hist. Rev.* 107 (2002): 798–820; and Richard White, "Environmental History: Watching a Historical Field Mature," *Pacific Historical Review* 70 (2001): 103–12.

shift within medical geography and environmental history to more place-centered approaches and more embodied local geographies of health and disease offers an important point of contact with historians of science similarly engaged with questions about the place-centered, situatedness of scientific knowledge.[8] We hope historians of science and medicine will continue to seek common ground with geographers and environmental historians well versed in studying the material, cultural, and social relations embedded in place, be it an immigrant neighborhood, a city, or a therapeutic landscape.

In the essays in this volume and in related scholarship, we believe a lively dialogue and fusion of differing methodologies, narrative strategies, and conceptual approaches to the subject of environment and health is underway. Drawing together approaches from different disciplines, this emergent history of environment and health is highlighting subjects that have largely fallen between the cracks of environmental history, geography, history of science, medical history, and science studies.[9] Productive tensions and approaches can be seen within the pairings of papers in this volume. Cutting across all the essays here are three major themes: scale, materiality, and uncertainty. Each theme has grounded the perspective of our authors in important ways; by the same token, each imposes its own array of questions and choices on this, as well as future, scholarship. By pulling out the interweaving of these themes throughout the papers, we mean to provide an underpinning for further dialogue across fields and for scholarly creativity and insight to come.

SCALE

The Asian Brown Cloud is but a visible sink of transnational material flows that bind economics, resources, people, and pollutants across molecular, local, regional, and global scales. In their UNEP Assessment Report, INDOEX scientists spoke of the "enormous range of scales" with which they had to grapple. "From the severe health effects caused by indoor air pollution, to those of urban and rural pollution, [to] the impact of the aerosol on regional and even global climate change," scientists found themselves moving between local factories and global climate changes, between ecological cycles and multinational economic projects, between the United Nations and commixing molecules. Inefficient coal-burning in Thailand, from power stations sold as cheap exports to it and to other developing countries, and forest fires in Indonesia, caused by the illegal clearing of land for palm oil plantations, linked local and regional

[8] Robin Kearns and Wilbert Gesler, *Putting Health into Place: Landscape, Identity, and Well-Being* (Syracuse, N.Y., 1998); Crosbie Smith and Jon Agar, *Making Space for Science: Territorial Themes in the Shaping of Knowledge* (New York, 1998); Adi Ophir, Steven Shapin, and Simon Schaffer, eds., *The Place of Knowledge: The Spatial Setting and Its Relation to the Production of Knowledge, Science in Context* 4 (1991): 3–218; David Livingstone, "The Spaces of Knowledge: Contributions Towards a Historical Geography of Science," *Environment and Planning D: Society and Space* 13 (1995): 5–34; Nicolaas Rupke, ed., *Medical Geography in Historical Perspective* (London, 2000); Gregg Mitman, "Hay Fever Holiday: Health, Leisure, and Place in Gilded Age America," *Bull. Hist. Med.* 77 (2003): 600–35; Susan Craddock, *City of Plagues: Disease, Poverty, and Deviance in San Francisco* (Minneapolis, Minn., 2000); and Isabel Dyck, Nancy Davis Lewis, and Sarah McLafferty, eds., *Geographies of Women's Health* (New York, 2001).

[9] Recent historiographic essays working across some of these "cracks" include Conevery Bolton Valenčius, "Histories of Medical Geography," in Rupke, *Medical Geography* (cit. n. 8), 3–28; Christopher Sellers, "Thoreau's Body: Towards an Embodied Environmental History," *Environmental History* 4 (1999): 487–514; and Warwick Anderson, "Introduction: Postcolonial Technoscience," *Soc. Stud. Sci.* 32 (2002): 643–58.

issues of environment and health to world economies. Just days before the World Summit in Johannesburg, Greenpeace activists from Europe grabbed news headlines when they landed a hot-air balloon reading "Save the Climate" on a coal plant in Lampang, Thailand. In West Kalimantan, Indonesia, street entrepreneurs sold facemasks to the choking citizenry, and CNN flashed pictures of the masked children to its audience. Onlookers were reminded that for global problems in environmental health, ecological and environmental inequities in different nations could not be readily separated either from each other or from the politics of class, as when workers in Delhi, India, participated in massive marches to protest the loss of their jobs due to an Indian Supreme Court ruling that ordered the closing and relocation of small industries polluting the air.[10] For this volume, phenomena such as the Asian Brown Cloud raise the question, How do we develop multisited analyses linking historical narratives of environment and health across multiple scales, from the local and regional to the transnational and global?

While the brown haze hovered across southern Asia, news of its reach extended across the global information networks of CNN.com, environmental list servers, and international governing bodies. On the eve of the World Summit, through virtual technology, the Asian Brown Cloud became a global environmental health problem, one that disregarded the borders of sovereign nations, against which UN officials called for countries of the world to unite. But the points of intervention recommended by UNEP scientists—enhanced technoscientific surveillance, international cooperation between scientists and government officials—were largely predetermined by the modes of representation through which the Brown Cloud had become an artifact of expert and elite concern. Framing this global problem through a technoscientific imaginary, the UNEP report engaged in a politics of scale; that is, its technoscientific production emphasized particular scales—molecular, ecosystem, planetary—in which environmental health problems as well as solutions were conceived to reside. "We have the initial findings and the technological and financial resources available," noted Klaus Toepfer. "Let's now develop the science and find the political and moral will to achieve this for the sake of Asia, for the sake of the world."[11]

Arguably, this particular construal of the global depended on a largely American model of the relationships between expertise and political order, a vision hinging on intergovernmental harmonization, technical assistance, and international cooperation in scientific research that emerged after the Second World War.[12] As Nick King observes in this volume, international public health campaigns of the 1990s that focused on emerging diseases, such as AIDS, mad cow disease, Ebola, and West Nile virus, took up a similarly scalar politics, which guided their choices of geographic frames and causal analyses of those diseases, as well as the interventions they proposed. Books such as *The Hot Zone* and Hollywood blockbusters such as *Outbreak* helped create a particular imperialist imaginary whose "global" pretenses belied the fact that it was located not so much in any abstract global space as in specific locales: media, biomedical research communities, and institutions of public health and national security, mostly based in the United States.

[10] UNEP and Center for Clouds, Chemistry, and Climate, *Asian Brown Cloud* (cit. n. 1), 7.
[11] UNEP, "Regional and Global Impacts" (cit. n. 1).
[12] See, e.g, Clark Miller, "Scientific Internationalism in American Foreign Policy: The Case of Meteorology, 1947–1961," in *Changing the Atmosphere: Expert Knowledge and Environmental Governance,* ed. Clark Miller, A. Miller, and Paul N. Edwards (Cambridge, Mass., 2001), 167–218.

How and why scientists, public health officials, and governing bodies circumscribe the multiple scales at work in the production of landscapes of exposure are questions raised by many of the essays here. They also ask, Who or what gets lost in the movements across divergent scales when scientists travel from global climate models to the molecular biology of viruses or from the biochemistry and physiology of radioiodine to U.S. census statistics? In his paper, Scott Kirsch demonstrates how the Atomic Energy Commission (AEC) used spatial representations of fallout data that excluded people to distance scientists from the human consequences of nuclear testing. An oppositional scientist within the AEC then drew on information from local ranchers and Utah public health scientists to produce new kinds of maps and models that reinserted people into the spaces of the Nevada nuclear test sites. Such analyses point to how the representations of environmental health problems depend upon a scale politics—a preordained investment in particular geographically-defined ways of seeing and intervening—in which they are embedded and which they help sustain.

Claims to describe the globe have rhetorical power yet, by their very nature, incline toward the grandiose. Most scholars here, as elsewhere, have opted for more modest claims on the transnational scale, choosing to follow the specific path of a colonial scientist as he is posted at imperial capitals, to track the ecological spread of an insect across a region, or to trace the reach of a multinational corporation as it finds resources in one place and manufacturers in another, creating a toxic plume that blows as far as the wind. Stretching across not just factories, farms, and homes, but national boundaries, the Asian Brown Cloud resembles many other phenomena in the history of environment and health. Like numerous scholars in the humanities and social sciences today, those studying environmental health have been devoting growing attention to transnational events and processes. For some, attending to transnational process involves examining precisely how people, pollutants, economic arrangements, and activism extend across national boundaries. In Giovanna Di Chiro's paper, the border between the United States and Mexico is not just a transfer point but also a concrete limit directly felt in people's lives and survived through transnational advocacy networks. Such networks are not new. Historians of slavery emancipation have drawn out the lively world of transatlantic abolitionism; in this volume, Harold Platt discusses the transatlantic traffic of reformers in the Progressive Era.[13]

What counts as a transnational, as opposed to a more local, process? In Chris Sellers's paper on fluoridated water, the groundwater movement of natural fluoride may traverse county, national, or other political borders, but these human-drawn political lines mean nothing to the nonhuman actors crossing them. The work of geographers such as Saskia Sassen and Cindi Katz, among others, has troubled the opposition of categories such as global-local and transnational-national.[14] Does local history have to be bounded by geographic proximity and state lines, or might an international community of disease ecologists be another form of the local? What defines the local across which so many processes flow? To what extent are neighborhoods actually more "local," produced against, as well as through, global processes? Instead of

[13] Paul Gilroy, *The Black Atlantic: Modernity and Double Consciousness* (Cambridge, Mass., 1995); Daniel Rogers, *Atlantic Crossings: Social Politics in a Progressive Age* (Cambridge, Mass., 1998).

[14] Saskia Sassen, "Spatialities and Temporalities of the Global Elements for a Theorization," *Public Culture* 12 (2000): 215–32; Cindi Katz, "On the Grounds of Globalization: A Topography for Feminist Political Engagement," *Signs* 26 (2001): 1213–34.

seeing space as a pregiven presence across which people, commodities, and natural processes surge, we would perhaps do better to see localness of place as constituted precisely by the movement, tensions, and connections it bears. For example, Gregg Mitman shows that as frontier towns such as Denver became linked to transnational markets through which drugs and capital flowed, new spaces of hope built through biomedical research were nevertheless shaped by Denver's history and landscape as a health resort. As other papers here show, from Susan Jones's contrast between urban and rural responses to a suggested human threat from bovine tuberculosis to Conevery Bolton Valenčius's juxtaposition of the comforting emotional environment of New England to that of the healing physical environment of Santa Fe, multisited analysis is an important methodology for revealing the work of scale and connection constituting so many issues of environment and health.

If economies of scale have been important in considering the historical relationships between environment and health, so, too, have ecologies of scale. Whether the community, region, population, or ecosystem, the ecological unit of analysis has greatly affected what is, and is not, seen by both historians and their subjects. Following reigning practices in environmental history, many papers in this volume center on health topics without mentioning ecological science per se; nevertheless they broach issues of complexity and relationality via historical imaginaries that are arguably "ecological."

New scholarship on the environmental dimensions of twentieth-century public health has raised new questions about the variety of "environmental" and "ecological" perspectives that emerged and the ways they have overlapped, intertwined, or clashed. While Anglo-American health scientists were more likely to style their work as "environmental," important visions of this work drew upon a relational sense defined by systems and material flows, with roots in both engineering and ecological science. In the United States, for example, Rachel Carson's classic *Silent Spring*, which helped make ecology a household word in the 1960s, drew upon the emerging field of ecosystem ecology, along with older balance of nature concepts and the public health discipline of environmental toxicology, to trace the health effects of pesticides on humans and wildlife. Parallel scientific developments of her time greatly aided her integrated perspective. For instance, among health scientists, a rising focus on the hazardous effects of the inorganic realm brought closer alignment with the work of Atomic Energy Commission ecologists, who were funded to help alleviate public fears about nuclear fallout and for whom the ecosystem served as the unit of analysis to investigate the flows of radioactive materials through soils, water, vegetation, and wildlife. At the same time, as Linda Nash observes in her essay, Carson's consumer-oriented focus on the ecology of suburban spaces and federal wildlife refuges—the habitats of professional ecologists—made the bodies and hazardous environments of farmworkers relatively invisible to her and her readers.[15] While historians such as Samuel Hays have suggested that public concern over environmental health emerged

[15] Thomas Dunlap, *DDT: Scientists, Citizens, and Public Policy* (Princeton, N.J., 1982); Linda Lears, *Rachel Carson: Witness for Nature* (New York, 1997); and Christopher Sellers, "Body, Place, and State: The Makings of an Environmentalist Imaginary in the Post-WWII U.S.," *Radical History Review* 74 (1999): 31–64. On the AEC, ecologists, and fallout concerns, see Stephen Bocking, "Ecosystems, Ecologists, and the Atom: Environmental Research at Oak Ridge National Laboratory," *Journal of the History of Biology* 28 (1995): 1–47; Joel B. Hagen, *An Entangled Bank: The Origins of Ecosystem Ecology* (New Brunswick, N.J., 1992), 100–21; and Chunglin Kwa, "Radiation Ecology, Systems Ecology, and the Management of the Environment," in *Science and Nature: Essays in the History of the Environmental Sciences,* ed. Michael Shortland (Oxford, 1993), 213–49.

in post–World War II America because of increased standards of living, the papers in this volume suggest that such a view is an artifact of the past historiography of U.S. environmentalism.[16] Attention has moved beyond the spaces of wilderness and middle-class suburbs into the city and factory, into war zones and colonial projects.

The material flows of energy and matter circulating the globe made visible through environmental sciences such as ecosystem ecology, meteorology, and environmental engineering have not been the only intellectual building blocks of the ecological and environmental imaginaries through which bodily threats are conceived. The recent popularity of ecological world history, as evidenced by best-selling books such as Jared Diamond's *Guns, Germs, and Steel,* reveals yet another ecology at play, one in which humans and microbes are seen as equal actors in a large-scale evolutionary drama. Diamond's book is indebted to a previous generation of scholarship in environmental history, one that includes William McNeill's *Plagues and People* and Alfred Crosby's *The Columbian Exchange,* which in turn borrowed heavily from the work of Australian parasitologist F. Macfarlane Burnet and American bacteriologist Rene Dubos in constructing global ecological narratives.[17] Both Warwick Anderson and Helen Tilley in this volume document the intellectual tradition of this research in tropical medicine and population ecology and its colonial and settler society contexts. Scaling up into global environmental problems, later experts fashioned sometimes apocalyptic ecological imaginaries that focused upon population, disease, and environment as the primary forces driving past and future human societies.[18]

Although the rise of laboratory medicine supposedly eclipsed a Hippocratic emphasis on airs, waters, and places by the early twentieth century, we find a persistence of Hippocratic concerns through this period and beyond. We can, in retrospect, identify many notions of regional disease and health as "ecological," whether informed by medical geography and climatotherapy in the nineteenth century or by self-identified "ecological" conceptions of community in the twentieth century. Environmental and medical historians have only recently begun to explore how scientific and popular practices as well as regional economies were shaped by ideas and experiences of the interplay between health and nature. Aldo Leopold's extensive references to the health of the land and conservation as the art of land doctoring offer one indication of the ways in which ecological conceptions of conservation and community borrowed heavily from Hippocratic ideals and experiential wisdom of the relationships between the health of body and place.[19] Throughout the volume, contributors find themselves in dialogue with a range of environmental sciences and lay visions of ecology, either in framing the scale of analysis or in exploring how

[16] Samuel Hays, *Beauty, Health, and Permanence: Environmental Politics in the United States, 1955–1985* (Cambridge, 1985).

[17] Jared Diamond, *Guns, Germs, and Steel: The Fate of Human Societies* (New York, 1997); William McNeill, *Plagues and Peoples* (Garden City, N.Y., 1976); Alfred Crosby, *The Columbian Exchange;* and idem, *Ecological Imperialism: The Biological Expansion of Europe, 900–1900* (Cambridge, 1986).

[18] Robert Kaplan, *The Coming Anarchy: Shattering the Dreams of the Post Cold War* (New York, 2000); Thomas Homer-Dixon, *Environment, Scarcity, and Violence* (Princeton, N.J., 1999).

[19] For an exploration of the ways in which notions of community ecology played into twentieth-century biomedicine, see Gregg Mitman, "Natural History and the Clinic: The Regional Ecology of Allergy in America," *Studies in the History and Philosophy of the Biological and Biomedical Sciences* 34 (2003): 491–510. Aldo Leopold, *A Sand County Almanac* (New York, 1949). On the persistence of aerial concerns, see Christopher Sellers, "The Dearth of the Clinic: Lead, Air, and Agency in Twentieth-Century America," *Journal of the History of Medicine and Allied Sciences* 58 (2003): 255–91.

ecological ideas and practices are themselves embedded in particular economic, political, social, and material relations.

MATERIALITY

Just who or what made the Asian Brown Cloud? Clearly, scientists were central to its transformation into a widely recognized menace to global public health. Just as clearly, however, what they christened the Brown Cloud comprised materials in whose production these scientists had no hand. The haze whose traces they captured in satellite photos and whose components they trapped on board ship and plane had other origins. The Brown Cloud was generated in and among other places, Thai power plants, Indonesian forest fires, chemical reactions within otherwise natural clouds, and the propulsion of soot and gas across vast distances. The diverse origins and dynamics of the Brown Cloud's nonhuman nature connect with human narratives of global capital, imperialism, and inequality.

Transnational events such as the Asian haze have helped inspire the most recent burst of interest in the history of environment and health among not only scientists and activists but also environmental historians and other scholars. Anthropologist Arjun Appadurai has proclaimed the circulations of people, technology, ideas, media, and money to be the "five dimensions of global cultural flows"; environmental historians, in particular, have made the case for a sixth—an ecological—dimension.[20] This dimension encompasses a wide range of material flows, from commodities to the inadvertent byproducts of industry and agriculture to more obscure and natural processes that have not always fallen within the analytic domain of human culture. A fine example of what environmental historians have been up to is J. R. McNeill's *Something New under the Sun,* which surveys the international streaming of goods and pollutants and what the author finds to be a dramatic broadening of this ecological dimension, of human intervention in, and exploitation of, the natural world in the twentieth century.[21]

The project of forging "ecological" versions of health history has involved spotlighting those material flows and processes many social theorists are apt to neglect. Such a project often starts with an assumption that these material flows are composed of relatively consistent entities—whether the fluoride in Sellers's paper, the tsetse fly in Tilley's, or pesticides in Nash's. These nonhuman natures retain enough constancy for the historian to follow their trajectory into times and places in which their existence was framed in terms far different from those used today. Whereas historians of science historicize their scientist's subjects by demonstrating the vast diversity of "science" across space and historical time, ecological approaches to history, by moving material actors to center stage, open the doors of inquiry into a widening variety of roles nonhuman actors have played in the human past, even when there were no recognizable scientists around to describe them. Often drawing upon the latest scientific findings to help guide the search for past ecological relationships, environmental history approaches here and elsewhere emphasize contingency and change along these material dimensions of the human past, especially as creatures or chemicals become caught up in the projects of human societies and economies. Much of the narrative and

[20] Arjun Appadurai, *Modernity at Large: Cultural Dimensions of Globalization* (Minneapolis, Minn., 1997), 33.
[21] J. R. McNeill, *Something New under the Sun: An Environmental History of the Twentieth-Century World* (New York, 2000).

analytical power in this approach derives from an assumption that nonhuman substances or organisms have concrete effects on history that we, as historians, can recognize, even if past actors saw them quite differently or not at all.

It is worth noting that many historians of science, as well as other scholars, prefer different ways of treating objects historically. They would rather historicize scientific conceptions such as "ecology" than use them as analytical frames.[22] That being said, many kinds of materialist approaches abound in the history of science and, like the materialist approaches of environmental history, inform many of the papers in this volume. The work of Michel Foucault and his attention to the effects of architecture and practice on epistemology and subject positions have influenced countless historians interested in tracing the process, not just the outcome, of epistemological regimes. In the field of history of science, scholars such as Lorraine Daston and Ian Hacking have built on Foucault to develop the term "historical ontology" to describe the historicity of things as formed in relation to epistemological traditions.[23] In science studies, the work of Bruno Latour and "actor network theory" have inspired inquiries into the social production of scientific objects as nonhuman agents, making the argument that only once germs had been produced in 1864 had they been there all along. Building on actor network theory, others have more recently opened up questions about ontological diversity, by showing how objects produced simultaneously by different domains of knowledge and practice can become imbued with conflicting, yet no less material, qualities and boundaries.[24] The materiality of instruments, scientific practices, and information in producing artifacts has also provided focal points for much scholarship in science studies over the past decade.[25] Additionally, Margaret Lock's concept of "local biologies" draws attention to the ways physical embodiment is produced through political and cultural circumstances.[26] A similar range of approaches to materiality characterizes this volume.

Scholarship wrestling with the intersection between environment and health almost inevitably confronts tangles of economy and flesh. Thus another important materialist tradition informing the work in this volume is historical-geographic materialism and its attention to the concrete effects of capitalism and political economy in

---

[22] Paolo Palladino, *Entomology, Ecology, and Agriculture: The Making of Scientific Careers in North America, 1885–1985* (Amsterdam, 1996); Gregg Mitman, *The State of Nature: Ecology, Community, and American Social Thought, 1900–1950* (Chicago, 1992); Peter J. Taylor and Ann S. Blum, "Ecosystems as Circuits: Diagrams and the Limits of Physical Analogies," *Biology and Philosophy* 6 (1991): 275–94; Gregg Mitman and Kevin Dann, "Essay Review: Exploring the Border of Environmental History and the History of Ecology," *J. Hist. Biol.* 30 (1997): 291–302; and Peter J. Taylor, "Technocratic Optimism: H. T. Odum and the Partial Transformation of Ecological Metaphor After World War 2," *J. Hist. Biol.* 21 (1988): 213–44.

[23] Lorraine Daston, "Historical Epistemology," in *Questions of Evidence: Proof, Practices, and Persuasion across the Disciplines,* ed. James Chandler, Arnold Davidson, and Harold Harootunian (Chicago, 1994), 282–89; Ian Hacking, *Historical Ontology* (Cambridge, Mass., 2002).

[24] Bruno Latour, *Pandora's Hope: Essays on the Reality of Science Studies* (Cambridge, Mass., 1999); Anne Marie Mol, *The Body Multiple: Ontology in Medical Practice* (Durham, N. C., 2002); and Michel Callon and John Law, "Agency and the Hybrid Collectif," *South Atlantic Quarterly* 94 (1995): 481–507.

[25] Andrew Pickering, *The Mangle of Practice: Time, Agency, and Science* (Chicago, 1995); Adele Clarke and Joan Fujimura, eds., *The Right Tools for the Job: At Work in Twentieth-Century Life Sciences* (Princeton, N.J., 1992); and N. Katherine Hayles, "The Materiality of Informatics," *Configurations* 1 (1992): 147–70.

[26] Margaret Lock, *Encounters with Aging: Menopause in Japan and North America* (Berkeley, Calif., 1995); Barbara Duden, *The Woman beneath the Skin: A Doctor's Patients in Eighteenth-Century Germany,* trans. Thomas Dunlap (Cambridge, Mass., 1991).

the production of body and place. Economic regimes of accumulation and production coalesce in particular locations and seep into the character, sensibility, and experience of everyday life. Sweeping notions such as Donald Worster's adaptation of the Marxian "modes of production" and David Harvey's "regimes of accumulation" designate large-scale historical patterns, the practices and cultures by which resources are extracted, manufactured into commodities, distributed, and consumed. Also including the patchwork of state regulations (or lack of them) and its own characteristic stratifications of labor, each "mode" or "regime" gives shape to the locales it encompasses.[27] The regime of colonial accumulation under the British Empire, for example, featured an escalating extraction of materials from African mines, whose byproducts generated scattered local epidemics of respiratory disease among black mine workers. Colonization also set in motion the spread of colonial public health administrations that saw the colonies as their own laboratories in which to develop Western science.[28]

The factories, mines, and shipyards of industrial capitalism, like the regimes of colonial accumulation, similarly transformed the earth, altering beyond recognition the landscapes and neighborhoods immediately surrounding industrial sites.[29] Inside the workplaces of shipyards, nuclear reactors, and manufacturers, industrial exposures intensified in flurries of asbestos, silica, lead, and radiation. Disasters and accidents, or purposeful acts of violence, scarred entire regions. In Arthur McIvor and Ronald Johnson's paper, the stark and heedless pathology of asbestosis provides a core reality they can map on to Glasgow itself, around which the awareness of their asbestos workers circles and lands, and with which it then struggles. From Glasgow shipyards to the Chernobyl plant that shadows Adriana Petryna's sufferers to the Bhopal chemical plant that Kim Fortun's scientists study, the modern industrial workplace produces not just commodities, chemicals, and uneven power relations but also its own varieties of disease. Michelle Murphy shows how the very office building in which EPA scientists worked had its own effects, insulating their work in such a way that they could reimagine their white-collar employment in labor union terms, while at the same time averting their eyes from the environmental plight of poor black communities, including those just beyond the agency's doors.

Scholars working at the intersection of environment and health feel free to take the factuality of many hazards for granted: that Chernobyl and Bhopal did indeed kill many people, that urban pollution exacerbates asthma, that tropical environments cause sleeping sickness through tsetse flies. Historically, however, as each of these new certainties has crystallized, a multitude of new uncertainties has sprouted around them. Much history of environment and health is about what we don't, or don't quite, know.

---

[27] Worster et al., "Roundtable" (cit. n. 7); David Harvey, *The Condition of Postmodernity* (New York, 1990); idem, *Justice, Nature, and the Geography of Difference* (Cambridge, Mass., 1996); and idem, *Spaces of Capital* (New York, 2001). For other historical accounts of how capitalism changed place, see William Cronon, *Nature's Metropolis: Chicago and the Great West* (New York, 1992); Mike Davis, *Late Victorian Holocausts: El Nino Famines and the Making of the Third World* (New York, 2001).

[28] Packard, *White Plague, Black Labor* (cit. n. 5); Megan Vaughan, *Curing Their Ills: Colonial Power and African Illness* (Stanford, Calif., 1991).

[29] See, e.g., Theodore Steinberg, *Nature Incorporated: Industrialization and the Waters of New England* (Cambridge, 1991). See also the following historiographic reviews: Stine and Tarr, "At the Intersections of Histories" (cit. n. 6); Chris Rosen and Chris Sellers, "The Nature of the Firm: Towards an Ecocultural Approach to Business History," *Business History Review* 73 (1999): 577–600.

UNCERTAINTY

Brown clouds of particulates hovering over a continent, poisons exhaling from strange lands, pesticides clinging to perfect fruits—the connection between place and bodies is often understood in terms of *exposure,* as contact between misplaced matter and flesh. Whereas most medical and public health history foregrounds disease itself, as manifested within individual bodies (pathology) or in populations (epidemics), most papers in this volume revolve around myriad historical forms of exposure. Privilege and violence are built into these forms: from the worrisome miasmas that stalked settler societies, to the stench and filth attributed to the "great unwashed" of the urban metropolis or the colonies, to the variety of perils made possible by industrial production in the West and elsewhere.[30] So, too, are competing notions of expertise. While biological exposures carried by wayward insects or human others were studied by sanitarians, epidemiologists, and disease ecologists, chemical and industrial exposures became the concern of industrial hygienists, engineers, and toxicologists. Although individual contributions to this volume concentrate on one side or the other of the resultant division between biological and nonbiological exposures, the collection as a whole points to how these differing kinds of expertise, and the uncertainties embedded within them, arose side by side.

A century of modern scientific scrutiny has made clear that chemically defined exposures often prove notoriously complex and, thus, difficult to address through public health policy. Seen through a late-twentieth-century toxicological or juridical lens, the demands for certainty about the chemical causes of illnesses are scandalously hard to meet. The lack of certitude reverberates here and in many other realms of the history of environment and health, for instance, realms in which presidents and prime ministers justify their refusal to ratify international environmental agreements, such as the Kyoto Protocol, by pointing to a lack of scientific consensus.

Uncertainty is perhaps the single most pervasive characteristic of the history of exposure collectively sketched by our volume. It inheres in the very material formation of many environmental health problems, starting with the time lag between exposure and possible health effect. But just as we understand perceptions of the natural world as historically constructed, situated, and conditioned, so we can approach uncertainty as a historical artifact, produced by particular ways of apprehending the world or by clashes between different versions of the world. Compelled to affirm the materiality of pesticides, viruses, or poisoned revolutionaries, historians are nevertheless confronted with contests over knowledge, both scientific and popular, within which cyclones of uncertainty swirl. Exploring the shape and tenor of uncertainty in different times and places, these papers contribute to a novel historical vein: the history of invisibility, imperceptibility, and doubt. If nonhuman actors have concrete effects in the history of environmental health, so, too, does the deployment of uncertainty—"geographies of unknowing" to Kirsch, "regimes of imperceptibility" to Murphy. The controversial nature of many environmental health problems raises the tricky question of how exposures remain

---

[30] See, e.g., Conevery Bolton Valenčius, *The Health of the Country: How American Settlers Understood Themselves and Their Land* (New York, 2002); Alain Corbin, *The Foul and the Fragrant: Odor and the French Social Imaginary,* trans. Miria Kochan (Cambridge, Mass., 1988); and Tom Lutz, *American Nervousness, 1903: An Anecdotal History* (Ithaca, N.Y., 1991).

unknown or come to seem less known, even as others become more known.[31] When, how, where, and under what modes of production do uncertainty and its frequent intellectual companions, such as "risk" and "complexity," accrete in modern cultures?

Changing modes of disclosure and information management provide important insights into the historical production of uncertainty. As Kirsch and Petryna show, knowledge about radiation was withheld, managed, and tailored in ways to calm public fears or thwart legal action in the name of national security during the cold war. Surveillance, a longstanding scientific response to environmental uncertainty, has been ratcheted up in intensity by the public health experts in King's paper, as the technocratic optimism of environmental sciences in the 1960s has yielded to recognition of the global scale and complexity of many disease threats. Today, as Fortun suggests in her account of EPA's digitalized "environmental subject," representational simplicity can dictate what appears known or unknown in the production of Web-based informational systems displaying regional toxic hazards.

Inquiries into uncertainty also build on a lively body of literature on postwar science and what Ulrich Beck labeled the "risk society."[32] This volume, however, also takes up the production of uncertainties that lurked in the practice of field scientists and laypersons before, or outside of, the formal enunciation of any risk calculus. Tilley's colonial medical officers, for example, working in the field as well as the lab, developed tropical medicine in ways that emphasized complex interactions, control rather than disease eradication, and the integration of human and nonhuman nature. Moreover, faced with a vast array of nonscientific ways of apprehending environmental health problems, scientific practice could often prove weak in its ability to link exposures to health effects and to induce political or social transformation. We learn from Nash's and Murphy's papers that twentieth-century environmental and occupational health scientists in the United States had trouble resolving questions about environmental pathology, hindered as they were by their techniques, by their personal distance from the "fields" they studied, and by the transient, variable nature of the exposures in question.

Fortifying clashes in values, the collision of multiple, intersecting, or competing epistemologies has been integral to the formation of uncertainty. The experts in controversies over environmental exposures covered by these papers often have not been the doctors or public health specialists featured in most medical history narratives, more or less committed to curing and preventing disease. A pesticide, for instance, is conceived and manufactured to be dangerous to certain kinds of life by chemists, entomologists, agri-business, the military, advertising firms, and state agencies. Health issues pertaining to pesticide exposure have involved the knowledge and actions of many additional groups: industrial hygienists, toxicologists, wildlife ecologists, lawyers, farmers, and citizens. While the history of environment and health encompasses a wide range of experts and their epistemological frames, it also builds on work, such as that of sociologists Phil Brown and Steve Epstein, exploring interac-

[31] Robert Proctor, *Cancer Wars: How Politics Shape What We Know and Don't Know About Cancer* (New York, 1995); Lisa Mitchell and Alberto Cambriosio, "The Invisible Topography of Power: Electromagnetic Fields, Bodies, and the Environment," *Soc. Stud. Sci.* 27 (1997): 221–71; and Michelle Murphy, "The Elsewhere within Here: Or How to Build Yourself a Body in Safe Space," *Configurations* 8 (2000): 87–120.
[32] Ulrich Beck, *Risk Society: Towards a New Modernity,* trans. Mark Ritter (London, 1992).

tions between professional and popular epistemology.[33] Signaling the centrality of this type of research in the field, at least half of the papers in this volume explore lay knowledge and activism.

With these emphases come methodological dilemmas. Among them: How do we, as historians, navigate alternative or conflicting accounts of a phenomenon when laypeople and experts, or different groups of experts, disagree? As individuals assemble knowledge out of diverse and intersecting epistemologies, each with its own complicated terrain of knowing and unknowing, historians confront a politics not just of epistemology but of ontology as well.[34] How should historians handle the ontological status of exposures when claims about them are so diverse at any given moment, controverted by different strands of science as well as by nonscientific epistemologies? Luise White's paper foregrounds this problem. Though poisoning presented the same signs and symptoms to all the groups involved in Zimbabwe's war of liberation, nearly incommensurable worldviews stirred controversies about the poisoning's external causes, in particular, about the existence of chemical warfare. Ambiguous expert conclusions unfolded alongside mercenary memoirs, rumors, and the agitation of witches working both sides of the war, giving chemical warfare a schizophrenic ontological purchase. On a historiographical level, attention to both the materiality of exposures and the multiple uncertainties that may be involved in their description raises interesting questions and challenges, since their terms of existence can prove so historically variable and so politically charged.

To historical actors themselves, uncertainties about relationships between bodies and places abound. Inhabiting landscapes where exposure is sustained by powerful interests and forces, individuals and communities worry about personal and family well-being, while often being forced to confront stark differences in scientific opinion. In the face of such uncertainties, they frequently raise difficult and troubling questions about identity and place—about who, as well as where, they are. Senses of self and belonging evolved through experiences of exposure and health as well as new encounters with related state, scientific, and economic regimes. Study of these changing relationships between health, identity, and place has rich precedents upon which to draw: anthropological and historical work on identity and nationalism, scholarship in feminist science studies on the relationship between bodies and epistemology, and explorations of place in the formation of community identities conducted by urban and environmental historians, as well as cultural geographers.[35] In this volume, Adriana Petryna develops the term "biological citizenship" to examine the ways individuals exposed to radiation around Chernobyl fashioned identities and lives anew at the intersection between medicine, unemployment, and state. McIvor and Johnston ask how masculinity shaped the way Scottish shipbuilders heard and articulated each other's testimonies of asbestos exposure, adding a new set of questions to the rapidly

---

[33] Phil Brown, *No Safe Place: Toxic Waste, Leukemia, and Community Action* (Berkeley, Calif., 1997); Steven Epstein, *Impure Science: Aids Activism and the Politics of Knowledge* (Berkeley, Calif., 1998); Steve Kroll-Smith, Phil Brown, and Valerie Gunter, eds., *Illness and The Environment: A Reader in Contested Medicine* (New York, 2000); and Brian Wynne, "Sheep Farming After Chernobyl: A Case Study in Communicating Scientific Information," in *Dirty Words: Writing on the History and Culture of Pollution,* ed. Hannah Bradby (London, 1990), 139–54.

[34] For an example of a recent work explicitly attending to the politics of clashing ontologies, see Helen Verran, *Science and an African Logic* (Chicago, 2002).

[35] See, e.g., Donna Haraway, "Situated Knowledge: The Science Question in Feminism and the Privilege of Partial Perspective," *Feminist Studies* 14 (1988): 575–99.

developing field of occupational health history. The subjectivity of scientists who enter into such controversies also needs interrogation. With its examination of state scientists in racially segregated regions and colonies, EPA investigators and policy makers in Washington, D.C., sanitary scientists and engineers hoping to influence Progressive Era politics, and others, this volume raises many questions about how scientists, doctors, and other experts have envisioned themselves and their epistemologies and practices in contexts of unequal power relations.[36]

These many examinations of identity and crises highlight the need for a wary, critical eye not only toward the inequitable distribution of hazards but also toward the power inequities at play in efforts to diagnose and solve environmental health problems. What historian Peggy Pascoe calls "relations of rescue" are often at work among the historical groups and contests depicted in this volume.[37] It matters immensely just who imagines themselves to be rescuing whom, whether they seek to save or cure the victims of exposure or to compensate or atone for the resultant health problems. In Nash's paper, state investigations into pesticides were motivated more by concerns over consumer safety than by the plight of migrant farmworkers, while in Jones's paper, public health veterinarians charged with the prevention of bovine tuberculosis balanced the protection of children against the economic security of the dairy industry. Historical perspectives encourage consideration of the subtle ways rescue projects were embedded within, not separate from, the generation of uneven distributions of health dangers. Through our methodological choices and commitment to particular actors or groups that our narratives aim to aid, we as scholars have staked out our own relations of rescue as well.

CONCLUSION

Whatever choices each of our authors or readers may make, the array of approaches and topics in this volume reflects the fertility of our collective search for new versions of the past that can help us to better understand the Brown Cloud in our present. As this volume suggests, the multitudinous exposures permeating our modern world have already sparked a proliferation of new scholarly insights into the broader origins of those exposures. These essays may augur greater historiographic changes to come. We anticipate it will become increasingly difficult to write the history of modern public health without asking many more questions about environment, ecology, and place. By the same token, histories of modern environments a few years hence may seem incomplete if they ignore a place's health implications, uncertainties, and impacts. Study of past expertise in environment and health, as it reaches out to encompass lay epistemology and ontology, may help lead a mutation of the "history of science" into a "history of knowledge."

As the first edited collection to strive after ways of bringing together histories of environment and illness long told asunder, this volume of *Osiris* treads into intellectual borderlands. Of these, as well as of future scholars, the borderlands demand a diffi-

---

[36] Warwick Anderson, "The Trespass Speaks: White Masculinity and Colonial Breakdown," *Amer. Hist. Rev.* 102 (1997): 1343–70; Jill Morawski, "White Experimenters, White Blood, and Other White Conditions: Locating the Psychologist's Race," in *Off White: Readings on Race, Power, and Society,* ed. M. Fine, L. Weis, L. Powell et al. (New York, 1997), 13–28.

[37] Peggy Pascoe, *The Search for Female Moral Authority in the American West, 1874–1939* (Oxford, 1990).

cult navigation. Coursing between certainty and uncertainty, between material flow and perception, between bodies and all manner of places, from the vast to the microscopic, the history of environment and health is spawning methods and messages as varied as the constituents of the Asian haze. These essays we see as only a beginning. We hope they will serve as stimulus for much further exploration by historians and other scholars into what it means to write history in the era of the Brown Cloud.

ECOLOGY AND INFECTION

Ecologies of Complexity:

Tropical Environments, African Trypanosomiasis, and the Science of Disease Control in British Colonial Africa, 1900–1940

By Helen Tilley*

ABSTRACT

Tropical Africa was one of the last regions of the world to experience formal European colonialism, a process that coincided with the advent of a range of new scientific specialties and research methods. The history of British attempts to understand and control African trypanosomiasis (sleeping sickness in humans and nagana in cattle), following the intense human epidemics that broke out between 1895 and 1910, reveals hitherto ignored scientific research in the fields of ecology, epidemiology, and tropical medicine that helped produce a new understanding of the "ecology of disease." Often generated within a transnational and interdisciplinary context, this knowledge increasingly assumed that vector-borne diseases in tropical environments were highly complex, dynamic, and interrelated phenomena. Thus while many people continued to hope that trypanosomiasis could be eradicated, research results made this prospect seem unlikely, if not impossible.

INTRODUCTION

In August 1939, Kenneth Morris, a young entomologist employed in Britain's Gold Coast medical department, wrote two articles for *The Times* (London) on his experiences "fighting a fly."[1] Morris had worked in West Africa since the late 1920s and was one of the first entomologists there to undertake systematic field investigations of the various tsetse flies in the region, adopting what was then called a "bionomic," or

* History Department, 129 Dickinson Hall, Princeton University, Princeton, NJ 08544; htilley@princeton.edu.

This essay is based on research conducted for my doctoral thesis and for a second project on the history of African trypanosomiasis. See Helen Tilley, "Africa as a 'Living Laboratory'—the African Research Survey and the British Colonial Empire: Consolidating Environmental, Medical, and Anthropological Debates, 1920–1940" (Ph.D. diss., Oxford Univ., 2001).

My thanks to William Beinart, Christopher Sellers, Linda Nash, Suman Seth, Eugene Cittadino, Gregg Mitman, Luise White, Warwick Anderson, Leo Slater, Michael Gordin, and the participants in the *Osiris* conference and the Princeton seminar in the History of Science for their comments on previous drafts. This research was funded in part by a grant from the Wellcome Trust (grant # 063276).

[1] Kenneth Morris, "Fighting a Fly: The Pest of the Tsetse, I. Its Meaning to Africa," *Times,* 7 Aug. 1939; and idem, "Fighting a Fly: The Tsetse in His Fortress, II. Some Victories of Science," *Times,* 8 Aug. 1939. Unless otherwise stated, all quotations are from the second article.

environmental, approach.[2] Morris's preoccupation with understanding life from the fly's point of view was evident in his articles. "This small grey insect . . . which dominates so much of Africa's most fertile land," he told his audience, "is an aristocrat of the insect world. . . . [H]e is quick and sly in his habits and is remarkable among insects in possessing so few natural enemies. And in the wild untouched bush of his natural home he exists in millions." Much as it was for aristocrats in Britain, one of the fly's main occupations was the hunt for prey.[3] At this point, Morris's admiration for the tsetse fly was superseded by his allegiance to his own species, for among the fly's prey were humans. "[E]very one of the tsetse is a potential carrier of trypanosomiasis," the parasitical disease that causes the fatal illnesses of sleeping sickness in humans and nagana in cattle. "Civilization," the entomologist concluded, "is incompatible with the presence of the tsetse fly."

For Morris, hope for future progress in tropical Africa lay "[o]nly in the quiet persistence of science in discovering and exploiting [the tsetse's] weaknesses." Researchers were subjecting to scientific scrutiny all facets of the fly's breeding grounds, including soil and climate, its means of securing blood meals, its digestive processes, its manner of infection by the trypanosomes, its affinity with particular flora and fauna, and even its "activity from hour to hour throughout the day." From his experience in the field, Morris felt that "tackling the problem from one aspect only, is bound to be a failure." What was needed was a comprehensive approach, which "combined the forces of entomology and medicine, veterinarians and agricultural experts," to reclaim land where humans, animals, and tsetse flies came into contact.[4] "This study of the fly's relationship to its environment, known as ecology, is the very keystone," assured Morris, "to an understanding of the problem of extermination. A full knowledge of the fly's habits and of the complexities of his environmental requirements may make it possible to determine how the alteration of some small factor will lead to the disappearance of the pest. . . . This highly involved study must be carried out on the spot, in the fly's own home."[5]

Luise White, in an innovative article on the social dimensions of African trypanosomiasis research and control in the interwar period, has argued that efforts in East and Central Africa "to study specific relationships—human to landscape, human to animal—that might cause or limit the disease gave way to the study of a vector alone, abstracted into the 'fly.'"[6] On one level, she is absolutely correct: British academics, colonial officials, and technical research officers often tended to organize their observations about trypanosomiasis under the heading of "the tsetse fly." Yet as Morris's comments should suggest, for the researchers themselves this was less an abstraction than a shorthand designation to represent the full spectrum of issues related to the disease they wanted to stem. By following the fly, they could consider particular dynamics both in isolation and in relation to one another: fly to climate, fly to

[2] Charles Tettey, "A Brief History of the Medical Research Institute and Laboratory Service of the Gold Coat [sic] (Ghana), 1908–1957," *West African Medical Journal* 9 (1960): 73–85. Morris's position was suspended in 1931 as a result of the economic depression; he returned in 1937.

[3] On aristocrats in colonial Africa, see Roderick P. Neumann, "Dukes, Earls, and Ersatz Edens: Aristocratic Nature Preservationists in Colonial Africa," *Society and Space* 14 (1996): 79–98.

[4] Quotations from first article (cit. n. 1), 7 Aug. 1939.

[5] Morris went on to author several important articles on the "ecology of sleeping sickness" in the 1950s.

[6] Luise White, "Tsetse Visions: Narratives of Blood and Bugs in Colonial Northern Rhodesia, 1931–9," *Journal of African History* 36 (1995): 219–45, on 226.

fauna, fly to flora, fly to parasite, fly to human. Yet the fly was never the sole point of entry for such studies. Researchers often chose to stress those realms for which their scientific training prepared them: parasitologists looked at trypanosomes, botanists studied vegetation, veterinarians explored cattle and game, physicians concentrated on humans, and entomologists, not surprisingly, dwelled on the fly.

The history of British efforts to understand African trypanosomiasis provides a useful lens to explore the dynamic interplay between ontology and epistemology, tropical and temperate environments, and colonies and the metropoles. In what ways was tropical Africa a significant site for ecological investigations regarding disease? How did studies undertaken there affect the thinking of practitioners in British institutions? Were there parallel trends elsewhere that fed into and reverberated with these new conceptual and methodological developments?[7]

A story this complicated—involving two continents, dozens of individuals, and multiple research institutes—requires a multifaceted approach, one that jumps back and forth between the local and individual, whether in England or Africa, and the international and institutional. What unites the narrative are a set of conditions—epidemics, colonial rule, and the fear that trypanosomiasis might pose permanent obstacles to the (European) development of Africa—and a range of ideas, about natural complexity, infectious disease, ecological interrelationships, and even disciplinary reconfigurations. Above all, the story explores some of the epistemological and epidemiological consequences of imperial conquest.

COLONIZATION, EPIDEMICS, AND THE GROWTH OF AFRICAN MEDICAL DEPARTMENTS

The intensive colonization and territorial demarcation of tropical Africa occurred between 1880 and 1920, coinciding with a period in human history marked by rapid changes in bioscientific theory. As disciplines proliferated so, too, did new ideas: the germ theory of disease, hypotheses concerning a genetic basis of life, the advent of ecological studies, and advances in, among other fields, immunology, bacteriology, psychology, and neurology.[8] In important respects, these changes were themselves an outgrowth of colonial encounters, yet their development and application across the European empires proved simultaneously uneven and sporadic. To capture this unevenness with respect to administrative concerns, imperial protagonists often likened their empires to organisms, suggesting both an "infinite variety of conditions" and an unpredictability with respect to their control.[9] As technical and scientific work took root across the African continent, such metaphors were extended to African environments and their human populations.

One of the most unsettling experiences the French, British, German, Belgian, and Portuguese powers faced in the first decade of the twentieth century, following the "Scramble for Africa," was the outbreak of sleeping sickness epidemics across the

[7] On this question, see especially Warwick Anderson, "Natural Histories of Infectious Disease: Ecological Vision in Twentieth-Century Biomedical Science" (this volume).

[8] For science and medicine at the turn of the twentieth century, see John Krige and Dominique Pestre, eds., *Science in the Twentieth Century* (London, 1997); and Roger Cooter and John Pickstone, eds., *Medicine in the Twentieth Century* (London, 2000).

[9] See, e.g., C. S. Goldman, ed., "The Imperial Organism," pt. 1 of *The Empire and the Century* (1905; reprinted London, 1998); and Charles Jeffries, *The Colonial Empire and Its Civil Service* (Cambridge, 1938), xvi.

continent's tropical territories. Although the actual mortality rates will never be known for certain, scholars conservatively estimate that between 750,000 and one million Africans died in this period.[10] Even before the scope of the catastrophe was understood, scientists and financiers in the various metropoles mobilized to send research commissions to the territories to determine the disease's etiology. Between 1901 and 1913, fifteen such commissions were completed (eight under the auspices of Britain).[11] While the immediate priority was understanding and stemming the epidemics, two underlying imperatives concerned interimperial competition and colonial development. No nation wanted to be trumped scientifically or otherwise by one of its rivals. More to the point, however, the colonial powers believed that, until the epidemics could be stopped, tropical Africa's economic future hung in the balance.[12]

From these on-the-ground investigations, scientists offered new explanations of the disease phenomena they observed. In a preliminary report from 1903, the second Sleeping Sickness Commission of the Royal Society reported two findings: the animal disease, known by its Zulu name *nagana,* was in fact related to sleeping sickness, and "trypanosoma fever" and sleeping sickness were considered to be the same disease in humans. Its research in Uganda, the commission concluded, offered "proof" that "the trypanosome when it finds its way into the cerebro-spinal fluid produces sleeping sickness, and that the parasite is carried to man by a biting fly which is a species of the same genus, *Glossina,* as that which carries the trypanosome of nagana to cattle."[13] Many questions remained, however, and these continued to preoccupy researchers in the years to come: Was there more than one species of trypanosome that caused sleeping sickness? What was the extent and duration of trypanosomal pathogenicity? What was the geographic distribution of the disease(s)? How was infection transferred between animals and humans? Did a kind of "racial" immunity to the disease exist? What types of tsetse fly species were responsible for carrying trypanosomes? What were the causes of different epidemic outbreaks? Finally, could the disease spread to other parts of the world, particularly British India?[14] Though a disease agent, the trypanosome, and a vector, the tsetse fly, had ostensibly been identified, many scientists remained skeptical that the relationship between the two was fully understood.[15]

---

[10] A. J. Duggan, "An Historical Perspective," in *The African Trypanosomiases,* ed. H. W. Mulligan (London, 1970), xliv, xlvi. Duggan estimates 500,000 deaths in the Congo Free State between 1896 and 1906 and another 200,000–300,000 in the Lake Victoria area between 1896 and 1908. These figures do not include deaths in French Equatorial or West Africa or in Portuguese Africa. Details on these areas can be found in Rita Headrick, *Colonialism, Health, and Illness in French Equatorial Africa, 1885–1935* (Atlanta, 1994); and Martin Shapiro, "Medicine in the Service of Colonialism: Medical Care in Portuguese Africa, 1885–1974" (unpublished Ph.D. diss., UCLA, 1983), chap. 5.

[11] Pieter de Raadt, "The History of Sleeping Sickness," in *Protozoal Diseases,* ed. H. M. Gilles (London, 1999), 253; and H. Harold Scott, *A History of Tropical Medicine* (London, 1939), 1:511–7.

[12] See, e.g., Headrick, *Colonialism, Health, and Illness* (cit. n. 10), 77.

[13] "The Etiology of Sleeping Sickness: The Trypanosome and Its Insect Carrier," *British Medical Journal* 2 (1903): 1343–50, on 1343.

[14] On these questions, see, e.g., Patrick Manson and C. W. Daniels, "Remarks on a Case of Trypanosomiasis," *British Medical Journal* 1 (1903): 1249–52; A. Laveran and F. Mesnil, *Trypanosomes and Trypanosomiasis,* trans. and enl. David Nabarro (London, 1907); Major W. B. Leishman, "On the Possibility of the Occurrence of Trypanosomiasis in India," *British Medical Journal* 1 (1903): 1252–4; and J. Everett Dutton, John L. Todd, and Cuthbert Christy, "Human Trypanosomiasis and its Relation to Congo Sleeping Sickness," *British Medical Journal* 2 (1904): 369–72.

[15] "Discussion on Trypanosomiasis," *British Medical Journal* 2 (1904): 367–79, on 378.

Scientific uncertainty notwithstanding, colonial administrations were forced to take action to control the epidemics and, as several historians have noted, often did so with haste and considerable coercion.[16] Colonial officials concentrated their attempts to manage epidemic outbreaks, with erratic success, on four methods: relocating populations from infested to noninfested areas; increasing population densities in lightly infested rural settlements; developing prophylaxis and chemotherapy treatments; and controlling, albeit on a variable scale, tsetse fly populations by clearing bush and vegetation, trapping flies, and killing game.[17]

The amount of attention devoted to describing these endeavors in Britain itself was sufficient to inspire the novelist John Masefield to write *Multitude and Solitude,* in which the protagonist decides to abandon his career in the arts for a career in science. "Art seemed to him to be taking stock of past achievement, science to be on the brink of new revelations. He knew so little of science that his thought of it was little more than a consideration of sleeping sickness." The disease was "an almost human thing, a pestilence walking in the noonday. . . . It fascinated him."[18] In this, he was not alone.

The period between the two world wars marked an important turning point in colonial governments' attempts to study and control diseases of all kinds. Not only were the new schools of tropical medicine and the newly established African medical departments augmenting and expanding their own capacities, but these changes were overseen and, in some respects, coordinated by Britain's Colonial and Foreign Offices and by the new intergovernmental Health Organisation of the League of Nations. The periodic international and interimperial conferences held in London, Paris, South Africa, and across the dependent territories (see Table 1) provided important vehicles for such coordination. In this way, with almost equal emphasis on African trypanosomiasis and general health care, the imperial powers rapidly built up an infrastructure for research, experimentation, and biomedical provision.

ECOLOGY INTO MEDICINE AND MEDICINE INTO ECOLOGY: DISCIPLINARY DEVELOPMENTS

In an important article on German and English epidemic theories in the interwar period, Andrew Mendelsohn opened with the question, "[W]here did the modern, ecological understanding of epidemic infectious disease come from?" While he writes that the "obvious answer would seem to be that it came from ecology," he rejects this explanation. "How indeed is one to imagine that the fledgling ideas and methods of upstart population ecology, or the premises of parasitology, which were of uncertain relevance to bacterial and viral disease . . . could have conquered bacteriology?" He goes on to argue that, in fact, epidemics "became complex," and epidemiology more syncretic, through intellectual currents within the discipline of

[16] Kirk Arden Hoppe, "Lords of the Fly: Colonial Visions and Revisions of African Sleeping Sickness Environments on Ugandan Lake Victoria," *Africa* 67 (1997): 86–105; Maryinez Lyons, *The Colonial Disease: A Social History of Sleeping Sickness in Northern Zaire, 1900–1940* (Cambridge, 1992).

[17] Michael Worboys, "The Comparative History of Sleeping Sickness in East and Central Africa, 1900–1914," *History of Science* 32 (1994): 89–102; Hoppe "Lords of the Fly" (cit. n. 16); Lyons, *Colonial Disease* (cit. n. 16); Headrick, *Colonialism, Health, and Illness* (cit. n. 10), and Shapiro, "Medicine in the Service of Colonialism" (cit. n. 10).

[18] John Masefield, *Multitude and Solitude* (London, 1909), 150.

Table 1: Interterritorial, Imperial, and International Conferences
and Commissions

1907–08	British Foreign Office—Int'l Conference on Common Sleeping Sickness Policy (led to establishment of Sleeping Sickness Bureau)
1920	Imperial Entomological Conference, London
1925	League of Nations—First Int'l Conference on Sleeping Sickness in Africa, London
1925	West African Medical Staff Conference of Senior Members, Accra
1925	Imperial Entomological Conference, London
1926–27	League of Nations—First Int'l Sleeping Sickness Commission in Entebbe, Uganda
1928	League of Nations—Second Int'l Conference on Sleeping Sickness in Africa, Paris
1930	Imperial Entomological Conference, London
1932	League of Nations—International Conference of Representatives of Health Services of African Territories and British India, South Africa
1932	Veterinary Services of Nigeria, Niger, and French Cameroons Conference in Nigeria
1933	Co-ordination of Tsetse & Tryps Research (Animal & Human) at Entebbe, Uganda
1933	Co-ordination of General Medical Research at Entebbe, Uganda
1934	Co-ordination of Veterinary Research at Kabete, Kenya
1935	League of Nations—Pan-African Health Conference, South Africa
1935	Imperial Entomological Conference, London
1936	Commonwealth Scientific Conference, London
1936	Co-ordination of Tsetse & Tryps (Animal & Human) Research at Entebbe, Uganda
1936	Co-ordination of General Medical & Veterinary Research in Nairobi, Kenya

SOURCE: Modified from Helen Tilley, "Africa as a 'Living Laboratory'—the African Research Survey and the British Colonial Empire: Consolidating Environmental, Medical, and Anthropological Debates, 1920–1940" (Ph.D. diss., Oxford Univ., 2001), 122–3.

bacteriology itself.[19] He locates this development in the decade following the First World War, a time when the influenza pandemic called into question existing explanations on the "causes and nature of [human] epidemics."[20]

The full story behind the question Mendelsohn posed, however, remains untold. As he acknowledged, in the interwar period there were a "diversity of sciences" affected by such concepts as "holism," "complexity," "equilibrium," "web of causes," and "system," suggesting there had been "a broad intellectual transformation" among many scientific disciplines and their adherents, not just one.[21] The evidence supports this interpretation and even pushes Mendelsohn's periodization back in time. Eco-

[19] J. Andrew Mendelsohn, "From Eradication to Equilibrium: How Epidemics Became Complex after World War I," in *Greater Than the Parts: Holism in Biomedicine, 1920–1950,* ed. Christopher Lawrence and George Weisz (London, 1998), 303–31, on 303–4.

[20] Ibid., 304.

[21] Ibid., 323.

logical ideas, principles, and research, fragmented though they may have been, clearly did have an effect on medical researchers, including epidemiologists, bacteriologists, and public health officials.[22] Yet the converse was also true: adherents to ecology, in particular those concerned with questions of disease and health, drew upon the work of medically trained researchers, not least because such arguments bolstered the formers' evidence. A strong case could be made that the "new epidemiology" being promoted in Britain, which supported the idea that epidemics should be examined from the "bird's eye view of all from an aeroplane," was in many respects an organic bedfellow to that other synthesizing and "aerial" science, ecology.[23]

There was a third dimension to this picture: tropical medicine. Historians have already made the point that this young discipline emerged equally from the biological and the medical sciences. "Structured around the life-cycle of parasites," writes Michael Worboys, "tropical medicine required detailed knowledge of the taxonomy of vector species and ecological management, which found application in the tropical environment."[24] However much the different European powers saw each other in a competitive light at the turn of the twentieth century, their experience of sleeping sickness epidemics was a shared one. All classified the disease as endemic to "that portion of the African continent lying between the Tropics of Cancer and Capricorn," and all devoted considerable resources to its study in their respective institutes of tropical medicine.[25] By the twentieth century's second decade, tropical medicine itself was undergoing key changes, shifting gradually away from a narrow and linear understanding of disease causation, in which microbes alone were targeted as the culprit, toward a more integrated and comprehensive approach. "[T]ropical medicine," noted Britain's Medical Research Council in 1925, "is no separate branch of medicine but touches all the fields of medicine and needs the services of all the medical sciences."[26]

[22] See, e.g., Clifford Allchin Gill, *The Genesis of Epidemics and the Natural History of Disease: An Introduction to the Science of Epidemiology Based upon the Study of Epidemics of Malaria, Influenza, and Plague* (London, 1928), especially pt. 4, "The Bionomics of Disease"; Sheldon F. Dudley, "Can Yellow Fever Spread into Asia? An Essay on the Ecology of Mosquito-Borne Disease," *Journal of Tropical Medicine and Hygiene* 37 (1934): 273–8; idem, "The Ecological Outlook on Epidemiology—President's Address," *Proceedings of the Royal Society of Medicine* 30 (1936): 57–70; Richard Strong, "The Importance of Ecology in Relation to Disease," *Science* 82 (1935): 307–17; Smith Ely Jelliffe, "The Ecological Principle in Medicine," *Journal of Abnormal and Social Psychology* 32 (1937): 100–21; L. W. Hackett, *Malaria in Europe: An Ecological Study* (London, 1937); F. M. Burnet, "Inapparent Virus Infections: With Special Reference to Australian Examples," *British Medical Journal* 1 (1936): 99–103; idem, *Biological Aspects of Infectious Disease* (Cambridge, 1940), especially chap. 1, "The Ecological Point of View"; idem, *Virus as Organism: Evolutionary and Ecological Aspects of Some Human Virus Diseases* (Cambridge, 1945); Alfred C. Reed, "Environmental Medicine," *Science* 82 (1935): 447–52; and Burton E. Livingstone, "Environments," *Science* 80 (1934): 569–76.

[23] Major Greenwood, "The Epidemiological Point of View," *British Medical Journal* 2 (1919): 405–07, on 405. Greenwood was the first professor of epidemiology and vital statistics at the London School of Hygiene and Tropical Medicine. Also partially quoted in Mendelsohn, "From Eradication to Equilibrium" (cit. n. 19), 321.

[24] Michael Worboys, "Manson, Ross, and Colonial Medical Policy: Tropical Medicine in London and Liverpool, 1899–1914," in *Disease, Medicine, and Empire: Perspectives on Western Medicine and the Experience of European Expansion,* ed. Roy MacLeod and Milton Lewis (London, 1988), 21–38, on 22.

[25] Andrew Balfour, E. Van Campenhout, Gustave Martin et al., *Interim Report on Tuberculosis and Sleeping-Sickness in Equatorial Africa* (Geneva, 1923), 7.

[26] From the obituary for Sir William Leishman, member, U.K. Medical Research Council, 1925, quoted in A. Landsborough Thomason, *The Programme of the Medical Research Council,* vol. 2 of *Half a Century of Medical Research* (London, 1975), 190; see also Michael Worboys, "Tropical Diseases," in *Companion Encyclopedia of the History of Medicine,* ed. W. F. Bynum and Roy Porter (London, 1993), 1:512–36, especially 523–4.

A unifying preoccupation of ecology, the "new epidemiology," and tropical medicine in this period was to discern how organisms functioned and interacted, whether in terms of animals, humans, plants, or parasites. Yet as one of the key protagonists, Julian Huxley, observed in this respect following on the heels of his research trip to East Africa: "organisms . . . have no biological meaning apart from their environment."[27] Thus whether in the field or the laboratory, practitioners began to emphasize "the inter relations of different parts": species diversity to climate, epidemic patterns to population levels, and disease distribution to geography.[28] Above all, these practitioners tended to share a fascination with the question of whether diseases could be permanently "eradicated." These preoccupations were by no means unique to the interwar decades, but it was only then that approaches in all three specialties began to converge around the analytical categories of "environment" and "disease."

The idea of the elimination of diseases can be traced back at least to the year 1801, when Edward Jenner pronounced that "vaccine inoculation" would result in "the annihilation of the Small Pox." Not until the late nineteenth century, however, was the term "eradication"—literally, "to pull out by the roots"—applied in a more systematic manner to infectious diseases. Some of the earliest attempts at "eradication" were directed toward diseases of domestic animals, such as bovine pleuropneumonia and rabies; these tended to be regional efforts, undertaken in places such as the United States, Britain, and South Africa, and were often effective on a limited geographic scale. Only in the early twentieth century were transnational disease "campaigns" first conceptualized under the rubric of eradication, the best known being the Rockefeller Foundation's Commission for the Eradication of Hookworm Disease (founded 1909; phased out beginning in 1922), which eventually sponsored projects, with limited success rates, in more than fifty countries. This campaign was followed, in 1915, by the Rockefeller Commission for the Eradication of Yellow Fever, which was frustrated by a lack of knowledge of the "jungle" variety of yellow fever and by measures that proved inadequate to reduce the mosquito vector (*anopheles aegypti*) in various centers of infection.[29] One of the questions arising from an examination of disciplinary developments in the first several decades of the twentieth century is how particular conceptual shifts affected national, imperial, and international confidence in "eradication" as a control strategy.

THE AFRICAN FIELD:

ECOLOGICAL APPROACHES TO TSETSE FLIES AND DISEASE CONTROL

Just after the First World War, at the invitation of the Portuguese administration, the Mozambique Company, Charles Swynnerton, game warden of Tanganyika, undertook a study of the tsetse fly problem in North Mossurise.[30] In the landmark article published as a result of this work—one of the first to take into account African ex-

---

[27] Julian Huxley, *Biology and Its Place in Native Education in East Africa* (London, 1930), 6.

[28] For general background, see Lise Wilkinson, "Epidemiology," in Bynum and Porter, *History of Medicine* (cit. n. 26), 2:1262–82, especially 1278–9.

[29] Fred Soper, "Rehabilitation of the Eradication Concept in Prevention of Communicable Diseases," *Public Health Reports* 80 (1965): 855–69; Marcos Cueto, "The Cycles of Eradication: The Rockefeller Foundation and Latin American Public Health, 1918–1940," in *International Health Organizations and Movements, 1918–1939*, ed. P. Weindling (Cambridge, 1995), 222–43.

[30] C. F. M. Swynnerton, "An Examination of the Tsetse Problem in North Mossurise, Portuguese East Africa," *Bulletin of Entomological Research* 11 (1921): 315–85.

perimental methods of tsetse control via bush burning—Swynnerton wrote that "the practical study of tsetse is a matter for the botanist and oecologist rather than the unaided entomologist." Some years later, he stressed, "[O]nly by an investigation on ecological lines, never applied to [tsetse research] previously, were these problems likely to be solved."[31] Just what were the ecological methods he had in mind, and how did they differ from previous traditions subsumed under the rubrics of natural history and bionomics?

Organized studies of the bionomics of tsetse fly species had begun in the twentieth century's second decade in association with the third Sleeping Sickness Commission of the Royal Society. Their primary author was Geoffrey Carpenter, a contemporary of Swynnerton's, who had received his M.D. in tropical medicine from the London School in 1910 and joined Uganda's medical service following the war. In addition to his scientific articles,[32] Carpenter published a popular account of his endeavors in 1920, *A Naturalist on Lake Victoria: With an Account of Sleeping Sickness and the Tse-Tse Fly*. Between 1911 and 1914, Carpenter had spent much of his time making observations on the recently depopulated islands of Lake Victoria in the hope that "the presence or absence of *Glossina* might be found to be correlated with definite factors."[33] He was concerned not just with the life history of the "tse-tse" or its "natural enemies" but also with "the relations between Tse-tse, Trypanosome, and the 'alternative hosts' of the latter from which the fly acquires it."[34] He thus explored tsetse feeding patterns, testing the blood of the islands' fauna, including antelopes, crocodiles, hippos, birds, lizards, tortoises, and even fish; the fly favored antelopes, he concluded, but never fed upon fish. He also correlated climate with tsetse population levels and tracked the distances the flies could travel at any one time. They tended, he discovered, to be less numerous during the hotter and humid months; the extent of their mobility remained unclear, though they could use fauna, including humans, for long-distance "travel." Given his findings among the faunal populations, Carpenter could conclude, at least for the islands, that to "exterminate Sleeping Sickness two animals must be kept from each other—the Situtunga antelope, from which the fly obtains the Trypanosome, and the fly, which inoculates the Situtunga with the Trypanosome. Each without the other is harmless."[35]

However, in reviewing this and other "eradication" experiments conducted by the Portuguese on the island of Principe off the West Coast of Africa, the authors of the East Africa Commission, who surveyed the territories in 1924, remarked: "[I]t is one thing to deal with an island and altogether another proposition to deal with a continent."[36] During their survey tour, commision members, William Ormsby-Gore (a Conservative M.P.), A. G. Church (a Labour M.P.), and F. C. Linfield (a Liberal

[31] Ibid., 317; Swynnerton, "The Tsetse Flies of East Africa: A First Study of Their Ecology, with a View to Their Control," *Transactions of the Royal Entomological Society of London* 84 (1936): 10. See also John MacKenzie, "Experts and Amateurs: Tsetse, Nagana, and Sleeping Sickness in East and Central Africa," in *Imperialism and the Natural World*, ed. John MacKenzie (Manchester, 1990), 187–212.

[32] Carpenter published five reports on the bionomics of tsetse species between 1912 and 1919 as part of the Sleeping Sickness Commissions.

[33] G. D. Hale Carpenter, *A Naturalist on Lake Victoria: With an Account of Sleeping Sickness and the Tse-Tse Fly* (New York, 1920), x.

[34] Ibid., 26.

[35] Ibid., 60–1.

[36] W. Ormsby-Gore, Major A. G. Church, and F. C. Linfield, *Report of the East Africa Commission* (London, 1925), 76.

M.P.), had met Swynnerton, Carpenter, and several East African colleagues. The commissioners were struck by the various methods at Britain's disposal to stem the disease, which seemed of "urgent necessity" given its role in obstructing economic development and the alarming reports claiming that tsetse fly populations were extending their territory. Not only were further bionomical studies and fly surveys necessary, but so, too, were field experiments aimed at exterminating the fly.[37] The commission's recommendations were considered on several fronts, and in 1925, when the British Parliament approved the creation of the Committee of Civil Research, one of the first subcommittees established was on the tsetse fly.[38]

The years between 1925 and 1929 marked a central shift in tsetse research in several respects. It was during this period that ecology superseded bionomics as the organizing concept for this work. While bionomics was, in principle, the study of organisms in their environments, in practice it was often much more, as Carpenter's far-reaching explorations should indicate. The perception of a close relationship between ecology and bionomics was reinforced by a Colonial Office administrator who remarked, after looking up the word "ecology" in the *Encyclopedia Brittanica* in 1936: "It appears to mean all . . . those factors which go to maintain or destroy animal and/or plant life and is apparently the modern jargon for our old pal 'bionomics'!"[39] Yet the semantic switch had wider implications and was accompanied by institutional and financial changes.[40] Shortly after the creation of the Tsetse Subcommittee of the Committee of Civil Research, the Colonial Office designated two territories, Tanganyika and Nigeria, then the two largest regions under its jurisdiction, as the coordinating centers for all research on African trypanosomiasis. As part of this process, the East African Loan Scheme provided Tanganyika with a five-year grant, and the Tsetse Subcommittee asked Swynnerton to direct efforts on tsetse research. Thus, in 1927, the Tsetse Research Department was founded.

Swynnerton had been conducting small-scale field experiments in tsetse control in regions across Tanganyika since 1923; with the creation of the department, he wanted to expand and decided to hire a "trained ecologist." He recruited a young South African, Dr. John Phillips, then a forest research officer in Knysna, South Africa, who had received his doctorate from Edinburgh in forestry and botany, with a strong emphasis on ecological methods. Phillips joined the department in late 1927 as deputy director and ecologist.[41] By 1930, the department had twenty-one staff members and rivaled in size many other technical departments across the African territories.[42] The driving vision behind the department was that ecological science could help untangle the problem of African trypanosomiasis, a problem Phillips considered "as bewilder-

---

[37] Ibid., 79.

[38] Roy MacLeod and Kay Andrews, "The Committee on Civil Research: Scientific Advice for Economic Development, 1925–1930," *Minerva* 7 (1969): 680–705, 691, 701.

[39] Colonial Office minute from J. E. W. Flood to D. O'Brien, 23 Oct. 1936, CO822/74/9, "Coordination of Research-Ecological," Public Records Office, London.

[40] Few studies discuss the history of bionomics explicitly; for an article on its development in the American context, which reinforces the idea that bionomics and ecology shared an interest in seeing "interactions in the natural world as complex and multi-causal," see Mark A. Largent, "Bionomics: Vernon Lyman Kellogg and the Defense of Darwinism," *Journal of the History of Biology* 32 (1999): 465–88, 482.

[41] Swynnerton, "Tsetse Flies of East Africa" (cit. n. 31), 10.

[42] Staff numbers were reduced to fifteen in 1934 following the Great Depression. Ibid., 17–8.

ing in its complexity, as comprehensive in its interrelations" as any problem in Africa then being confronted.[43]

In large part because of his initial experiences in Tanganyika, Phillips wrote a proposal in which he encouraged the British territories to undertake a systematic ecological survey of South, Central, and East Africa. Such a survey, he felt, would help place the challenges of agriculture, forestry, soil fertility, and trypanosomiasis in their widest possible context. One of the obstacles, however, was that "the teaching of dynamic ecology" had been relatively neglected in British and overseas universities. "[T]he need," Phillips wished to emphasize, was to train many more ecologists and to provide them with skills

> in such subjects as agriculture, forestry, veterinary or medical science, for the very good reason that without one or the other of these (or in special instances all) ecologists cannot be expected to realise and understand the great practical problems set them in the field, nor can they hope to work in the fullest and most intelligent co-operation with officers engaged in these particular professions. . . . The necessary basis is a dynamic ecological survey in which the co-operative team-work is one of the inspiring forces.[44]

Work in Nigeria, in the meantime, had begun with a Sleeping Sickness Service. To compare methods across the different territories, W. B. Johnson, recently appointed director of the Medical Services in Nigeria, undertook a tour in 1928 "to observe general medical organisation and methods of trypanosomiasis control."[45] Of all the research efforts Johnson observed, those in Tanganyika received his highest praise. In his forty-page report covering the Belgian Congo, the Sudan, Uganda, Tanganyika, and French Equatorial Africa, he devoted eleven pages to a discussion "in support of the research concept and programme" in Tanganyika. The "Tsetse Research Team," Johnson concluded, had undertaken a program that was "objective, dynamic, comprehensive, balanced, the outcome of the vision sufficiently broad, sufficiently penetrating."

> [A]ny attempt to lessen the conceptional scope of the investigation, would result in a general weakening of the possibility of the team's ultimate success. . . . A concept embracing [a] purely pathological, histological, veterinary, medical, or entomological basis of research it is believed, would fail to achieve more than a fraction of that knowledge which the biological-ecological concept outlined in this communication should lay before us.[46]

As a result of his trip, Johnson would recruit to Nigeria's Medical Department one of Tanganyika's young ecological entomologists, Thomas Nash. Nash would remain in that territory conducting research on tsetse flies and trypanosomiasis for the next twenty-six years, ultimately as the director of the West African Institute of Trypanosomiasis Research.[47]

---

[43] John F. V. Phillips, "The Application of Ecological Research Methods to the Tsetse (*Glossina* SPP.) Problem in Tanganyika Territory: A Preliminary Account," *Ecology* 11 (1930): 713–33, on 713.

[44] John F. V. Phillips, "Ecological Investigation in South, Central, and East Africa: Outline of a Progressive Scheme," *Journal of Ecology* 19 (1931): 474–82, on 477 and 482.

[45] W. B. Johnson, *Notes upon a Journey through Certain Belgian, French, and British African Dependencies to Observe General Medical Organisation and Methods of Trypanosomiasis Control* (Lagos, 1929).

[46] Ibid., 38 (italics in original).

[47] Thomas Nash, *A Zoo without Bars: Life in the East Africa Bush, 1927–1932* (Tunbridge Wells, 1984), 75, 147.

This very brief discussion has been meant to illustrate that ecological methods, as embraced in Tanganyika in particular, were in many respects a direct response not only to the heterogeneity of Africa's environments but also to the sheer intricacy of the attempts to control such environments. Swynnerton and other researchers made this point repeatedly: the variety of tsetse species, the different strains of try-panosomes, the diverse flora and fauna, the changing settlement patterns of human populations, all these factors brought to light "the magnitude and complexity of the problems that a tsetse investigation has to solve."[48] Such complexity tempered optimism about the ease with which African trypanosomiasis might be either controlled or eradicated. Britain's colonial undersecretary, William Ormsby-Gore, remarked upon just these points at the Second International Conference on Sleeping Sickness in 1928:

> Let me at once emphasise . . . that the problem of trypanosomiasis is very varied, but must be regarded as a whole. The fact that the human and the animal diseases are conveyed by the same family of blood-sucking flies—the glossinae—and that the protozoan organisms which the flies carry are so similar, makes it absolutely essential that both the medical and veterinary problems should be discussed in their widest ecological bearings.

Referring explicitly to the "continued work of Mr. Swynnerton," Ormsby-Gore went on to remark that "[o]ne thing is clear from the work that he has carried on for the last five years, namely, that no one form of attack upon the tsetse fly is universally practicable. . . . [N]o one method [of trypanosomiasis control] can be universally applied, but each administration faced with the problem has much to learn from the experiments carried out by its neighbors."[49] With the substitution of ecology for bionomics, an explicitly interdisciplinary and site-specific method of scientific research was instituted in British Africa, one that considerably expanded the scope of analysis.

THE UNITED KINGDOM:
ANIMAL ECOLOGY, PARASITES, AND THE "ECOLOGY OF DISEASE"

To examine in more detail how scientists in Great Britain took up these issues, it will help to turn to the work of Julian Huxley and Charles Elton, two individuals whose interest in, and attention to, colonial Africa has often been overlooked.[50] Huxley and Elton were pivotal players in the interwar effort to establish the field of ecology on a firm footing: Huxley as a mentor, popularizer, and booster for the field; Elton as a researcher, editor (*Journal of Animal Ecology*), and founder of institutions (Bureau of Animal Populations, Oxford). Their social and intellectual networks were, of course, extensive; numerous other individuals, including the demographer and social scientist Alexander Carr-Saunders and the statistician and eugenicist Ronald A. Fisher, provided important stimulus along the way. Carr-Saunders, for instance, author of the widely influential book *The Population Problem: A Study in Human Evolution* (1922), was one of Elton's supervisors at the University of Oxford and helped coor-

---

[48] Swynnerton, "Tsetse Flies of East Africa" (cit. n. 31), 6.

[49] William Ormsby-Gore, opening address, *Report of the Second International Conference on Sleeping Sickness—Paris, November 5–7, 1928* (Geneva, 1928), 14–9, on 15, 18–9.

[50] An important exception is Peder Anker's recent book, *Imperial Ecology: Environmental Order in the British Empire, 1895–1945* (Cambridge, 2001), which includes an astute discussion of Huxley, Elton, and H. G. Wells.

dinate, with Huxley, the scientific side of the first expedition to Spitsbergen Island, in 1921, on which Elton served.[51] Likewise, Fisher was author of *Statistical Methods for Research Workers* (1925) and *The Genetical Theory of Evolution* (1930) and helped train or mentor several research workers bound for the Colonial Service, including Charles Jackson and Thomas Nash, who went to work in the Tsetse Research Department in Tanganyika.[52] What all these men shared was an interest in relating population patterns—densities, growth, and fluctuations—to wider environmental changes.

One of the first indications of Huxley's interest in relationships between ecology and disease emerged in a short review of a utopian novel by H. G. Wells, *Men Like Gods*. Wells was, of course, the preeminent science fiction writer of his generation, but he was also a science popularizer and shared with Huxley—even before they met—a preoccupation with the liberating possibilities of scientific application.[53] Seizing upon Wells's optimism regarding the idea of disease eradication, Huxley wrote, "The triumphs of parasitology and the rise of ecology have set him thinking; and he believes that, given a real knowledge of the life-histories and inter-relations of organisms, man could successfully proceed to wholesale elimination of a multitude of noxious bacteria, parasitic worms, insects, and carnivores."[54] The idea that one could predict and control the natural world to such an extent was not new, but the conceptual tools of ecology were, and these gave Wells and Huxley some cause to temper their utopian visions. Understanding species interrelations was no simple affair; already scientists and land "managers" were learning that the precise kinds of controls being described could sometimes have unpredictable consequences. However, wrote Huxley, "Mr. Wells does not need to be reminded [of this,] . . . his Utopians proceed with exemplary precautions."[55]

Interest in biological relationships continued to preoccupy the two men, and in 1927, they, along with Wells's son George, teamed up to write what became a three-volume study titled *The Science of Life*. In the chapters "The Science of Ecology" and "Life under Control," they elaborated on the challenges involved in understanding, as well as predicting, dynamics in the natural world and drew attention to important distinctions between temperate and tropical environments.[56] "To work out this web of interrelations in detail for a whole community is all but impossible even in our temperate regions—let alone in the richer tropics." Because biological interrelations were so unstable, they argued, to tamper with the "web's weaving, whereby a twitch on one life-thread alters the whole fabric, . . . it behoves the would-be benefactor of humanity to proceed with caution; if he is not careful, he will do infinitely more harm than good."[57]

[51] On the Oxford expeditions to Spitsbergen, see ibid., 89–97.

[52] See, in particular, C. H. N. Jackson, "Some New Methods in the Study of Glossina Morsitans," *Proceedings of the Zoological Society, London* (1937): 811–96.

[53] For Huxley's early essays on these subjects, see Huxley, *Essays of a Biologist* (London, 1923); and idem, *Essays in Popular Science* (1926; reprinted London, 1937).

[54] The review was written in 1923 and republished in Julian Huxley, "Biology in Utopia," in *Essays in Popular Science* (cit. n. 53), 65; cited in Anker, *Imperial Ecology* (cit. n. 50), 111. Anker offers an analysis of the philosophical and ideological underpinnings of Wells and Huxley's collaborative writings, which I do not explore.

[55] Huxley, "Biology in Utopia" (cit. n. 54), 65.

[56] Also expressed by Huxley in University of Oxford, ed., *Scientific Results of the First Oxford University Expedition to Spitsbergen (1921)* (London, 1925), 1:v.

[57] H. G. Wells, Julian Huxley, and G. P. Wells, *The Science of Life* (London, 1930), 3:664, 677, 679.

The effect of *The Science of Life* on lay and science audiences alike was far-reaching. At least one young medic, Macfarlane Burnet, who later received the Nobel Prize for his work on immunological tolerance, claimed that it "had a major impact on my thinking. It made me particularly interested in the ecological aspects of microbiology and epidemic disease, the field I was then just entering as a laboratory worker."[58] In an article written for the *British Medical Journal* in 1936, Burnet wrote that since the First World War, "the most characteristic development of epidemiology has been the adoption of what may be called a more oecological point of view, in which the activities of the two organisms concerned—man and the pathogenic microorganism—are both considered from the point of view of survival of the species."[59] That same year, Burnet began to write his book *Biological Aspects of Infectious Disease,* and titled the first section "The Ecological Viewpoint," which owed both its framework and many of its examples to *The Science of Life.*[60]

The other individual Burnet acknowledged as having a formative influence on him was, not surprisingly, Charles Elton. In 1927, at Huxley's urging, Elton had written one of the first full-length textbooks on animal ecology, in which he devoted an entire chapter to a discussion of the ecology of parasites and their relationships to "complex food chains." Like animal populations, parasites, in Elton's schema, were treated as dynamic organisms, which shaped and were shaped by their relations with other organisms and their environments. In this context, epidemics were defined as a consequence of parasites' "power of multiplication" creating a situation in which they were able to overwhelm their hosts in the food chain. Unlike some of his colleagues at the time, Elton took issue with the idea that any simple "balance of nature" existed with respect to the regulation of animal and parasite populations; his objection stemmed from field studies in which he realized that in examining "the dynamics of a whole community," it was extremely difficult "to predict the effects of variation in numbers of one species upon those of other ones in the same community."[61] This statement, he believed, held for parasites as well.

Behind the discussions of prediction in both *The Science of Life* and *Animal Ecology* lay a deeper issue: the way in which disease was to be defined. If one considered disease the result of population dynamics among organisms and their environments, dynamics which were themselves fluctuating and unpredictable, it called into question two assumptions: first, the idea that either linear or simple relationships between species and their environments or among species was actually the norm; second, that it was realistic, in the case of diseases that demonstrated nonlinear and complex dynamics, to believe they could be "eradicated." By challenging these assumptions, Elton argued that epidemic disease, whether among humans or animals, should not be considered an aberration but "a normal and frequent phenomenon in nature."[62] As a

---

[58] F. M. Burnet, *Changing Patterns: An Atypical Autobiography* (London, 1968), 23; a deeper analysis of Burnet's ecological approach to disease can be found in Anderson, "Natural Histories of Infectious Disease" (cit. n. 7).

[59] Burnet, "Inapparent Virus Infections" (cit. n. 22), 100.

[60] In the preface, Burnet wrote that he "should like to think that this book expresses the same general point of view that runs through Wells, Huxley, and Wells' *Science of Life.*" Burnet, *Biological Aspects of Infectious Disease* (cit. n. 22), ix.

[61] Charles Elton, *Animal Ecology* (London, 1927), 101.

[62] Charles Elton, "The Study of Epidemic Diseases among Wild Animals," *Journal of Hygiene,* 31 (Oct. 1931): 435–56, on 436; see also idem, "Plague and the Regulation of Numbers," *Journal of Hygiene* 24 (1925): 138.

result of this redefinition, he believed that "our general attitude towards health in the human population . . . [and] existing theories as to the origin of human disease in history may have to be reconsidered."[63] Humans were themselves organisms and interacted dynamically with other organisms in their environments; the phenomena of disease in human populations needed to be viewed in terms of interrelations among organisms rather than as a simple chain of cause and effect.[64] In the editor's introduction to *Animal Ecology,* Huxley drew attention to the implications of these findings for medicine and disease control:

> [U]nder the magic of the germ-theory and its spectacular triumphs, medical research on disease was largely concentrated upon the discovery of specific "germs" and their eradication. But as work progressed, the limitations of [this] mode of attack were seen. *Disease was envisaged more and more as a phenomenon of general biology, into whose causation the constitution and physiology of the patient and the effects of the environment entered as importantly as did the specific parasites.* . . . The discovery of the tubercle bacillus, [for instance], has not led to the eradication of tuberculosis: indeed it looks much more likely that this will be effected through hygienic reform than through bacteriological knowledge. . . . In other words, a particular pest may be a symptom rather than a cause; and consequently over-specialisation in special branches of applied biology may give a false optimism, and lead to waste of time and money through directing attention to the wrong point of attack.[65]

It was in this context that Elton and Huxley began to discuss the creation of a "Bureau of Ecology for Africa" and the importance to this project of research on "medical ecology in relation to disease."[66] Huxley was the prime mover behind this shift in attention, which was the direct result of his 1929–1930 tour of East Africa on behalf of the Colonial Office Advisory Committee on Native Education. Like the East Africa Commission before him, he had been captivated by the issue of sleeping sickness and African trypanosomiasis; he wrote several articles for *The Times* on his "first impressions," which later became the substance of his book *Africa View.*[67]

Huxley had already observed that "especially in the tropics . . . climate gives such an initial advantage to [humans'] cold blooded rivals, the plant pest and, most of all, the insect."[68] Since eradication was increasingly seen as a problematic concept, the alternative would be to research "the ecological status" of vector species and parasites in relation to their hosts and environments. Such a program would form the basis of the Bureau of African Ecology that he envisaged:

> The fluctuations in the population of wild animals, which occasionally culminate in epidemics, have been little studied in tropical countries. However they have important bearings. In large animals they are of direct interest to game departments. Indirectly through rinderpest outbreaks, they concern veterinary departments, and through game acting as

---

[63] Elton, "Study of Epidemic Diseases" (cit. n. 62), 436.

[64] See also Elton's monograph, *The Ecology of Animals* (London, 1933), 78–9.

[65] Julian Huxley, "Editor's Introduction," in Elton, *Animal Ecology* (cit. n. 61), xiv-xv (my italics).

[66] Julian Huxley, "A Proposal for a Bureau of Ecology for Africa," 29 April 1931, Folder 3 16/26d, Royal Institute of International Affairs, London; Charles Elton in "Report of the Science Committee on African Problems," 28 Feb. 1930, File 2792, Rhodes Trust Archives, Oxford.

[67] Julian Huxley, "Aspects of Africa: I. Some First Impressions," "II. White and Black," "III. The Tsetse Fly," "IV. The Linked Tale," *Times,* 6, 7, 8, and 9 Jan. 1930; and idem, *Africa View* (London, 1931).

[68] Huxley, "Editor's Introduction" (cit. n. 65), xiv.

a host to tsetse fly, to medical departments as well. In smaller animals like rodents they are of direct concern to human health; e.g. the dependence of plague on rodent cycles. . . . [T]he tsetse problem is broadly speaking an ecological one. The abundance and movements of game, tsetse-fly habits, the ecological peculiarities of different kinds of bush, the effects of burning, grazing and planting, soil-erosion, and afforestation and even bee keeping all enter the ecological nexus. . . . [Soil erosion] is immediately seen to be connected with tsetse research on one side, forestry on another, and native customs ("human ecology") on a third.[69]

Huxley's understanding and conceptualization of ecological and disease patterns was clearly far-reaching and inclusive. He positioned humans within nature and saw human, medical, and animal ecology as intimately related. Elton took a similar view and was equally concerned about breaking down some of the disciplinary and institutional boundaries that seemed to prevent important insights from coming to light. During his and Huxley's several-year correspondence about the Bureau of African Ecology, Elton had begun to investigate more carefully some of the research on trypanosomiasis in the Lake Victoria region of Kenya, Uganda, and Tanganyika. In a letter to Huxley in 1932, he wrote, "In my recent delvings on the subject of Africa I have been astonished at the lack of coordination of the information already in existence." The work he examined, including that of Carpenter, he said, was "all in itself first-rate" and the researchers had "their organizations, bureaus, etc.," but this still seemed insufficient. His primary interest was to correlate tsetse fly populations with climatic cycles and epidemic outbreaks of the disease. The evidence he reviewed only confirmed his concerns about plans to develop the region economically:

[I]t follows that there is a [cyclical] fluctuation in potential sleeping sickness conditions and that any plans . . . to control the lake waters [for development] will have to reckon with this fact. We therefore have a) medical, b) meteorological c) engineering people, all more or less aware of each other's work, but unconscious of its significance. This seems to be the role of ecology in the next fifty years: super-coordination.[70]

The ecological perspective on disease advanced by Huxley and Elton, among others, directly challenged the idea that a vector-borne disease could be eradicated. Huxley might have referred to this attitude as an instance of "false optimism." The crux of the issue rested on distinctions between eradication and control. Imagining all of nature as a dynamic and interrelated "web of life" did not preclude the idea that one could understand, and in turn control, that web, but it did make matters of control infinitely more complicated.[71] The challenges posed in any effort to literally "eradicate" a species seemed in many respects overwhelming, particularly in the face of admitted ignorance regarding relationships among parasite strains, vectors, hosts, and their environments. Even when sufficient information was thought to exist, financial, institutional, and sociological factors were often presented as significant obstacles. As it was

[69] Julian Huxley, "Proposal for a Bureau of Ecology" (cit. n. 66).

[70] Charles Elton to Julian Huxley, 9 March 1932, Box 2, File 4, Joseph Oldham Papers, Rhodes House, Oxford.

[71] See, e.g., Huxley's rather confident assertion: "There is no reason why diseases like malaria and plague should not disappear from Africa as thoroughly as they have from northern Europe. . . . [T]he prime cause of malaria's diminution has always been and will continue to be the general raising of the human standard of life, which results in better houses, less contact with mosquitoes, more drainage, greater resistance, readiness to call in a doctor, and readiness to take more trouble about sanitary matters in general." Huxley, *Africa View* (cit. n. 67), 289–90.

increasingly recognized that vectors and disease patterns were highly localized, that very fact was being incorporated into the fabric of the new science of ecology.

There were two other challenges to the concept of eradication, stemming from developments in bacteriology and the emerging science of immunology. The first dealt with adaptations that vulnerable organisms, including humans, underwent to cause them to be partially or completely protected from infections. The second concerned changes produced in the strains of disease or the vectors themselves that consequently gave them unstable characteristics. *The Science of Life* dealt with both features at some length.[72] These developments had a direct bearing on the activities of researchers in the African medical services in the interwar period, many of whom adopted an extremely cautious approach to the idea of eradication. While this was not always the case in urban environments or with nonendemic diseases, it certainly was true with respect to malaria and trypanosomiasis, two widely endemic diseases with extremely complex vector-parasite-host relationships. In these instances, many researchers often advocated methods of control and careful experimentation, rather than either vector or disease eradication.[73] In this respect, the sciences of ecology and epidemiology came together quite literally in the field. As one medical officer doing research on sleeping sickness in Uganda expressed it:

> For the effective transmission of sleeping sickness in a community, a minimum breadth of contact between fly and man is necessary. Breadth of contact is essentially dependent on numerous biological factors which vary not only in different localities, but at different times in the same locality. The most essential of these factors are human and fly density, the domestic and economic activities of the human population, and fly activity in relation to its human host. . . . These factors are closely interwoven with each other, and are to a certain extent, mutually interdependent. . . . Their sum total in a given area results in what may be termed an epidemic potential the degree of which determines the nature and extent of an outbreak of sleeping sickness should a strain of trypanosome of a certain inherent transmissibility be introduced. . . . This, a knowledge of local conditions, is a most important point in the epidemiology of the disease, and one to which every consideration should be given in estimating the probability and extent of an outbreak occurring in any given time.[74]

Thus issues of endemicity, social organization, population densities (of both vectors and humans), habitat changes, and individual susceptibilities were increasingly brought into a single analytical framework.[75] This is not to say that control efforts themselves reflected this level of integration; such a goal would have been in many respects impossible, in part because of the sheer regional heterogeneity. But it is to stress that the intellectual milieu of the period, which had been facilitated largely by research coordinating

---

[72] Wells, Huxley, and Wells, *Science of Life* (cit. n. 57), 3:694–5.

[73] The Tsetse Research Department in Tanganyika offers an excellent example of this. See Swynnerton, "Tsetse Flies of East Africa" (cit. n. 31), 28–30.

[74] R. E. Barrett, "Notes on the Epidemiology of Sleeping Sickness with Special Reference to Conditions in the West Nile District of Uganda," *East African Medical Journal* 11 (1934–1935): 20–28, on 20. Barrett cites Swynnerton and Carpenter among his primary references. See also P. Granville Edge, "The Incidence and Distribution of Human Trypanosomiasis in British Tropical Africa," *Tropical Diseases Bulletin* 35 (1938): 3–18. Edge, for instance, writes of the "curiously localized incidence" of the disease; his article is a summary of medical department reports for 1936.

[75] Charles Wilcocks, J. F. Corson, and R. L. Sheppard, *A Survey of Recent Work on Trypanosomiasis and Tsetse Flies—Based on Reports and Papers Published During the Period 1932–1944* (London, 1946), especially 54–69; and Colonial Office, *A Note on Some of the Scientific Studies Undertaken by Members of the Colonial Medical Service during the Period 1930–47* (London, 1949).

conferences and concomitant training programs that stressed interdisciplinary team-work and local specificity, enabled a more comprehensive analysis to emerge.

CONCLUSION

It comes, then, as little surprise that the mid-1930s marked the moment when "the importance of ecology in relation to disease" began to find a wider audience.[76] It is also fitting that this occurred in the context of vector-borne and infectious diseases often relating to the fields of tropical medicine and epidemiology. By using an ecological framework to synthesize specialties as well as ideas, practitioners in this period were simultaneously reconfiguring disciplinary boundaries. "Epidemiology," declared Sheldon Dudley during his presidential address to the Royal Medical Society's Epidemiology Section in 1936, "has to draw on the data and use the methods of many other sciences, and the epidemiologist has to try to keep all the factors involved in the origin and distribution of epidemic and endemic disease in view at the same time, and take care that each factor is seen in proper perspective. In fact scientific epidemiology is medical ecology."[77] The extent to which research on sleeping sickness and nagana—diseases limited to the African continent—played a role in catalyzing these discussions helps highlight the transnational and geographically interdependent nature of knowledge production. Behind such fluid dynamics, however, also lurked the concrete, messy, and politically charged terrain of imperial affairs.

A careful and detailed historical study of research on African trypanosomiasis awaits an author. This cursory overview says little, for instance, of the disputes among scientists, of the impact of their work in specific territories, or of the wider colonial context in which they operated. Nor does it explain how ecological methods could become increasingly pervasive and yet have, at times, a seemingly negligible impact on wider medical treatment and disease control programs. What it does demonstrate, however, is that alternative understandings of disease causation, based on analyses of populations and their interactions with their environments, emerged in part from investigations of the disease patterns of British Africa. John Phillips, describing his experience studying the "complex biological problem of tsetse-fly and its control" in Tanganyika between 1927 and 1931, made a similar point: "Habitat, vegetation, game and other forms of animal life and the vector of the trypanosome responsible for human and animal trypanosomiasis . . . provided a series of relationships requiring analysis, interpretation and the synoptic view of the ecologist."[78] This new approach, bridging tropical medicine, epidemiology, and ecological science, allowed those interested in the study of the "ecology of diseases" to embark, albeit unevenly and sporadically, on a course of research that has only in the past few decades genuinely flourished.[79]

[76] This phrase comes from Strong, "Ecology in Relation to Disease" (cit. n. 22); Strong's discussion of sleeping sickness is on 314–7.

[77] Dudley, "Ecological Outlook on Epidemiology" (cit. n. 22), 57–8.

[78] John Phillips, "A Tribute to Frederic E. Clements and His Concepts in Ecology," *Ecology* 35 (1954): 114–5, on 114.

[79] For examples of such recent research as well as historical commentary, see Roy Anderson and Robert May, *Infectious Diseases of Humans: Dynamics and Control* (London, 1991); Mervyn Susser, "Does Risk Factor Epidemiology Put Epidemiology at Risk? Peering into the Future," *Journal of Epidemiology and Community Health* 52 (1998): 608–11; and the five responses to Susser's article on 612–8. See also Eli Chernin, "The Early British and American Journals of Tropical Medicine and Hygiene: An Informal Survey," *Medical History* 36 (1992): 70–83.

Natural Histories of Infectious Disease:
Ecological Vision in Twentieth-Century Biomedical Science

By Warwick Anderson*

ABSTRACT

During the twentieth century, disease ecology emerged as a distinct disciplinary network within infectious diseases research. The key figures were Theobald Smith, F. Macfarlane Burnet, René Dubos, and Frank Fenner. They all drew on Darwinian evolutionism to fashion an integrative (but rarely holistic) understanding of disease processes, distinguishing themselves from reductionist "chemists" and mere "microbe hunters." They sought a more complex, biologically informed epidemiology. Their emphasis on competition and mutualism in the animated environment differed from the physical determinism that prevailed in much medical geography and environmental health research. Disease ecology derived in part from studies of the interaction of organisms—micro and macro—in tropical medicine, veterinary pathology, and immunology. It developed in postcolonial settler societies. Once a minority interest, disease ecology has attracted more attention since the 1980s for its explanations of disease emergence, antibiotic resistance, bioterrorism, and the health impacts of climate change.

INTRODUCTION

The end of the twentieth century found Joshua Lederberg reflecting on the history of infectious diseases research. "During the early acme of microbe hunting, from about 1880 to 1940," he wrote, "microbes were all but ignored by mainstream biologists." Moreover, "medical microbiology had a life of its own, but it was almost totally divorced from general biological studies." Bacteriologists "had scarcely heard of the conceptual revolutions in genetic and evolutionary theory." Although germs had long been "recognized as living entities . . . the realization that they must inexorably be evolving and changing" was slow to penetrate public health and medical practice.[1] Yet Lederberg could identify a few conceptual bridges linking bacteriology and general

* Department of Medical History and Bioethics, University of Wisconsin Medical School, 1440 Medical Sciences Center, 1300 University Ave., Madison, WI 53706–1532; whanderson@med.wisc.edu.

I would like to thank Gregg Mitman, Michelle Murphy, and Chris Sellers for asking me to write this essay for the conference on "Environment, Place, and Health," at the University of Wisconsin at Madison. Susan Craddock provided helpful comments at the meeting. Barbara Gutmann Rosenkrantz and Mark Veitch have from the start guided my work on disease ecology. I am also grateful to David Abernathy, Nick King, Linda Nash, Charles Rosenberg, and Conevery Bolton Valenčius for their advice on earlier versions of this essay. Frank Fenner, Joshua Lederberg, Steve Boyden, and Tony McMichael cheerfully corrected some misconceptions, though they may not agree with all that remains.

[1] Joshua Lederberg, "Infectious History," *Science* 288 (2000): 287–93, on 288, 291.

biology (most of them, admittedly, constructed after 1940). Once O. T. Avery, at the Rockefeller Institute, discovered that nucleic acid was the transmissible factor responsible for pneumococcal transformation, bacteria and bacterial viruses had "quickly supplanted fruit flies as the test-bed for many of the subsequent developments of molecular genetics and biotechnology."[2] Indeed, in 1956 Lederberg himself had been awarded a Nobel Prize for his work in microbial genetics. He noted, too, that the synoptic texts of Macfarlane Burnet and René Dubos had described the nexus of microbiology and general biology; their pioneering integrative work characterized the relations of disease, environment, and evolutionary processes in newly fashionable ecological terms.[3]

Increasing confidence in antibiotic and vaccine development during the 1960s and 1970s, however, led to the neglect of such ecological interpretations of infectious disease. In the 1980s, nature struck back. Emergent diseases, such as AIDS, and problems of microbial resistance to antibiotics, prompted "widespread re-examination of our cohabitation with microbes." It was time, Lederberg wrote in the year 2000, for us to abandon the old metaphor of a war between germs and humans, replacing it with "a more ecologically informed metaphor, which includes the germ's-eye view of infection." Above all, he concluded, we need more "research into the microbial ecology of our own bodies."[4]

Historians generally have neglected ecological traditions in biomedical science. Like most of the scientists and physicians they study, historians have chosen instead to emphasize the development during the twentieth century of simplified laboratory models for complex pathophysiological mechanisms. Moreover, diagnosis and prevention are commonly framed in terms of "microbe hunting," and treatment in terms of "magic bullets." Sometimes the story will conclude with a monitory account of the pitfalls of progress: technically biomedicine is doing better than ever, but at the cost of its interpretive or exegetical power, its ability to define and represent our place in nature. Medical science, in these accounts, has concentrated on elucidating mechanisms of disease, abandoning the older efforts—frequently associated with the names of Hippocrates and Sydenham—to make sense of life forms and their relations to the environment. It is easy to find examples of such cautionary tales, with their typical mixture of satisfaction with contemporary achievement and nostalgia for a more integrative, or holistic, worldview.[5] When Charles-Edward Amory Winslow came to write his history of the "conquest" of epidemic disease, he admitted that "the practical triumphs of bacteriology did indeed tend to over-simplify the problem and to cause medical men for nearly half a century to ignore the true many-sidedness of disease."[6] Following a similar line of reasoning, Mirko Grmek has

---

[2] Ibid., 288. See René J. Dubos. *The Professor, The Institute, and DNA* (New York, 1976).

[3] F. M. Burnet, *Biological Aspects of Infectious Disease* (New York, 1940); and René J. Dubos, *Man Adapting* (New Haven, Conn., 1965). See also Joshua Lederberg, "J. B. S. Haldane (1949) on Infectious Disease and Evolution," *Genetics* 153 (1999): 1–3.

[4] Lederberg, "Infectious History" (cit. n. 1), 289, 290, 293. Even in 1988, Lederberg, citing Burnet, had urged us "to come to grips with the realities of our place in nature" ("Medical Science, Infectious Disease, and the Unity of Humankind," *Journal of the American Medical Association* 260 [1988]: 684–5, on 684).

[5] For my own contribution to this literature, see Warwick Anderson, "Disease and Its Meanings," *Lancet 2000* 354 (1999): SIV49.

[6] C.-E. A. Winslow, *The Conquest of Epidemic Disease: A Chapter in the History of Ideas* (1943; reprinted Madison, Wis., 1980), 335.

pointed out that since Pasteur and Koch, "much more importance is attached to investigations of the biology of germs than to knowledge of the influence of milieu."[7] Recently, Michael Worboys has argued that even if germ theories did not lead physicians simply to switch from holism to reductionism, interest in social and environmental influences gradually declined, and increasingly "disease was constituted in relations between bacteria and individual bodies."[8] The pace may be slower, but the destination is the same: medical science eventually becomes triumphantly, yet perhaps meretriciously, reductionist. Anyone would think that modern biomedicine is just a matter of culturing germs in the laboratory, identifying their physicochemical properties, and tracking them in the community—that is, little more than microbe hunting.

In this essay, however, I want to review the history of infectious diseases research in the twentieth century to recover various emerging forms of ecological understanding from what has sometimes seemed an arid waste of reductionism. In particular, I would like to sketch out personal connections between the major advocates of the ecology—or "natural history"—of disease, to describe an international research network, and to explore the various institutional and social niches these scientists occupied. Reconstructing this social ecology of ecological knowledge in medical science is no easy task.[9] For one thing, each of the major pioneers of a broader biological conception of disease processes tended to represent himself as singular, as the sole author of the idea, and rarely cited others, even those linked to him by education and friendship. For many early proponents, this rhetoric of singularity and marginality was a crucial aspect of the argument and part of their own prophetic self-fashioning. Yet the training and the career paths of key figures such as Theobald Smith, F. Macfarlane Burnet, and René Dubos structured an intricate network of influence, counsel, and criticism. This is not to say that a uniform, well-defined school of disease ecology existed in the twentieth century. The differences between many of these scientists can at times be as great as any similarity: Dubos, for example, was by the end of his career a significant outlier from this cluster of scholars. But an emphasis on singularity, and on variation in laboratory work, or in argument and appeal, often disguised a fundamental similarity in approach, a family resemblance in theory, perspective, and career. These scientists had common points of reference and a shared rhetorical recourse to ecological and other broadly integrative approaches to understanding disease. That is, the intellectual interactions of these and other scientists forged a recognizable subdiscipline, a shared conceptual framework and rhetoric, within infectious diseases research, even if their own local commitments and laboratory style sometimes differed.

It is important to distinguish this assertion of an "environmental" perspective in medical science from an earlier concern with medical geography, the role of the physical

[7] Mirko Grmek, "Géographie médicale et histoire des civilisations," *Annales: Economies, Sociétés, Civilisations* 18 (1963): 1071–89, on 1085. For similar statements, see Fielding H. Garrison, "The Newer Epidemiology," *Military Surgeon* 53 (1923): 1–14, 10; and Erwin H. Ackerknecht, *History and Geography of the Most Important Diseases* (New York, 1965), 1.

[8] Michael Worboys, *Spreading Germs: Disease Theories and Medical Practice in Britain, 1865–1900* (Cambridge, 2000), 285.

[9] Charles E. Rosenberg, "Toward an Ecology of Knowledge: On Discipline, Context, and History," in *The Organization of Knowledge in Modern America, 1860–1920*, ed. Alexandra Oleson and John Voss (Baltimore, 1979), 440–55.

milieu in causing disease.[10] The work of Burnet, in particular, was structured more around a biologically mediated environment and derived, in part, from the parasitological tradition in tropical medicine and veterinary pathology, which was preadapted to ecological explanations of this sort, not from older Hippocratic notions of direct environmental determinism. Unlike most medical geography—which persisted in the twentieth century in medical history and in geography itself—disease ecology postulated an evolutionary time scale, models that were integrative and interactive, and a global scope. Moreover, disease ecology was less explicitly racial in its arguments, though a concern with population quality undoubtedly persisted. In general, the spatial imaginary of disease ecology was more abstract and biologically animated than medical geography, and the processes it described usually were visible only to experts, not readily discerned or experienced by the general public. That is, the fine pattern of microbial interaction was generally less evident than a change in season or a shift in the wind direction. But the ecological understanding of the global as a site of infectious disease emergence could nonetheless be compelling: it was, after all, Dubos who coined the slogans "Only One Earth," and "Think Globally, Act Locally." Furthermore, it would be disease ecology that provided the most plausible explanation for the emergence of "new" diseases in the 1980s.

The disease ecologists whose careers I trace here were not, of course, the only ones seeking to reinterpret the relations of health and environment in the twentieth century. There were researchers, for example, in toxicology and occupation medicine, who, as Christopher Sellers has pointed out, transformed those fields into "environmental health" after World War II, a metamorphosis most forcefully expressed, perhaps, in studies of the etiology of cancer and represented popularly in Rachel Carson's *Silent Spring*.[11] Although often not explicitly "ecological"—the term is used only twice in *Silent Spring*—studies of the effect of the physical environment on human health could, at times, draw extensively on work in animal ecology, and the emphasis on the balance of nature and the dark side of modernity conveyed a broadly ecological tone.[12] It is remarkable, then, just how rarely the proponents of an ecology of infectious diseases and the experts on the health impact of the physical environment refer to one another. They may have shared intellectual interests, even political concerns, but their social networks, career trajectories, and institutional niches seem very different. Such basic similarity allied with self-styled difference attests to both the broad salience and the interpretive flexibility of notions of "environment" and "ecology" in twentieth-century explanations of health and disease.

The relationship of disease ecology to medical holism, or constitutional medicine, and to theories of the pathogenic potential of civilization is just as ambiguous. Both Burnet and Dubos severely criticized what they saw as an exclusively reductionist

---

[10] See Grmek, "Géographie médicale" (cit. n. 7); Caroline Hannaway, "Environment and Miasmata," in *Companion Encyclopedia of the History of Medicine,* ed. W. F. Bynum and Roy Porter (London, 1993), 292–308; Conevery Bolton Valenčius, "Histories of Medical Geography," in *Medical Geography in Historical Perspective,* ed. Nicolaas A. Rupke (London, 2000), 3–29; and Frank A. Barrett, *Disease and Geography: The History of an Idea* (Toronto, 2000).

[11] Christopher C. Sellers, *Hazards of the Job: From Industrial Disease to Environmental Health Science* (Chapel Hill, N. C., 1997).

[12] I am grateful to Gregg Mitman and Chris Sellers for insisting on this. Carson observes that disease is "a problem of ecology, of interrelationships, of interdependence" (*Silent Spring* [1963; reprinted Harmondsworth, 1974], 169). But as Sellers suggests, "[E]ven as [Carson's] own ecological habits of mind proved crucial to her synthesis, not ecology itself but the sciences of human health supplied the core of her argument" (*Hazards of the Job* [cit. n. 11], 2).

trend in modern biomedicine, but they did not display much interest in holism at the level of the human organism, and their ecological models always contained some key reductionist elements.[13] Rather, they distrusted unnaturally narrow and overly simplified accounts of any sort, especially those revealing a preoccupation with physicochemical mechanism, promoting instead a more inclusive and integrated understanding of nature and society. Their annoyance with reductionism and dismay at the emerging character of modernity were not unrelated.[14] Certainly Dubos became a famous critic of the dangers of industrial capitalism and environmentally insensitive modernity. By the end of his career, condemnation of a destructive civilization, and its associated alienated rationalism, dominated his writing, and his arguments came to assume a more traditionally holistic, humanistic, and even mystical, cast. Burnet, too, displayed ambivalence toward modernity, pointing to the dangers of overpopulation, biological warfare, antibiotic resistance, and environmental degradation, all, in his opinion, the fruits of a narrowly reductionist—even obscurantist—worldview.

As Charles E. Rosenberg has suggested, anxieties about the risk of modern ways of life were, during the course of the twentieth century, "explained increasingly in terms not of the city as a pathogenic environment, but of evolutionary and global ecological realities."[15] It is the purpose of this essay to explain how one part of this complex explanatory framework was assembled and eventually made popular.

THE NEWER EPIDEMIOLOGY AND HOST-PARASITE INTERACTIONS

Writing in 1940, Iago Galdston, a New York psychiatrist and historian of medicine, remarked on the crude reductionism of most "post-Pasteurian" medicine.[16] Epidemiology, the statistical study of the character and cause of disease, had once provided a richly textured, multifactorial accounting of the environmental concomitants, and the social and moral circumstances, of epidemics that ravaged nineteenth-century Europe and North America. Since the demise of medical geography and the general acceptance of germ theories, however, an intricate and exacting calculus had dwindled into mere microbe hunting. This lamentation had become commonplace and, by this time, was perhaps more a necessary ritual than a realistic description. It was certainly convenient for reformers to represent the status quo as a collection of narrow-minded

---

[13] More generally, on holism in the twentieth century, see Charles E. Rosenberg, "Holism in Twentieth-Century Medicine," in *Greater than the Parts: Holism in Biomedicine, 1920–1950,* ed. Christopher Lawrence and George Weisz (New York, 1998), 335–55.

[14] Nor are they unrelated to the concerns that prompted social medicine, though the styles of analysis, and policy implications, were often quite different. Yet in his earlier work, John A. Ryle could claim that "there are better inspirations to thoughtful medicine to be found in the *Origin of Species* than in a modern textbook of bacteriology" (*The Natural History of Disease* [1936; reprinted London, 1988], 382). See also Iago Galdston, *The Meaning of Social Medicine* (Cambridge, Mass., 1954); George Rosen, "What Is Social Medicine? A Genetic Analysis of the Concept," *Bulletin of the History of Medicine* 21 (1947): 674–733; and Dorothy Porter, "Changing Disciplines: John Ryle and the Making of Social Medicine in Twentieth-Century Britain," *History of Science* 30 (1992): 119–47.

[15] Charles E. Rosenberg, "Pathologies of Progress: The Idea of Civilization as Risk," *Bull. Hist. Med.* 72 (1998): 714–30, on 723. Rosenberg describes "a more expansive global perspective in which inclusive and ecological styles of analysis have become increasingly pervasive" (726), citing Burnet. For a helpful discussion of the epidemiological tropes of "contamination" and "configuration" (i.e., a more ecological understanding of epidemics), see Rosenberg, "Explaining Epidemics," in *Explaining Epidemics, and Other Studies in the History of Medicine* (Cambridge, 1992), 293–304.

[16] Iago Galdston, *Progress in Medicine: A Critical Review of the Last One Hundred Years* (New York, 1940).

technicians, to represent the old guard as cloistered laboratory scientists dutifully cultivating their germs or as public health officials glibly documenting disease outbreaks, tracing germs through the community. The benefits of complexity and integrative analysis surely would stand out most clearly against such alleged routinism and artlessness.

But those who studied tropical medicine and veterinary pathology—and, indeed, most field epidemiologists—understood that tracking the microbial cause of a disease outbreak rarely proved as easy as one might hope. Other biological factors had to be taken into account. The development of the notion of "carrier status" during the 1890s—the idea that susceptibility to germs might vary in a human population, allowing some individuals to spread a germ without succumbing to it—had already presented a challenge to more simplistic efforts to track microbes.[17] The full complexity of host-parasite relations, however, was perhaps better revealed in studies of tropical medicine and veterinary pathology, in which various life forms were commonly involved in disease processes. Just before World War I, Ronald Ross had tried to get his colleagues in tropical medicine to accept a sophisticated dynamic-equilibrium model of disease transmission, but by then few of them took his work seriously.[18] It was instead Theobald Smith, a comparative pathologist at Harvard, who became the major advocate of the study of disease as a general biological problem.

Smith, the codiscoverer of the role of a tick in transmitting the parasite that causes Texas cattle fever, was fond of emphasizing the mutual dependence of host and microorganism, whether in health or disease. In an address at the 1904 St. Louis Universal Exposition, he had claimed that the "social and industrial movement of the human race is continually leading to disturbances of equilibrium in nature, one of whose direct or indirect manifestations is augmentation of disease."[19] Later, Smith would argue that the study of disease was no longer "in the hands of professional mystics" because "disease has become a biological problem. . . . [I]t is ranging itself among natural phenomena in our mind." He also positioned his comparative work in pathology and immunology, with its concern for dynamic biological processes, against the work of investigators trying to dig toward the "more fundamental concepts embodied in physics and chemistry."[20] Smith described health and disease as consequences of a struggle for existence between living things, predatory and parasitic. He reported on the life cycle of parasites, host-parasite conflict, symbiosis and mutualism, cell parasitism and phagocytosis, and variation and mutation among parasites. "Parasitism

---

[17] On the "healthy carrier," see Winslow, *Conquest of Epidemic Disease* (cit. n. 6); and Judith Walzer Leavitt, *Typhoid Mary: Captive to the Public's Health* (Boston, 1996).

[18] Ronald Ross, "Some Quantitative Studies in Epidemiology," *Nature* 87 (1911): 466–7. See also J. Andrew Mendelsohn, "From Eradication to Equilibrium: How Epidemics Became Complex after World War I," in Lawrence and Weisz, *Greater than the Parts* (cit. n. 13), 303–31. On the practice of disease ecology in African tropical medicine during this period, see Helen Tilley, "Ecologies of Complexity: Tropical Environments, African Trypanosomiasis, and the Science of Disease Control in British Colonial Africa, 1900–1940" (this volume).

[19] Theobald Smith, "Some Problems in the Life History of Pathogenic Mircoorganisms," *Science,* n.s., 22 (1904): 817–32, on 817. See also his "Parasitism as a Factor in Disease," *Science,* n.s., 54 (1921): 99–108. Smith became director of the Division of Animal Pathology of the Rockefeller Institute, in Princeton, New Jersey, in 1914. See Barbara Gutmann Rosenkrantz, "Theobald Smith," *American National Biography Online,* Feb. 2000, http://www.anb.org/articles/12/12–00861.html; and Hans Zinsser, "Biographical Memoir of Theobald Smith, 1859–1934," *Biographical Memoirs* (National Academy of Sciences) 17 (1936): 261–303.

[20] Theobald Smith, *Parasitism and Disease* (Princeton, 1934), viii, x.

may be regarded," Smith wrote in 1934, "not as a pathological manifestation, but as a normal condition having its roots in the interdependence of all living organisms." The prevailing modus vivendi was always temporary, always evolving: the "face of nature and of civilization is steadily changing and thereby changing the host-parasite relations."[21] Here was a naturalistic and evolutionary understanding of the interactions of organisms—micro and macro—in which human disease was decentered and environment, or milieu, became animated. Ultimately, "all that can be postulated is the universal struggle of living things to survive, and in this struggle the fundamental biological reactions gradually range themselves by natural selection under . . . categories of offence and defence." The environment that mattered most was alive, and any effect of climate or topography would be mediated through the interactions of organisms. Smith reflected on the difference between his dynamic modeling and the work of earlier medical geographers:

> Sanitarians looked to the variations in atmospheric moisture and temperature, the rise and fall of the water in subsoil, great fluctuations in temperatures, as favoring causes of epidemics. Today we are inclined to narrow them down to the human and animal world, their intercourse, migrations, the continual fluctuations in habits and modes of life, but especially in the increasing susceptibility of populations during the disease-free periods.[22]

If "the human race is in a rather delicate, unstable relation to its environment," as Smith contended, then a more complex epidemiology was required. Hans Zinsser, in his obituary of Smith, saw his friend and colleague struggling against a tendency "for bacteriological investigation to segregate isolated fractions of a problem for analysis, with frequent neglect of the correlation of results with the problem as a whole."[23] Complex, integrative statistical studies of disease were necessary.

The relation of illness to population mobilization and host resistance, evident during World War I and in the subsequent influenza pandemic, had already attuned many epidemiologists to such appeals for a more intricate and naturalistic accounting of disease patterns. Major Greenwood, a statistician at the Ministry of Health in Britain, and later the first professor of epidemiology at the London School of Hygiene and Tropical Medicine, claimed that in the nineteenth century William Farr and John Simon had recognized that "the conditions of human life—both, as we now say, environmental and eugenic—were quite as important to the epidemiologist as the materia morbi."[24] Since then, however, epidemiology had become oversimplified, too close to a mere addendum on bacteriology. Greenwood wanted more complex and evolutionary studies of disease in populations, not mere microbe hunting. Epidemics, he believed, were the consequence of a disturbance in biological equilibrium and thus not reducible to invasion stories.[25] Simon Flexner, director of the Rockefeller Institute and an admirer of Smith's work, also observed a reaction against a facile bacteriological epidemiology during the 1920s. He noted that "each generation receives its

[21] Ibid., 2, xi.
[22] Ibid., 162.
[23] Zinsser, "Theobald Smith" (cit. n. 19), 284.
[24] Major Greenwood, "The Epidemiological Point of View," *British Medical Journal* (hereafter *BMJ*) 2 (1919): 405–7. For more on Greenwood, see J. Rosser Matthews, *Quantification and the Quest for Medical Certainty* (Princeton, 1995), chap. 5; and Mendelsohn, "From Eradication to Equilibrium" (cit. n. 18).
[25] Major Greenwood, *Epidemics and Crowd Diseases: An Introduction to the Study of Epidemiology* (London, 1935).

particular impression of epidemic diseases" and that the postwar generation was making an effort to "define epidemiology in terms wider than those of the microbic incitants of disease alone."[26] According to Flexner, this would mean developing an experimental epidemiology in which the host-parasite interactions Smith had sketched out might be explored further.

Studies of the biological complexities of host-parasite interaction soon reached a more popular readership, which wanted to learn about the latest scientific theories on man's place in nature. In *Rats, Lice, and History,* Hans Zinsser recounted the "biography" of typhus fever, tracing the impact of the disease on the rise and fall of civilizations. Zinsser, a professor of bacteriology and immunology at Columbia and later at Harvard, was respected for his investigations of leucocyte function and the precipitin reaction, and he wrote extensively on allergy and the immune response. Having trained initially as a biologist—with E. B. Wilson at Columbia—Zinsser knew that "physicochemical analysis will never give the final clue to life processes" and that Theobald Smith's study of "complex systems" was required instead. "Infectious disease is one of the great tragedies of living things," Zinsser wrote, "the struggle for existence between different forms of life."[27] A similarly tragic biological view could be seen in Percy Ashburn's *The Ranks of Death,* a popular book on the contribution of disease to the European conquest of America. Influenced by Zinsser, Ashburn evoked European migration to the Western Hemisphere as "the greatest mobilization of disease, of its introduction to new and susceptible peoples, the most striking example of the influence of disease upon history, of which we can speak with any certainty." The native peoples of the Americas had little immunity to the diseases Europeans brought with them: the preexisting biological equilibrium was upset, with devastating human consequences. Ashburn, who had served in the Philippines, worried that whites might yet meet their match if exposed to the "yellow man," who was even more immunologically competent, the result of "the general dearth of sanitation in Asia, and the unrivaled opportunities for ingesting a neighbour's dung." Ashburn's understanding of the natural history of infectious disease led him to wonder if, in the coming centuries, it would be "machinery or science or immunity to disease that will most influence racial dominance?"[28]

Disease history, usually shorn of its racial links, would continue to appeal to biologically inclined historians, at first through the *Annales* school and then, in the 1970s, through American historians seeking a broad conceptual framework for ambitious surveys. In *The Columbian Exchange,* for example, Alfred W. Crosby Jr. expressed a desire to understand man above all as a "biological entity." Inspired by Ashburn and Zinsser, and citing Henry Sigerist and Galdston, Crosby described how European invaders had disrupted the ecological stability of the New World, spreading disease to vulnerable populations. He charted the transfer of plants, animals, and germs between Europe and the Americas, arguing that this "Columbian exchange has left us with not

[26] Simon Flexner, "Experimental Epidemiology," *Journal of Experimental Medicine* 36 (1922): 9–14, on 9, 11.

[27] Hans Zinsser, *Rats, Lice, and History: A Study in Biography* (1934; reprinted Boston, 1963), 16, 7. During World War I, Zinsser had been a member of the Red Cross Sanitary Commission in Serbia, organized to control typhus by Richard P. Strong, later professor of tropical medicine at Harvard. Strong was a minor contributor to the development of disease ecology at Harvard. See Tilley, "Ecologies of Complexity" (cit. n. 18).

[28] P. M. Ashburn, *The Ranks of Death: A Medical History of the Conquest of America* (New York, 1947), 5, 211, 212. Most of the book was finished before 1937.

a richer but a more impoverished genetic pool."[29] A few years later, William H. Mc-Neill also tried to apply the theories of Smith and Zinsser to the historical study of disease outbreaks. McNeill was concerned about human persistence in "tampering with complex ecological relationships." Since World War II, a new generation of disease ecologists, in particular Macfarlane Burnet and René Dubos, had been warning of the biological dangers of population growth, biological warfare, and environmental degradation. Like them, McNeill feared that "a sequence of sharp alterations and abrupt oscillations in existing balances between microparasitism and macroparasitism can therefore be expected in the near future as in the recent past."[30]

MACFARLANE BURNET AND
THE *NATURAL HISTORY OF INFECTIOUS DISEASE*

In Melbourne, Australia, during the 1930s, Macfarlane Burnet, a young virologist, was reading the works of Theobald Smith and Hans Zinsser with great interest. Growing up in rural Victoria, Burnet had taken delight in natural history, which would remain a source of pleasure for the rest of his life. In his youth, he became an avid collector of beetles and participated enthusiastically in the local nature study movement.[31] While a medical student at the University of Melbourne, he continued his beetling and further developed his biological inclinations under the tutelage of Wilfred E. Agar, the professor of zoology.

Agar, an expert on the role of chromosomes in heredity, had become increasingly preoccupied with wide-ranging biological speculation. Inspired by his reading of Alfred North Whitehead, the professor had fashioned himself as an opponent of physicochemical reductionism and an advocate of vitalism and holism. "Surely it is not scientific," he pleaded in 1926, "to insist that biologists must interpret all their phenomena in the same terms as those found sufficient in physics and chemistry."[32] Familiar with Gestalt psychology and new work in ecology, Agar transmitted his enthusiasms to many of his students, including Burnet. For Agar, as for Burnet, all organisms, even microbes, were "links in the causal process which is the course of nature itself."[33]

As a young doctor, beginning his research on animal viruses and bacteriophages, Burnet also read Julian Huxley's *Essays of a Biologist,* and all of *The Science of Life,* by H. G. Wells, Huxley, and G. P. Wells.[34] "In one way or another," Burnet later

---

[29] Alfred W. Crosby Jr., *The Columbian Exchange: Biological and Cultural Consequences of 1492* (Westport, Conn., 1972), xiii, 219.

[30] William H. McNeill, *Plagues and Peoples* (New York, 1976), 254, 257.

[31] Tom Griffiths, *Hunters and Collectors: The Antiquarian Imagination in Australia* (Melbourne, 1996).

[32] W. E. Agar, "Some Problems of Evolution and Genetics," Presidential Address, Section D.—Zoology," in *Report of the 17th Meeting of the Australasian Association for the Advancement of Science, Adelaide, 1924* (Adelaide, 1926), 347–58, on 358. Whitehead's arguments for creative force, interrelations, and process in nature influenced a number of ecologists in the 1930s: see his *Process and Reality: An Essay in Cosmology* (New York, 1929), 127–97.

[33] W. E. Agar, *A Contribution to the Theory of the Living Organism* (Melbourne, 1943). Agar repeatedly cites Raymond Pearl, D'A. W. Thompson, and C. M. Child. Agar's interest in population policy and eugenics also infected Burnet. See Agar, "Some Eugenic Aspects of Australian Population Problems," in *The Peopling of Australia,* ed. P. D. Phillips and G. L. Wood (Melbourne, 1928), 128–44.

[34] Julian Huxley, *Essays of a Biologist* (London, 1923); and H. G. Wells, Julian Huxley, and G. P. Wells, *The Science of Life,* 2 vols. (1929; reprinted New York, 1931). See F. M. Burnet, *Changing Patterns: An Atypical Autobiography* (Melbourne, 1968); F. J. Fenner, "Frank Macfarlane Burnet, 1899–1985," *Biographical Memoirs of Fellows of the Royal Society* 33 (1987): 100–62; and Christopher Sexton, *The Seeds of Time: The Life of Sir Macfarlane Burnet* (Melbourne, 1991). Burnet also recalled (*Changing Patterns,* 75–6) the influence of Sinclair Lewis, *Arrowsmith* (London, 1925).

recalled, Wells, Huxley, and Wells "brought me into biological science and probably gave that ecological slant to the study of human disease which I think is characteristic of what I have written."[35] H. G. Wells and his coauthors provided an extensive account of the new science of ecology, which they defined as "the observation of animals in their proper surroundings and of their normal interaction and ways of life, and what one may call field physiology."[36] They described the gradual transition on Krakatoa from devastation to the "richness of tropical climax vegetation," the importance of parasite chains and food chains, the life cycle of lemmings, the biological control of prickly pear in Australia, and the proliferation of introduced deer in New Zealand. According to *The Science of Life,* "unrestrained breeding, for man and animals alike, whether they are mice, lemmings, locusts, Italians, Hindoos, or Chinamen, is biologically a thoroughly evil thing." Burnet would take to heart explanations of the "struggle for Life," "the tangled web of interrelationships," and the dangers of overcrowding and human mobility.[37] Wells warned that

> freedom of intercourse and communication stimulates both trade and thought; but it gives disease-germs new facilities for rapid spreading. . . . Thus the growth of civilization has been marked by a trail of plagues, more explosive and more widespread than anything which primitive man can have experienced.[38]

Yet in its treatment of health and disease, *The Science of Life* mostly presented a conventional version of bacteriology. Burnet must have been disappointed, but the book would stimulate him to think more about the ecology of the viruses he was cultivating in the Walter and Eliza Hall Institute during the 1930s.

At the Hall Institute in Melbourne, Burnet completed his studies of the interactions of bacteria and bacteriophage, stimulated during an earlier visit to the Lister Institute in London, and examined the behavior of viruses, especially the influenza virus, in the chick embryo. In 1935, an opportunity to investigate psittacosis confirmed his growing interest in biological aspects of infectious disease. Burnet was able to demonstrate that asymptomatic infection with psittacosis was common among wild parrots, but when the birds were confined and stressed, the illness would become manifest.[39] Such latent, inapparent infection drew Burnet into a broader consideration of disease ecology and immunity. The small scale of the medical research enterprise in Australia would frequently compel the Hall Institute to take on a microbiological service role for government, much of which involved field studies of disease outbreaks such as psittacosis. As a consequence, Burnet, though a brilliant laboratory experimentalist, was never able to limit his research to bench work. His presence in a "peripheral" settler society—a place preoccupied with population problems and its fragile environment—and the occasional need to

---

[35] F. M. Burnet, "Life's Complexities: Misgivings about Models," *Australian Annals of Medicine* 4 (1969): 363–7, on 364. William B. Provine argues that "Huxley's discussion of evolution was the single most encompassing presentation of a neo-Darwinian viewpoint available in 1930," and he suggests that "the influence of *The Science of Life* on scientists and the educated public deserves careful study by historians" ("England," in *The Evolutionary Synthesis: Perspectives on the Unification of Biology,* ed. Ernst Mayr and William B. Provine [Cambridge, Mass., 1980], 332).

[36] Wells, Huxley, and Wells, *Science of Life* (cit. n. 34), 1:22.

[37] Ibid., 2:974, 1012, 1010, 1012.

[38] Ibid., 2:1016.

[39] F. M. Burnet, "Enzootic Psittacosis amongst Wild Australian Parrots," *Journal of Hygiene* 35 (1935): 412–20.

undertake field microbiology led him repeatedly toward an ecological perspective on disease.

In 1937, Burnet began to write *Biological Aspects of Infectious Disease,* a book designed for the nonspecialist reader. He wanted to demonstrate that "infectious disease can be thought of with profit along ecological lines as a struggle for existence between man and micro-organisms." That is, he would argue that there was a natural

> conflict between man and his parasites which, in a constant environment, would tend to result in a virtual equilibrium, in which both species would survive indefinitely. Man, however, lives in an environment constantly being changed by his own activities, and few of his diseases have attained such an equilibrium.[40]

Burnet regarded his work as a combination of epidemiology and immunology viewed from a wide biological perspective. Examining a number of common diseases, he sought to provide an evolutionary explanation of the relations between human populations and their parasites. Infectious disease was, he claimed, nothing more or less than "a manifestation of the interaction of living beings" in a changing environment. Processes such as migration, urbanization, and general population increase thus would lead to redistribution of old diseases and the emergence of "new" diseases. "Wars, internal and external, financial depressions and labour troubles," he wrote, "are all breeders of infectious disease, and the future of disease will depend on the essentially fortuitous circumstances which will let loose or withhold these calamities."[41] In conclusion, Burnet warned (perhaps presciently) of the dangers of biological warfare, the deployment of germs such as anthrax.

Invited to give the Edward K. Dunham Lectures at Harvard in 1944, Burnet distinguished his recent work from those disciplines "which deal essentially with the organism in isolation and which are chiefly concerned with its structure, chemical and morphological, and with its functioning as a single unit." Rather, he associated his own research interests with "the sciences whose object is to interpret and control the phenomena of living things as they are, in such environments as the presence of nature and the activities of mankind have allotted them." Burnet wanted to understand how organisms are "distributed in space and time" and the "long-term historical aspects of the interaction between organism and environment."[42] He had begun to view himself as a contributor to the new evolutionary synthesis that Julian Huxley described in the early 1940s. The synthesis was initially a combination of Mendelism, chromosome theory, mathematical studies of heredity in populations, and natural selectionism; a consequence of the revival of Darwinism in the 1930s, it brought experimental geneticists together with naturalists who studied evolution in the field.[43] In a sense, then, Burnet was extending the synthesis to microbiology and immunology.

---

[40] Burnet, *Biological Aspects of Infectious Disease* (cit. n. 3), 3, 23. See also idem, "Changes in Twenty-Five Years in Outlook on Infectious Disease," *The Medical Journal of Australia* (hereafter *MJA*) 2 (1939): 23–8; and idem, "Charles Mackay Lecture: Biological Approaches to Infectious Disease," *MJA* 2 (1941): 607–12.

[41] Burnet, *Biological Aspects of Infectious Disease* (cit. n. 3), 4, 307.

[42] F. M. Burnet, *Virus as Organism: Evolutionary and Ecological Aspects of Some Human Virus Diseases* (Cambridge, Mass., 1945), 3.

[43] Mayr and Provine, *Evolutionary Synthesis* (cit. n. 35). See Julian Huxley, *Evolution: The Modern Synthesis* (London, 1942); R. A. Fisher, *The Genetical Theory of Natural Selection* (1930; reprinted New York, 1958); and Theodosius Dobzhansky, *Genetics and the Origin of Species* (1937; reprinted New York, 1951).

Huxley recognized Burnet's contribution in the second edition of *Evolution: The Modern Synthesis,* though he emphasized mostly the research on microbial genetics and antibody production rather than the more speculative ecological modeling of disease processes.[44]

Burnet's antipathy to facile reductionism and his discomfort with industrial modernity became more pronounced in later editions of his survey of disease ecology, retitled the *Natural History of Infectious Disease.* In 1953, he lamented that "the older generation of bacteriologists who were mostly trained as medical men has now almost been replaced by workers trained primarily as biochemists." The new generation had little or no interest in ecological aspects of infectious disease. The advent of antibiotics had further narrowed the focus of medical research, and simplified the clinical encounter. In this and later editions, Burnet, an admirer of Malthus, amplified his warnings of an increase in population numbers, and a decline in population quality, as infectious disease became a less significant social factor. He was also more worried during the cold war that the great powers would resort to biological warfare. The prospect of germ warfare was, he wrote, "a strange and gloomy ending of this account of the natural history of infectious disease."[45]

In 1972, David O. White, the professor of microbiology at the University of Melbourne, joined Burnet as the author of the fourth edition of the *Natural History of Infectious Disease,* but the conceptual framework hardly altered. This last edition included some additional introductory chapters on susceptibility and resistance, the transmission of infection, evolution and the survival of host and parasite, antibiotics, and hospital infections and iatrogenic disease. The antireductionist theme continued, along with the disparagement of genetics and the new molecular biology. "The fascination with molecular biology and its implications at all levels of biology has persisted," Burnet and White wrote, "and amongst bacteriologists and virologists there has been a trend away from what may be called the ecological aspects of infectious disease." In the fourth edition, ecology was again defined as "the interaction of organisms with their environment and especially with other organisms of their own or different species in that environment."[46] More than ever before, the text was strewn with terms such as "climax state," "niche," "virgin-soil epidemics," and "ecosystem." Thus "every organism is itself a product of evolution and ecologically related to almost every other organism in the ecosystem." Urbanization, overcrowding, and human mobility were causing ever-greater disruption to such ecosystems. All the same, antibiotics seemed in the 1970s to have controlled infectious diseases, whose future looked dull. "There may be some wholly unexpected emergence of a new and dangerous infectious disease, but nothing of the sort has marked the last fifty years."[47] Burnet and White remained concerned, however, that nuclear or bacteriological warfare was imminent and that modern civilization was environmentally unsustainable. "There is unease everywhere amongst research microbiologists with a social con-

[44] Julian Huxley, *Evolution: The Modern Synthesis,* 2d ed. (London, 1963), xxx–xxxi, l.

[45] F. M. Burnet, *Natural History of Infectious Disease,* 2d ed. (Cambridge, 1953), 3, 351. The third edition, published in 1962, was little altered.

[46] F. M. Burnet and David O. White, *Natural History of Infectious Disease,* 4th ed. (Cambridge, 1972), 3, 4.

[47] Ibid., 138, 263. During the 1960s, Burnet also explained that "communicable disease has more than a human significance; it is also a manifestation of the way of life of another organism, the responsible pathogen" ("The Natural History of Infectious Disease," in *The Theory and Practice of Public Health,* 2d. ed., ed. W. Hobson [London, 1965], 121).

science," they noted. It seemed that "the advance of science is allowing the development of wholly unnatural modes of human domination and mass destruction."[48]

Burnet and White pointed to immunology as one of the few branches of "medical" research to embrace ecological approaches. "The medical aspects of immunity have moved away from the centre of scientific interest as in the last ten years immunology has begun to develop its full status as part of fundamental biology."[49] Apparently, the stimulus to this development was Burnet's clonal selection theory of antibody formation, and his explanation of immunological tolerance, which had earned him and Sir Peter Medawar the 1960 Nobel Prize for physiology or medicine. In the late 1950s, Burnet had encountered Niels Jerne's "selective" hypothesis, which stated that an antigen would combine with the best fit among the organism's diverse natural globulins and transport it to antibody-producing cells, ready to make multiple copies of the presented globulin. Burnet reconfigured this idea, arguing that it would make sense if cells produced a pattern of globulin for genetic reasons and the arrival of the corresponding antigenic determinant caused the proliferation of the appropriate globulin. The theory implied a genetic polymorphism of the lymphocytes, with a variety of surface receptors and coding for antibody globulins arising from somatic mutation or from some other process occurring in differentiation and development.[50] This "biological" concept replaced Linus Pauling's earlier "instructive" hypothesis, which suggested that the antigen acted directly as a template for antibody production—this idea had been far too "chemical" for Burnet to countenance it. The clonal selection theory would come to represent the culmination of Burnet's "ecological" thinking.

"In more senses than one I have always been a human biologist," Burnet declared in 1966, "and I know that to look at man as a product of evolution, as part of the whole living world, has given me some useful insight into my own characteristics and difficulties, and added a special intensity to my appreciation of the world as it is."[51] In Burnet's lexicon "biological," or "ecological," signaled attention to an integrative, Darwinian conceptual framework; a biological approach did not therefore imply the abandonment of reductionist methods so much as their integration into a dynamic and interactive model of the relations of organism and environment. Burnet generally used his interpretation of prewar ecology as a metaphoric resource rather than an analytic tool. After Burnet read Wells and Huxley and studied with Agar, his understanding of ecology had acquired a more or less permanent form. Despite the occasional reference to "ecosystem" in the fourth edition of the *Natural History of Infectious Disease*, he

---

[48] Burnet and White, *Natural History of Infectious Disease,* 4th ed. (cit. n. 46), 266, 267. Burnet in particular hoped that ethology might provide some insight into human aggression and suggest a means of channeling it into more constructive activities. He communicated extensively with Konrad Lorenz and frequently visited him.

[49] Burnet and White, *Natural History of Infectious Disease* (cit. n. 46), 31.

[50] F. M. Burnet, "A Modification of Jerne's Theory of Antibody Production Using the Concept of Clonal Selection," *Australian Journal of Science* 20 (1957): 67–9; and idem, *The Clonal Selection Theory of Acquired Immunity* (Nashville, 1959). For an earlier version of the theory, see F. M. Burnet and F. J. Fenner, *The Production of Antibodies,* 2d ed. (Melbourne, 1949). See also Joshua Lederberg, "The Ontogeny of the Clonal Selection Theory of Antibody Formation: Reflections on Darwin and Ehrlich," *Annals of the New York Academy of Sciences* 546 (1988): 175–87; Arthur Silverstein, *History of Immunology* (San Diego, 1989); Anne-Marie Moulin, *Le dernier langage de la médecine: Histoire de l'immunologie de Pasteur au Sida* (Paris, 1991); and Eileen Crust and A. I. Tauber, "Selfhood, Immunity, and the Biological Imagination: The Thought of Frank Macfarlane Burnet," *Biology and Philosophy* 15 (2000): 509–33.

[51] F. M. Burnet, *Biology and the Appreciation of Life.* Boyer Lectures 1966 (Sydney, 1966), 1.

appears to have disregarded the mathematical modeling and systems theory of post-war ecology.[52] Indeed, although many of the scholars who shaped twentieth-century ecology worked in Australia, there is no evidence that Burnet interacted with them.[53] For Burnet, the term "ecology" continued to suggest a general appreciation of natural complexity and evolutionary processes, the transcendence of mere physicochemical mechanism; he would use some older ecological insights, but he did not embark on systematic ecological research.

RENÉ DUBOS AND *MAN ADAPTING*

In 1943, René Dubos, the professor of comparative pathology at Harvard, had recommended Burnet as the next Dunham lecturer at the medical school. Dubos, though he had not met the Australian scientist, was interested in Burnet's views on disease ecology.[54] After doctoral studies in bacteriology at Rutgers with Selman Waksman, Dubos had investigated the properties of soil microorganisms, concentrating on the role of competition, and its biochemical mediators, in limiting populations.[55] Later, settled at the Rockefeller Institute, he extracted a bacteriostatic substance, gramicidin, from *Bacillus brevis,* and demonstrated its use in the treatment of external infections. Such investigations of soil microbiology and bacterial competition quickly attuned Dubos to an interactive and ecological vision of human health and disease.[56]

Surprisingly, Dubos even found support for his emerging ecological views from O. T. Avery, his mentor at the Rockefeller Institute, who otherwise preferred to conduct cloistered laboratory studies of immunochemistry and bacterial transformation. Dubos thought that Avery's work had little influence on medical practice; rather, he contributed to "the understanding of biological phenomena." "When Avery became interested in a biological phenomenon, he first observed it for the sheer fun of it, as a naturalist."[57] The study of microbiology should be more than a mere medical instrument; it should also help to illuminate a wide range of biological problems. The scientist should therefore acquire a broad biological perspective on disease processes. According to Dubos, his senior colleague was convinced that "the interplay between the life processes of the host and those of the parasite was the way of the future."[58]

---

[52] See, e.g., W. C. Allee et al., *Principles of Animal Ecology* (Philadelphia, 1949); and Eugene Odum, *Fundamentals of Ecology* (Philadelphia, 1953). For more on the history of ecology, see Gregg Mitman, *The State of Nature: Ecology, Community, and American Social Thought, 1900–1950* (Chicago, 1992).

[53] See A. J. Nicholson, "The Balance of Animal Populations," *Journal of Animal Ecology* 2 (1933): 551–98; and H. G. Andrewartha and L. C. Birch, *The Distribution and Abundance of Animals* (Chicago, 1954). The leading Australian ecologists are discussed in Sharon E. Kingsland, *Modeling Nature: Episodes in the History of Population Ecology* (Chicago, 1985), 116–23, 171–2; and Martin Mulligan and Stuart Hill, *Ecological Pioneers: A Social History of Australian Ecological Thought and Action* (Cambridge, 2001), chap. 7.

[54] Alexander N. Zabusky, "Ecological Odyssey: The Intellectual Development of René J. Dubos" (senior honors thesis, Harvard University, 1986). Dubos was at Harvard only between 1942 and 1944; otherwise he worked at the Rockefeller Institute.

[55] Bernard D. Davis describes Waksman, who was awarded a Nobel Prize for his discovery of streptomycin, as "primarily a natural historian of the soil, cataloguing the micro-organisms found there, and focusing on their taxonomy and their ecological effects" ("Two Perspectives: On René Dubos, and on Antibiotic Actions," *Perspectives in Biology and Medicine* 35 [1991]: 37–48, on 40).

[56] Barbara Gutmann Rosenkrantz, "René Jules Dubos," *American National Biography Online,* Feb. 2000, http://www.anb.org/articles/12/12–01795.html.

[57] Dubos, *The Professor, the Institute, and DNA* (cit. n. 2), 89, 113.

[58] Ibid., 100.

In the early 1940s, Dubos began to formulate a more complex biological account of host-parasite relationships, drawing certainly on Theobald Smith's previous work and probably on Burnet's recent monograph. In *The Bacterial Cell,* published in 1945, Dubos described

> the organization of those molecular groupings which, on account of their chemical activity, condition the behavior of the cell both as an independent functioning unit and in its relation to the environment. The ultimate understanding of the natural history of infectious diseases, and the rational development of methods for their control, depend upon this knowledge.[59]

It was an effort to place bacteria in an evolutionary time frame, report on bacterial variability and mutation, and indicate the biochemical manifestations of organismic competition. Dubos claimed that in the past, microbiologists had narrowly studied "bacteriological 'events' rather than bacteria themselves." These researchers were "almost uninfluenced by the doctrines and methods of classical biology." Dubos sought to remedy this deficiency. "It is necessary," he declared, "to abandon the anthropomorphic attitude which characterized earlier efforts; bacteria must be studied, not only in the effects which they have on practical human problems, but also for what they are and what they do as independent living organisms."[60]

Dubos, however, was gradually moving away from laboratory investigation and refashioning himself as a popular writer and commentator. In a series of books, he would argue for a more integrative social and biological understanding of human disease.[61] His dominant theme was that "the states of health and disease are the expressions of the success or failure of the organism in its efforts to respond adaptively to environmental challenges." Like Burnet, he believed "organismic and environmental biology" needed as much attention as "physicochemical biology." He wrote, "In comparison with the enormous effort devoted to the components of the body machine, living as a process has hardly been studied by scientific methods."[62] In books such as *Mirage of Health* and *Man Adapting,* Dubos sought to explore "the complex inter-relationships between man and his physicochemical and biological environment."[63] Many of his arguments would have been familiar to readers of Burnet's work: the "interplay" between organisms reaching an equilibrium, the balance between parasitism and predation, the determinants of bacterial virulence and host resistance, the impact of increasing population density, and the general evolution of microbial diseases. However, Dubos, unlike Burnet, would increasingly focus on more direct physical influences on human health, referring back to Hippocrates and medical geography and pointing to the dangers of environmental pollution.

[59] René J. Dubos, *The Bacterial Cell: In its Relation to Problems of Virulence, Immunity, and Chemotherapy* (Cambridge, Mass., 1945), 17. The book was based on a series of lectures delivered in 1944 at the Lowell Institute in Boston.

[60] Ibid., 339, 342. See also René J. Dubos, "Utilization of Selective Microbial Agents in the Study of Biological Problems," *Bulletin of the N.Y. Academy of Medicine* 17 (1941): 405–22; idem, "Microbiology," *Annual Review of Biochemistry* 11 (1942): 659–78; and idem, "Trends in the Study and Control of Infectious Diseases," *Proceedings of the American Philosophical Society* 88 (1944): 208–13.

[61] The first of these was René Dubos and Jean Dubos, *The White Plague: Tuberculosis, Man, and Society* (1952; reprinted New Brunswick, N.J., 1987), which describes the impact of poverty and war on the incidence of this "social disease."

[62] Dubos, *Man Adapting* (cit. n. 3), xvii, xix, xx, 333.

[63] Ibid., xxi. See also René J. Dubos, *Mirage of Health: Utopias, Progress, and Biological Change* (London, 1959).

Although some of Dubos's and Burnet's interests and approaches continued to overlap, their paths began to diverge as well, as Dubos emphasized a critique of capitalism and civilization. In his work, monitory comment often overcame ecological scruples. His message was becoming ever-more traditional:

> To a surprising extent, modern man has retained unaltered the bodily constitution, physiological responses, and emotional drives which he has inherited from his Paleolithic ancestors. Yet he lives in a mechanized, air-conditioned, and regimented world radically different from the one in which he evolved.[64]

Dubos's condemnation of "the pathology of urban and suburban life, of antiphysiological leisure in a mechanized, automated and crowded environment," resonated with the sensibility of the 1960s protest movements, and he soon became a prophet of the "counterculture."[65] He began to talk more about the moral cost of adaptation to degraded conditions. "All too often," he lamented, "the biological and social changes that enable mankind to overcome the threats posed by the modern world must be eventually paid for at a cruel price in terms of human values." Adaptability, therefore, was "often a passive acceptance of conditions which really are not desirable for mankind." This led Dubos to argue that "the biological view of adaptation is inadequate for human life"—a statement that Burnet was unlikely to utter.[66]

Dubos continued in much the same vein throughout the 1960s and the 1970s. In 1968, he warned, "[M]an will ultimately destroy himself if he thoughtlessly eliminates the organisms that constitute essential links in the complex and delicate web of life of which he is a part." Increasingly, he represented himself as an heir of the Hippocratic tradition: "today, as in Hippocrates' time, good medical care implies attention not only to the body but to the whole person and to his total environment."[67] Although he was now calling attention to his "holistic" understanding of disease, Dubos still argued that this general conceptual framework had to be informed by precise laboratory knowledge. He tried to resist, not always successfully, environmental determinism and to hold on to a more interactive, ecological model. "All natural phenomena," he wrote, "are the result of complex inter-relationships; all manifestations of human disease are the consequences of the interplay between body, mind, and environment."[68] But increasingly, he also wanted to condemn the damage industrial capitalism was doing to "human values," counterposing radical humanism to his ecological sensibility. "Medical problems posed by the environmental stimuli and insults of modern civilization have acquired a critical urgency," he asserted. Indeed, his main worry had become "the threat to mankind posed by technologies derived from modern physicochemical and biological sciences."[69]

FRANK FENNER AND THE ECOLOGY OF VIRUS ERADICATION

Frank Fenner, an Australian microbiologist, worked in Dubos's laboratory at the Rockefeller Institute in 1948 and 1949. There he studied the production of BCG, since

---

[64] Dubos, *Man Adapting* (cit. n. 3), xviii. See Rosenberg, "Civilization as Risk" (cit. n. 15).
[65] Dubos, *Man Adapting* (cit. n. 3), 252.
[66] Ibid., 275, 279.
[67] René J. Dubos, *Man, Medicine, and Environment* (New York, 1968), 9, 61.
[68] Ibid., 61.
[69] Ibid., 88, 111.

Dubos was thinking more generally about tuberculosis at the time.[70] Just after World War II, Fenner had gone to the Hall Institute in Melbourne to investigate the experimental epidemiology of infectious ectromelia of mice and to conduct research more generally on poxviruses. It was Burnet who suggested that Fenner might appreciate working with Dubos. Fenner was, in a sense, preadapted to this intellectual terrain. While studying medicine at the University of Adelaide, he attended the lectures of J. B. Cleland, the professor of pathology and later author of an essay on the naturalist in medicine, and C. Stanton Hicks, the professor of physiology and a pioneer ecologist.[71] As a medical student, Fenner even attached himself to the South Australian Museum as its first and last "craniologist"—a joke title, he later recalled. Working in an older tradition of the natural history of man, Fenner had used his skull measurements to challenge prevailing theories of Aboriginal homogeneity, postulating instead a hybrid origin. As there was no future in craniology, during the war he retrained as a microbiologist, completing a diploma of tropical medicine, and then treating malaria, scrub typhus, and dengue in Papua New Guinea. Throughout his subsequent career, Fenner would retain his interest in natural history and in the relations of humans and their environment, especially the animated environment.[72]

In 1949, Fenner took up the foundation chair of microbiology at the new Australian National University, though he continued for a few years to work at the Hall Institute on *Mycobacterium ulcerans* and its peculiar terrestrial relations. In the 1950s, as myxomatosis, a poxvirus, was introduced into the rabbit population of southern Australia, Fenner decided to investigate evolutionary change in the organism's virulence and in the resistance of rabbits to it. It was, he believed, "the best natural experiment on the co-evolution of viral virulence and host resistance available for a disease of vertebrates."[73] At first, 99 percent of rabbits died from infection with the myxoma virus. The virus, however, acted as a powerful selection pressure and the next generation of rabbits was far less susceptible. In this research project, Fenner collaborated with Francis Ratcliffe, a zoologist at the CSIRO. Ratcliffe had trained with Charles Elton and Julian Huxley in England before venturing to Australia where initially he studied the distribution of the fruit-eating "flying-fox" and later the problem of soil erosion in the outback.[74] Fenner's ecological orientation was confirmed and enriched through his association with Ratcliffe, who since the 1940s had been seeking biological means to control rabbit populations.[75]

[70] René J. Dubos, F. J. Fenner, and C. H. Pierce, "Properties of a Culture of BCG Grown in Liquid Media Containing Tween 80 and the Filtrate of Heated Serum," *American Review of Tuberculosis* 61 (1950): 66–76; and F. J. Fenner, "Bacteriological and Immunological Aspects of BCG Vaccination," *Advances in Tuberculosis Research* 4 (1951): 112–86.

[71] On the environmentalist and sometimes neo-Lamarckian cast of the Adelaide medical school, see Warwick Anderson, *The Cultivation of Whiteness: Science, Health, and Racial Destiny in Australia* (New York, 2003), chap. 7. See also J. B. Cleland, "The Naturalist in Medicine, with Particular Reference to Australia," *MJA* 1 (1950): 549–65; and C. Stanton Hicks, *Soil, Food and Life* (Melbourne, 1945). Fenner's father, Charles, a geographer who earned his living as superintendent of education in South Australia, was perhaps a more important influence.

[72] F. J. Fenner, "Nature, Nurture, and My Experience with Smallpox Eradication: A Career Influenced by Chance Events," *MJA* 171 (1999): 638–41. When he was at the Hall Institute, Fenner became the junior author, with Burnet, of the second edition of *The Production of Antibodies* (cit n. 50).

[73] Fenner, "Nature, Nurture and My Experience with Smallpox Eradication" (cit. n. 72), 639. See also idem, ed., *History of Microbiology in Australia* (Canberra, 1990).

[74] F. N. Ratcliffe, *Flying Fox and Drifting Sand: The Adventures of a Biologist in Australia* (London, 1938). Julian Huxley wrote the introduction to this book.

[75] F. J. Fenner and F. N. Ratcliffe, *Myxomatosis* (Cambridge, 1965).

During the 1960s, the writing of books and accumulating administrative duties took Fenner away from the laboratory bench.[76] *The Biology of Animal Viruses,* a major revision of Burnet's *Principles of Animal Virology,* was Fenner's first attempt to describe in detail the "broader biological principles of animal virology."[77] In a chapter on the transmission of viruses, he argued that "the epidemiology of infectious diseases is a branch of ecology concerned with the spread and survival of infectious agents in nature." He reported on routes of entry into the body and exit from it, the differences between single-host and multiple-host viruses, variations in immunity, the influence of weather conditions on disease manifestation, and the importance of experimental epidemiology (for which he cited Greenwood). "Animal viruses," Fenner concluded, "survive in nature as part of an ecosystem."[78] In a later part of the same chapter, he considered the evolution of virulence and host response. Focusing on myxomatosis, influenza, and dengue, he emphasized the processes of natural selection, adaptation, and genetic drift.

Even in 1968, Fenner was concerned that increasing human mobility and environmental alteration would lead to the emergence of new diseases. His warnings were amplified in the second edition of *The Biology of Animal Viruses,* published in 1974. "Man's disturbance of the biosphere, and some medical innovations," wrote Fenner and his coauthors, "have created 'new' viral diseases for man himself and for his domesticated animals." It was clear by then that "the great increase in the numbers of human beings, their crowding into ever larger cities, and the increasing communication by men between these cities all over the world, are tending to make the 'human' world into a single ecological unit."[79] Rare, or previously unknown, microbes might proliferate with rapidity in these circumstances. In an especially prescient passage—in view of the outbreak of bovine spongiform encephalopathy in Britain a decade later—Fenner and his colleagues suggested that "the great concentrations of livestock that characterize 'industrial agriculture' in some Western countries, and the widespread shipment and aggregation of large numbers of animals for fattening, provide greatly increased opportunities for the spread of 'rare' viruses through large numbers of animals and for the emergence of viral diseases of livestock."[80] In the 1970s, Burnet may have thought that the outlook for infectious disease was still reassuringly "dull," but Fenner was increasingly pessimistic.

The prospects for smallpox eradication, however, did seem to improve. Dubos's distaste for human intervention in the environment had led him to dismiss efforts to eradicate diseases; he urged, rather, their control, by which he meant a "proper, skilful handling of the ecological situation." He asserted in 1965: "Eradication of microbial disease is a will-o'-the-wisp; pursuing it leads into a morass of hazy biological concepts and half-truths." This vain pursuit encouraged "the illusion that man can

---

[76] In 1967 Fenner became dean of the John Curtin School of Medical Research at the Australian National University. From 1973 until 1979, he was director of the Centre for Resource and Environment Studies there.

[77] F. J. Fenner, *The Biology of Animal Viruses,* 2 vols. (New York, 1968), 2:v. See also F. M. Burnet, *Principles of Animal Virology,* 2d ed. (New York, 1960). Elsewhere, Fenner claimed that Burnet's *Biological Aspects of Infectious Disease* "shares with Theobald Smith's *Parasitism and Disease* the distinction of being the first attempt to apply ecological principles to infectious diseases" ("Brahma, Shiva, and Vishnu: Three Faces of Science," *Australian Annals of Medicine* 4 [1969]: 351–60, on 351).

[78] Fenner, *Biology of Animal Viruses* (cit. n. 77), 2:760, 758.

[79] F. J. Fenner, B. R. McAuslan, C. A. Mims et al., *The Biology of Animal Viruses,* 2d ed. (New York, 1974), 618, 635.

[80] Ibid., 635.

control his responses to stimuli and can make adjustments to new ways of life without having to pay for these adaptive changes."[81] Having studied biological efforts to eradicate rabbits, Fenner was not so sure that complete failure was inevitable, though he, too, could be doubtful. Because the germs of malaria and yellow fever had multiple hosts, Fenner believed any attempt to eradicate these diseases was "biologically unsound." He could, however, conceive of the biological possibility of eradicating smallpox, since the virus was monotypic and had no animal host, the disease was readily identified, immunity after infection was life long, and vaccination was relatively easy. In 1968, the year after the launch of the Intensified Smallpox Eradication Campaign, he still thought "technical difficulties" stood in the way of the complete eradication of smallpox, but by 1974 he was more hopeful.[82] His expertise in poxvirus research, as well as his understanding of the ecological dynamics of vaccination, had attracted the attention of D. A. Henderson, the chief of the Smallpox Eradication Unit of the World Health Organization, who invited Fenner to serve on the committee overseeing the eradication program. Later, as chairman of the Global Commission for the Certification of Smallpox Eradication, Fenner, in 1979, was able to confirm the disappearance of the disease since 1977.[83]

In 1968, S. V. Boyden, an immunologist who had worked with Dubos and then in Fenner's department at the ANU, organized a conference in Canberra on civilization's impact on human biology. The meeting brought Burnet, Fenner, and Dubos together again. Boyden, the head of a new "urban biology" group, had developed an interest in the social and biological causes of human maladjustment. It was not surprising that he joined Dubos in condemning the costs of cultural adaptation to an unnatural civilization. According to Boyden, "[W]hen the conditions of life of an animal population deviate from those to which it has become, through natural selection, genetically adapted, some signs of biological maladjustment are almost inevitable."[84] Predictably, Dubos asserted that "in applying the concept of adaptation to the human species, it is . . . necessary to use criteria different from those used in Darwinian population theory." Instead, he believed, humans needed to "decide what kinds of adaptation are compatible with the maintenance of desirable human values."[85] Burnet, in his introduction to the symposium, avoided such talk of "human values," focusing on the changing distribution of human genotypes secondary to civilization. His old enthusiasm for eugenics never far below the surface, Burnet worried that as populations increased their quality was declining; the "overbreeding" of American "Negroes" especially dismayed him.[86]

Fenner provided what was perhaps the most thoroughly "ecological," or least anthropocentric, of the papers. "From the point of view of infectious diseases," he wrote, "the most important features of man's cultural development are the size of the

[81] Dubos, *Man Adapting* (cit. n. 3), 381. See Socrates Litsios, "René Dubos and Fred L. Soper: Their Contrasting Views on Vector and Disease Eradication," *Perspect. Biol. Med.* 41 (1997): 138–49. Soper, with the Rockefeller Foundation, led the campaign to eradicate malaria in the 1950s and 1960s.

[82] Fenner, *Biology of Animal Viruses* (cit. n. 77), 2:784.

[83] F. J. Fenner, D. A. Henderson, I. Arita et al., *Smallpox and its Eradication* (Geneva, 1988).

[84] S. V. Boyden, introduction to *The Impact of Civilisation of the Biology of Man,* ed. S. V. Boyden (Toronto, 1970), xiii–xiv, on xiii.

[85] René J. Dubos, "The Biology of Civilisation—With Emphasis on Perinatal Influences," in Boyden, *Impact of Civilisation* (cit. n. 84), 219–29, on 220, 222.

[86] F. M. Burnet, "Human Biology as the Study of Human Differences," in Boyden, *Impact of Civilisation* (cit. n. 84), xv–xx.

individual communities of men, the number and proximity of such communities, and the extent of movement and interchange between them."[87] Fenner discussed changes in host-parasite interactions in malaria, salmonellosis, cholera, measles, smallpox, yellow fever, and poliomyelitis. New viral diseases had been "recognized relatively commonly during this century: most but not all of these have been due to human intervention in some natural situation, or to changes in the social habits of man." Urbanization, human colonization, and air travel seemed especially to promote the spread of disease. In Fenner's opinion, the best solution to the problem of disease was "the eradication of the sources of infection by the elimination of poverty."[88] In discussing Fenner's paper, S. D. Rubbo, the professor of microbiology at Melbourne, warned of selection for antibiotic resistance among microbes. "The appearance of these new strains of bacteria in the affluent societies illustrates an unusual effect of the changing social organization on the infectious diseases of man."[89] In response, Fenner drew his audience's attention back to the emergence of "new" diseases. "I am sure we will witness the appearance of antigenically novel viruses in the next fifty years, and I do not mean only influenza viruses; possibly we may also witness the appearance of viruses of novel pathogenic potential."[90]

CONCLUSION

While some other medical investigators in the twentieth century had turned to sociology to gain a wider perspective on health and disease, advocates of disease ecology drew on their knowledge of biological processes. For them, even the social might become a figment of biology. Many of these scientists had trained in tropical medicine, veterinary pathology, or agricultural science and were thus preadapted to seeing the broader biological determinants of apparently specific or idiosyncratic pathological events.[91] Commonly, they had acquired in youth an enthusiasm for natural history, which continued to inflect their medical studies; and the problems they investigated— Texas cattle fever, psittacosis, soil microbiology, myxomatosis—tended to confirm their ecological inclinations. Resident in settler societies such as the United States and Australia, pioneer disease ecologists were especially attuned to the persisting impact of colonial development policies, to the lasting effects of agricultural change and human resettlement. Burnet, in particular, never lost his interest in the quality and quantity of populations, a eugenic preoccupation fostered by the intense debate in the 1920s over Australian population policy—or "carrying capacity."[92] Disease ecology

[87] F. J. Fenner, "The Effects of Changing Social Organisation on the Infectious Diseases of Man," in Boyden, *Impact of Civilisation* (cit. n. 84), 48–76, on 48.

[88] Ibid., on 63, 66.

[89] S. D. Rubbo, quoted in ibid., 69.

[90] Ibid., 76.

[91] Experience of tropical practice clearly influenced the ideas of Jacques M. May, a disease ecologist who mixed mostly with geographers but was also on the margins of the group I describe. In the 1940s and 1950s, with the support of the American Geographical Society, May tried to refashion medical geography as disease ecology. See his "Medical Geography: Its Methods and Objectives," *Geographical Review* 40 (1950): 9–41. On May's tropical career, see Jacques M. May, *Siam Doctor* (Garden City, N.Y., 1949).

[92] I discuss this debate in *The Cultivation of Whiteness* (cit. n. 71). Interestingly, Burnet seems to have followed a trajectory similar to that of Raymond Pearl. In Burnet's case, it led him away from the "Australia Unlimited" partisans of the Melbourne medical school, toward a concern with "overpopulation." On Pearl, see Garland E. Allen, "Old Wine in New Bottles: From Eugenics to Population Control in the Work of Raymond Pearl," in *The Expansion of American Biology,* ed. Keith R. Benson, Jane Maienschein, and Ronald Rainger (New Brunswick, N.J., 1991), 231–61.

thus emerged as a legacy of settler colonial anxieties.[93] For Burnet and others it would provide a cogent explanation of disease patterns in a world shaped by human mobility; it also allowed them to express their ambivalence toward the modernity colonialism had apparently made global.

In the 1980s, further changes in the natural and conceptual environments would provide a larger niche for ecological reasoning in medical science. Above all, the emergence of a global epidemic of HIV/AIDS shook those previously "nonecological" scientists who had become complacent about infectious disease in the developed world. Just as epidemiologists such as Flexner and Greenwood had called for a more complex understanding of disease transmission in the wake of the influenza pandemic of 1918–1919, many microbiologists now sought a more integrative explanation for the emergence of AIDS and other diseases, including legionnaires' disease, Lyme disease, dengue hemorrhagic fever, and new variant Creutzfeldt-Jacob disease. Of course, disease "emergence," usually a matter of old diseases conquering new territories, was hardly a novel occurrence, as Fenner could have told them. Since the late nineteenth century, many "new" diseases had spread widely, among them nontyphoid salmonellosis, poliomyelitis, kuru, and even the pandemic form of influenza. All of these emergent conditions had at the time stimulated speculation on their underlying biological or social causes, but AIDS would exert an even more profound impact on biomedical science, as it struck after a long period of complacency, when the future of infectious diseases, in Burnet's opinion, looked exceedingly dull in the developed world. In the developing world, however, there had never been much cause for contentment, for there the impact of economic development on microbial abundance and distribution was still demonstrated daily.[94] Ecological insight was rarely absent from tropical medicine; thus, in a sense, "mainstream" biomedical science was simply catching up, recognizing that disease even in Europe and North America might be the outcome of dynamic processes in a global ecosystem. The complacency shattered in the 1980s was not so much overconfidence in the global control of infectious disease, as the conventional assumption that Europe and North America had managed somehow to remove themselves from the natural processes that disease ecologists described.

[93] This echoes Richard Grove's argument for the emergence of a general environmental consciousness from concerns about colonial environmental degradation: see his *Green Imperialism: Colonial Expansion, Tropical Island Edens, and the Origins of Environmentalism, 1600–1860* (Cambridge, 1995). See also Libby Robin, "Ecology: A Science of Empire," in *Ecology and Empire: Environmental History of Settler Societies,* ed. Tom Griffiths and Libby Robin (Melbourne, 1997), 63–75; Thomas R. Dunlap, "Ecology and Environmentalism in the Anglo Settler Colonies," in Griffiths and Robin, *Ecology and Empire,* 76–86; John Mackenzie, "Empire and the Ecological Apocalypse: The Historiography of the Imperial Environment," in Griffiths and Robin, *Ecology and Empire,* 215–28; and Peder Anker, *Imperial Ecology: Environmental Order in the British Empire, 1895–1945* (Cambridge, Mass., 2001). By "British empire," Anker means Oxford and South Africa.

[94] Much of this work has focused on schistosomiasis and trypanosomiasis in Africa. See Charles C. Hughes and John M. Hunter, "Disease and 'Development' in Africa," *Social Science and Medicine* 3 (1970): 443–93; and Duncan Pedersen, "Disease Ecology at a Crossroads: Man-Made Environments, Human Rights, and Perpetual Development Utopias," *Social Science and Medicine* 43 (1996): 745–58. Helen Tilley has discussed the persisting ecological perspective of African tropical medicine in more detail in "Ecologies of Complexity" (cit. n. 18). See also Stephen J. Kunitz, *Disease and Social Diversity: The European Impact on the Health of Non-Europeans* (New York, 1994). In the 1960s and 1970s, medical anthropology frequently demonstrated an "eco-social" approach to understanding disease in the "Third World." See Marcia C. Inhorn and Peter J. Brown, *The Anthropology of Infectious Disease: International Health Perspectives* (Amsterdam, 1997).

Fenner's prediction in the 1960s of the emergence of "new" diseases in the developed world was fulfilled in the 1980s. Joshua Lederberg was among those who finally read the ecological lesson. A colleague of Dubos's at the Rockefeller University, Lederberg had worked with Burnet at the Hall Institute in the 1950s, so their work was familiar to him. In 1993, though, Lederberg complained that "the historiography of epidemic disease is one of the last refuges of the concept of special creationism"—pioneering ecological approaches had largely been ignored.[95] Now, however, many scientists were prepared to admit that evolutionary processes operating on a global scale were responsible for the emergence of "new" diseases. As environments changed, as urbanization, deforestation, and human mobility increased, so, too, did disease patterns alter, with natural selection promoting the proliferation of microbes in new niches. But pessimism should not overwhelm us. "We recall that since Frank Macfarlane Burnet, Theobald Smith, and others," wrote Lederberg, "we have understood that evolutionary equilibrium favors mutualistic rather than parasitic or unilaterally destructive interactions. Natural selection, in the long run, favors host resistance, on the one hand, and temperate virulence and immunogenic masking on the parasite's part on the other." All the same, Lederberg, echoing Dubos, remained concerned that too good a human adaptation to an increasingly degraded environment might yet be detrimental to human values. "In a biological sense," he mused, "we may achieve new genomic equilibria with these radically altered environments; but the price of natural selection is so high that I doubt we would find it ethically acceptable: it conflicts violently with the nominally infinite worth that we place on every individual."[96]

During the 1990s, amplified concern about emerging infectious diseases, along with fears of increasing antibiotic resistance and the health effects of climate change, would boost interest in disease ecology. Stephen S. Morse, a virologist and immunologist at the Rockefeller University, joined Lederberg in arguing that since "most 'new' or 'emerging' viruses are the result of changes in traffic patterns that give viruses new highways," we need "a science of traffic patterns, part biology and part social science."[97] Interest in the emergence of "new" diseases soon led to a proliferation of conferences and symposia; it gave rise to numerous reports and to popular books, such as Laurie Garrett's *The Coming Plague*.[98] The journals *Emerging Infectious Diseases* and *Ecosystem Health* were launched in the mid-1990s.[99] At the end of the century, clinicians and scientists were also coming to recognize antibiotic resistance as a growing problem, and

[95] Joshua Lederberg, "Viruses and Human Kind: Intracellular Symbiosis and Evolutionary Competition," in *Emerging Viruses,* ed. Stephen S. Morse (New York, 1993), 3–9, on 3. This book derived from a 1989 conference on emerging viruses supported by the Rockefeller University and the National Institutes of Health, and attended by D. A. Henderson, Frank Fenner, William H. McNeill, and Robert M. May. See also Joshua Lederberg, R. E. Shope, and S. C. Oaks Jr., eds., *Emerging Infections: Microbial Threats to Health in the United States* (Washington, D.C., 1992). Like Burnet a generation earlier, Lederberg was profoundly shaped by Wells and Huxley: "*The Science of Life* was the most influential source of my perspective on biology and man's place in the cosmos, seen as evolutionary drama" ("Genetic Recombination in Bacteria: A Discovery Account," in *The Excitement and Fascination of Science,* ed. Joshua Lederberg [Palo Alto, Calif., 1990], vol. 3, part 1, 893–915, on 895).

[96] Lederberg, "Viruses and Human Kind" (cit. n. 95), on 8, 4.

[97] Stephen S. Morse, "Emerging Viruses: Defining the Rules for Viral Traffic," *Perspect. Biol. Med.* 34 (1991): 387–409, on 388, 404. Morse thanks Frank Fenner for discussing this paper with him.

[98] Laurie Garrett, *The Coming Plague: Newly Emerging Diseases in a World Out of Balance* (London, 1994).

[99] See Nicolas B. King, "The Scale Politics of Emerging Diseases" (this volume). King relates fears of biological warfare—a problem that haunted Burnet even in 1940—to the growing interest in emerging diseases.

they attributed it to evolutionary processes. According to S. B. Levy, a microbiologist at Tufts, profligate antibiotic use had delivered a selection pressure on microbes "unprecedented in the history of evolution."[100] "We must somehow," he wrote, "find a means to reverse the ecological imbalance that has occurred in terms of resistant and susceptible strains."[101] Others saw us reaping the ecological whirlwind of climate change. Alterations in the abundance and distribution of microorganisms and their vectors might, to a large extent, mediate the influence of climate change and other physical transfigurations. Thus as mosquitoes extended their range so, too, would malaria, dengue and other supposedly "tropical" pathogens. In *Human Frontiers, Environments, and Disease,* A. J. McMichael, who occupied Greenwood's chair at the London School of Hygiene and Tropical Medicine, declared that "as human intervention in the global environment and its life processes intensifies, we need better understanding of the potential consequences of . . . ecological disruptions for health and disease."[102]

In form and function, the understanding of disease ecology emerging at the end of the twentieth century often departed quite radically from the antecedent theories, and metaphoric borrowings, of Smith, Burnet, and Dubos. However, there was still a common interest in mitigating facile reductionism in medical research, a desire to assemble a more complex and integrative explanatory framework for disease patterns. Proponents of an ecological perspective on infectious diseases sought a means to relate *micro*biological processes to larger environmental or biological forces, a way to describe the interactive, dynamic relationships between host and parasite and physical milieu. In so doing, they conventionally invoked an evolutionary time frame and a global compass. Indeed, it is hard to imagine the development of disease ecology without the concomitant economic and political globalization occurring during the twentieth century. Now, as before, the leading instigators of wide-ranging ecological approaches argue that their views are marginal in biomedical research. Yet they continue to occupy key positions at elite research institutions, and they continue to acquire Nobel Prizes at an enviable rate. Despite claims of marginality, speculation on disease ecology has often functioned to distinguish scientists from mere technicians or from the alleged reductionism and routinism of junior colleagues and competitors. But theoretical speculation of this sort, once little more than a mark of intellectual distinction, a flashy bit of plumage, would eventually become a major selection advantage in the rapidly changing, and perplexing, natural and conceptual environments of the late twentieth century.

---

[100] S. B. Levy, "Antibiotic Resistance: An Ecological Imbalance," in *Antibiotic Resistance: Origins, Evolution, Selection and Spread,* Ciba Foundation Symposium 207 (Chichester, U.K,, 1997), 1–14, on 2. See also idem, *The Antibiotic Paradox: How Miracle Drugs Are Destroying the Miracle* (New York, 1992). Dubos predicted selection for antibiotic resistance as early as 1942. See Carol L. Moberg, "René Dubos: A Harbinger of Microbial Resistance to Antibiotics," *Perspect. Biol. Med.* 42 (1999): 559–80.

[101] Levy, "Antibiotic Resistance" (cit. n. 100), 8.

[102] A. J. McMichael, *Human Frontiers, Environments, and Disease: Past Patterns, Uncertain Futures* (Cambridge, 2001), xiv. The WHO opened an office to study the health implications of climate change in 1990. It is also important to recognize the contribution since the late 1970s of mathematical modeling of host-parasite interactions and population dynamics. See R. M. May, "Ecology and the Evolution of Host-Virus Interactions," in Morse, *Emerging Viruses* (cit. n. 95), 58–68; R. M. Anderson and R. M. May, "Population Biology of Infectious Diseases," *Nature* 280 (1979): 361–7, 455–61; and R. M. May and R. M. Anderson, "The Transmission Dynamics of Human Immunodeficiency Virus," in *The Epidemiology and Ecology of Infectious Disease Agents,* ed. R. M. Anderson and J. M. Thresh (London, 1988), 239–81. May regarded Frank Fenner as the "real hero" of the biological analysis of disease because of his classic mathematical analysis of myxomatosis (May, "Ecology and the Evolution of Host-Virus Interactions," 63).

The Scale Politics of
Emerging Diseases

By Nicholas B. King*

ABSTRACT

The concept of scale politics offers historians a useful framework for analyzing the connections between environment and health. This essay examines the public health campaign around emerging diseases during the 1990s, particularly the ways in which different actors employed scale in geographic and political representations; how they configured cause, consequence, and intervention at different scales; and the moments at which they shifted between different scales in the presentation of their arguments. Biomedical scientists, the mass media, and public health and national security experts contributed to this campaign, exploiting Americans' ambivalence about globalization and the role of modernity in the production of new risks, framing them in terms that made particular interventions appear necessary, logical, or practical.

"DISEASE KNOWS NO BORDERS"

In 1994, international health expert Milton Roemer wrote, "[I]n the modern world, the claim that 'disease knows no borders' has become a cliché that no mature health leader repeats." Roemer described a new approach to international health:

> Towards the end of the twentieth century . . . inequities became a major concern of international health work—a far cry from the original narrow focus on border quarantine. . . . The goal today is now to assure within all countries—rich and poor, large or small—the full health benefits of modern science and civilization. "One world" should mean not merely to eliminate the need for border quarantine, but to endow each country with the resources and strategies to achieve maximum health for all its people.[1]

Roemer was optimistic about the likelihood of success for this new approach—with good reason. In 1974, an international convention at Alma Ata in the former

* Center for Social Epidemiology and Population Health, 1214 S. University, 2d Floor, Ann Arbor, MI 48104; nbking@umich.edu.

The author thanks Jennifer Fishman, Michelle Murphy, Gregg Mitman, and all of the participants in the "Environment, Health, and Place in Global Perspective" conference at the University of Wisconsin–Madison for their comments and guidance.

[1] Milton I. Roemer, "Internationalism in Medicine and Public Health," in *The History of Public Health and the Modern State,* ed. Dorothy Porter (Atlanta, Ga., 1994), 421. Roemer had a distinguished career in public health research and practice, serving in the New Jersey State Department of Health and briefly as chief of the Social and Occupational Health Section of the newly formed World Health Organization, and publishing the comprehensive survey *National Health Systems of the World.* Among other awards, he received the American Public Health Association's Sedgwick Memorial Medal for Distinguished Service in Public Health and the Lifetime Achievement Award of the association's International Health Section.

Soviet Union declared that primary health care and social and economic justice should be the foundations of international public health. This declaration and documents such as the Lalonde Report on Canadian health and the Ottawa Charter for Health Promotion have been cited as evidence of a "third public health revolution." Instead of investment in health care and disease prevention efforts aimed at border control and modification of individual behaviors, "health promotion" and "population health" approaches emphasized research into the upstream determinants of health, provision of primary health care and political empowerment, and a self-consciously global strategy of "capacity for building health" in developed and developing nations.[2]

Yet even as Roemer proclaimed the demise of the truism that "disease knows no borders," a public health campaign that readily embraced it was well underway. This campaign argued that recent political, economic, and technological changes were giving rise to "emerging diseases": newly discovered pathogens such as the Ebola and Hantaviruses, and newly resistant strains of bacteria such as *Mycobacterium tuberculosis* and *Streptococcus pneumoniae*. With increasing international commerce and travel, emerging diseases could be rapidly transmitted from one country to another, thus constituting a global threat that demanded immediate response. As journalist Laurie Garrett observed in 1996, under the subheading "Diseases without Borders," "[G]eographic sequestration was crucial in all postwar health planning, but diseases can no longer be expected to remain in their country or region of origin."[3]

In this essay, I will sketch a brief history of the emergence of this campaign, which, despite some similarities, represented a pragmatic alternative to Roemer's idealistic vision of "one world." I will pay close attention to the *scale politics* of this campaign, particularly the ways in which different actors used scale as a resource: how they employed scale in geographic and political representations; how they configured *cause, consequence,* and *intervention* at different scales; and the moments at which they shifted between different scales in the presentation of their arguments.

In focusing on scale politics, I will be following the work of geographers Erik Swyngedouw and Neil Smith, who argue that scale should not be regarded as an ontologically given geographic territory or a priori unit of analysis. Instead, it is the outcome of a historically contingent political process, in which actors construct *scalar narratives* that invoke places and spaces at different geographic scales to explain events, enlist allies, and attract attention and funding.[4] Both Roemer's vision and the truism that "disease knows no borders" are forms of scalar narratives.

Historians and sociologists of science have also drawn attention to this category. Bruno Latour has argued that manipulation of scale was an integral part of Louis Pasteur's ability to secure alliances among his contemporaries and to convince others that in order to solve their problems, they must first "pass through" his laboratory. Part of

---

[2] See the articles in the March 2003 issue (vol. 93) of the *American Journal of Public Health,* especially I. Kickbusch, "The Contribution of the World Health Organization to a New Public Health and Health Promotion" (383–8); and D. Kindig and G. L. Stoddart, "What is Population Health?" (380–3).

[3] Laurie Garrett, "The Return of Infectious Disease," *Foreign Affairs* 75 (Jan./Feb. 1996): 69.

[4] Erik Swyngedouw, "Neither Global nor Local: Glocalisation and the Politics of Scale," in *Spaces of Globalization: Reasserting the Power of the Local* (New York, 1996), 137–66; and Neil Smith, "Geography, Difference, and the Politics of Scale," in *Postmodernism and the Social Sciences,* ed. Joe Doherty, Elspeth Graham, and Mo Malek (New York, 1992). See also Andrew E. G. Jonas, "The Scale Politics of Spaciality," *Environment and Planning D: Society and Space* 12 (1994): 257–64.

the social power of science lies in its ability to work at a small scale and to convince others that this scale is simple, efficient, and effective.[5] In a review of the literature on "Big Science," James Capshew and Karen Rader have urged historians to avoid the romanticized "drama of scale," which contrasts "huge machines, large organizations, and massive expenditures found in some contemporary research projects with the stereotyped lone investigator of the past."[6] Instead, Capshew and Rader advocate a synchronic investigation of the epistemological, institutional, and political conditions that make changes in scale possible.

This essay briefly reviews some of these conditions and concludes with reflections on the importance of scale in historical analysis. The emerging diseases campaign employed a strategic and historically resonant scale politics, making it attractive to journalists, biomedical researchers, activists, politicians, and public health and national security experts. Campaigners' identification of causes and consequences at particular scales were a means of marketing risk to specific audiences and thereby securing alliances; their recommendations for intervention at particular scales were a means of ensuring that those alliances ultimately benefited specific interests. Through their scalar narratives, campaigners exploited Americans' ambivalence about globalization and the role of modernity in the production of new risks and framed them in terms that made certain interventions appear necessary, logical, or practical. The truism that "disease knows no borders" was not just a convenient cliché for describing microbial transgression; it was also linguistic shorthand for the political production of scale.

STEPHEN MORSE AND THE EMERGENCE OF "EMERGENCE"

Concerns over the appearance of new diseases are centuries old, and the term "emerging diseases" can be identified in the medical literature at least as far back as the 1960s.[7] However, not until the 1990s did emerging diseases appear as a coherent concept and the intellectual kernel of a broad public health campaign. The person generally credited with originating it is Rockefeller University virologist Stephen S. Morse, who chaired the 1989 conference "Emerging Viruses: The Evolution of Viruses and Viral Disease" and published several articles on the topic during the early 1990s.[8]

Writing in the shadow of HIV/AIDS, an epidemic that had taken the biomedical

---

[5] Bruno Latour, "Give Me a Laboratory and I Will Raise the World," in *Science Observed: Perspectives on the Social Study of Science,* ed. Karin Knorr-Cetina and Michael Mulkay (London, 1983), 141–70. I disagree with Latour's assertion that smaller scales are inevitably simpler. No such a priori case can be made—intervening at the molecular level is not inherently simpler than intervening at the level of global political economy. Such distinctions are politically determined, not ontologically given.

[6] James H. Capshew and Karen A. Rader, "Big Science: Price to the Present," *Osiris* 7 (1992): 19.

[7] All the following publications foreshadowed contemporary concerns, the latter two identifying (ironically enough) chronic diseases as "emerging": Alfred S. Evans, "The Instant-Distant Infections," *Journal of the American Medical Women's Association* 21 (1966): 210–6; Richard M. Krause, *The Restless Tide: The Persistent Challenge of the Microbial World* (Washington, D.C., 1981); Eric Cassell and Wilson G. Smillie, "New and Emergent Diseases," in *Human Ecology and Public Health,* ed. Edwin D. Kilbourne (New York, 1969); and Paul F. Basch, *International Health* (New York, 1978), 221–2. I thank David Jones for bringing the Cassell-Smillie reference to my attention.

[8] Stephen S. Morse, "Emerging Viruses: The Evolution of Viruses and Viral Diseases," *Journal of Infectious Diseases* 162 (1990): 1–7; idem, "Regulating Viral Traffic," *Issues in Science and Technology* 7 (fall 1990): 81–4; idem, "Emerging Viruses: Defining the Rules for Viral Traffic," *Perspectives in Biology and Medicine* 34 (1991): 387–409; and idem, "Global Microbial Traffic and the Interchange of Disease," *American Journal of Public Health* 82 (1992): 1326–7. Papers from the 1989 conference were collected in Morse, ed., *Emerging Viruses* (New York, 1993).

sciences by surprise and resisted effective intervention for more than a decade, Morse focused on the appearance of apparently novel viruses. He was careful to distinguish between "truly new" viruses resulting from major evolutionary changes via mutation or recombination and existing viruses transferred unchanged or with slight variations to the human population. These latter "emerged" through a two-step process, first crossing over from an animal host to humans, then infecting and spreading within the human population.

Morse argued that while it was almost impossible to predict the evolution of new viruses, tracking and anticipating the emergence of pathogens was feasible given a proper understanding of the basic mechanisms of interspecies transfer and viral spread. He called these the "rules of viral traffic" and advocated a program of "viral traffic studies," incorporating virology, molecular genetics, field and evolutionary biology, ecology, and the social sciences. This synthetic approach would provide a comprehensive map of the biological, social, and ecological forces that govern the appearance and circulation of new viruses in human populations.

Morse's description of viral emergence established direct causal links between the largest and smallest scalar extremes. Urbanization, environmental degradation, war, migration, and commerce on a global scale had profound impacts on the microbial level, altering the evolution of microorganisms and the patterns of viral traffic. As these activities increased, viral emergence accelerated, rendering a comprehensive research program on emerging viruses urgent: "Like every other kind of traffic, viral traffic is increasing. . . . As deforestation progresses worldwide, as human activities continue to alter the environment, as population influx into Third World cities continues unabated, as every part of the world becomes more accessible, one would expect disease emergence to accelerate."[9]

For Morse, proper intervention against emerging diseases depended upon understanding the universal molecular and ecological laws underpinning the interaction between humans and diseases. In the introduction to his edited volume on emerging viruses, he called for a comprehensive program with virology at its core:

> What are the resources needed for anticipating and controlling emerging diseases? Most of all, we will need trained people, and active laboratory facilities and research programs in which training can take place. . . . These areas include viral ecology, viral traffic analysis, driving forces and constraints in viral evolution, technologies for detection, and increased understanding of how viruses cause disease and how they interact with their hosts and with host cells (viral pathogenesis and immunology).[10]

As Morse acknowledged in his early essays, and the chapters in this volume ably demonstrate, the concept of emergence had intellectual roots in older understandings of environmental and disease ecology.[11] In this respect, it resembled the antimodernist sentiment Charles Rosenberg has characterized as "the idea of civilization as risk."[12]

[9] Morse, "Defining the Rules" (cit. n. 8), 405.

[10] Stephen S. Morse, "Examining the Origins of Emerging Viruses," in *Emerging Viruses* (cit. n. 8), 25.

[11] See Warwick Anderson, "Natural Histories of Infectious Disease: Ecological Vision in Twentieth-Century Biomedical Science"; and Helen Tilley, "Ecologies of Complexity: Tropical Environments, African Trypanosomiasis, and the Science of Disease Control in British Colonial Africa, 1900–1940." (Both this volume.)

[12] Indeed, an editorial by Joshua Lederberg contended that "many aspects of emerging infections can be viewed as diseases of civilization." Lederberg, "Infection Emergent," *Journal of the American Medical Association* 275 (1996): 244.

Rosenberg argues that this narrative draws its enduring social power from its fluidity, its holistic insistence on considering both biology and society in explaining disease, and its emphasis on the dynamic balance between humans and their environment.[13] Like this narrative, the concept of disease emergence was immensely flexible and would appeal to a wide variety of analysts and advocates in turn-of-the-century America. It, too, emphasized the interaction between biology and society, and it, too, insisted that disease can only be tamed if dynamic ecological balance is reached. A long history of holistic and ecological inquiry laid the groundwork for the emergence of a scalar narrative linking large-scale environmental and economic change, microbial evolution, and disease outbreaks in specific locations.

However, Morse's work was notable in its identification of "civilization" not only as the *cause* of new risks but also as the source of their *solutions*.[14] In this respect, he presented a scalar narrative, invoking different scales to describe the causes, consequences, and proper points of intervention in viral emergence. By altering viral traffic patterns, the introduction of modern agricultural or industrial technologies in one location—"local" causes—might produce an international epidemic or pandemic—"global" effects. However, since the "laws of viral traffic" were universal, monitoring and intervening need not be bound to the same scales as either cause or consequence. Addressing "global" risks entailed making ecological changes legible to laboratory investigation or information processing in multiple locations, often far removed from the specific site of disease outbreaks.

INSTITUTIONALIZING EMERGENCE:
THE 1992 INSTITUTE OF MEDICINE REPORT

Morse's call did not go unheeded. In 1991, the National Academy of Science's Institute of Medicine (IOM) convened a committee of scientists and public health experts, which included Morse and was cochaired by his colleague Joshua Lederberg, a geneticist and microbiologist. Their report, *Emerging Infections: Microbial Threats to Health in the United States* (1992), was the most comprehensive and influential statement of the effects of global change on American health and security in the 1990s. This report transformed Morse's ideas into a civic advocacy campaign, distilling a complex constellation of ideas into a coherent yet flexible discourse intended to convince policy makers of the national consequences of global change.[15]

While Morse's essays gave the concept of emerging diseases intellectual coherence, the IOM report supplied a political rationale and a blueprint for building a network of institutions to study them. Written while the public health community, infectious disease researchers in particular, was reeling from decades of budget cuts, the report identified a novel threat that justified funding, and provided specific recommendations for how those funds should be spent. Like Morse's work, it sought to - draw lessons from the sudden appearance and tenacity of HIV/AIDS and linked the global and microbial scales in a description of the *causes* of disease emergence.

[13] Charles Rosenberg, "Pathologies of Progress: The Idea of Civilization as Risk," *Bulletin of the History of Medicine* 72 (1998): 728–30.

[14] Ulrich Beck, *Risk Society: Towards a New Modernity* (London, 1992).

[15] I borrow the term "civic advocacy" from Paul Rutherford, *Endless Propaganda: The Advertising of Public Goods* (Toronto, 2000).

However, the IOM report focused more narrowly on the *consequences* at a national scale, targeting American policymakers and framing its arguments in terms of American public health and national security. It began:

> As the human immunodeficiency virus (HIV) pandemic surely should have taught us, in the context of infectious diseases, there is nowhere in the world from which we are remote and no one from whom we are disconnected. Consequently, some infectious diseases that now affect people in other parts of the world represent potential threats to the United States because of global interdependence, modern transportation, trade, and changing social and cultural patterns.[16]

Defining emerging diseases as "clinically distinct conditions whose incidence in humans has increased," the report identified more than fifty emerging viruses, bacteria, and other microorganisms. It also discussed a broad range of "factors in emergence" that were responsible for the appearance of new pathogens and the growing problem of antimicrobial resistance. These factors included population growth and migration; changes in individual behaviors such as sexual activity and substance use; new medical technologies such as radiation therapy, immunosuppressive drugs, and antibiotics; economic development and changes in land use such as dam building, deforestation, and global warming; international travel and commerce; microbial adaptation and change; and the breakdown of public health measures such as sanitation, immunization, and vector control.[17]

The report also addressed interventions. Recommendations fell into four broad categories: surveillance, training and research, vaccine and drug development, and behavioral change. Arguing that "the key to recognizing new or emerging infectious diseases, and to tracking the prevalence of more established infectious diseases, is surveillance," the IOM committee recommended the development of a comprehensive, computerized, global epidemiologic surveillance network.[18] This network would depend on four components: the training of clinicians in standardized reporting guidelines, so data would be uniformly reliable; a network of laboratory facilities capable of identifying specific strains of individual pathogens; sophisticated information-processing systems, so data could be efficiently analyzed; and a reliable international communications network, so the data could then be rapidly disseminated.

Like Morse, the IOM stressed the importance of research and training, although its recommendations were more specific than his, focusing on scientific research into the biology, pathogenesis, evolution, and epidemiology of infectious agents. It devoted the most space to discussing the importance of pharmaceuticals, recommending the development of stockpiles of selected vaccines and "surge" capacity for rapid development and production in case of emergency; a commitment to developing new vaccines and antimicrobial drugs, which could include public financing and expedited approval of privately developed pharmaceuticals; and a commitment to the development of pesticides useful in suppressing vector-borne diseases. Finally, the report included a brief discussion of the need to support public health

---

[16] Joshua Lederberg, Robert E. Shope, and Stanley C. Oaks Jr., *Emerging Infections: Microbial Threats to Health in the United States* (Washington, D.C., 1992), v.

[17] Ibid., 34 (definition), 34–112 (list of factors in emergence is condensed).

[18] Ibid, 113.

education and community health measures, although it did not specify the mechanism for doing so.

During the next ten years, the IOM's vision of emerging diseases became the centerpiece of a broad campaign as its advocates engaged in activities that historians of science recognize as the early stages of discipline-building: issuing reports and other publications reiterating their themes; holding conferences; persuading existing institutions to adopt its framework of risk and response; developing independent institutes and funding streams; and establishing a journal dedicated to the topic.[19] The Centers for Disease Control and Prevention (CDC) and Cabinet-level National Science and Technology Council (NSTC) issued reports repeating the IOM's analysis with little modification.[20] In 1995 alone, meetings of the IOM and the New York Academy of Medicine were devoted to the topic; the CDC launched the online journal *Emerging Infectious Diseases;* and the World Health Organization (WHO) established a Division of Emerging and Other Communicable Diseases Surveillance and Control.[21] In October, the U.S. Senate convened a hearing on the topic, during which Lederberg warned that "the microbe which felled one child in a distant continent yesterday can reach your child today and seed a global pandemic tomorrow," and the WHO's James LeDuc stated that "infectious diseases, like the environment, do not recognize national boundaries."[22]

The following year, at the behest of the editors of the *Western Journal of Medicine,* the *Journal of the Norwegian Medical Association,* and the *Journal of the American Medical Association,* thirty-six medical journals in twenty-one countries devoted all or part of their issues to emerging diseases. In one of the lead editorials for this "global theme issue," Lederberg presented a pragmatic argument for funding emerging diseases prevention efforts, obliquely contrasting it to the approaches stressed by proponents of the "third public health revolution":

> World health is indivisible, [and] we cannot satisfy our most parochial needs without attending to the health conditions of all the globe. One line of social thought would argue that the only answer is a fundamental convergence on population and poverty. Even were the will to do so to exist, and that will needs every encouragement, the history of social experiment in the 20th century would leave one in despair. Health is also a precondition to economic development, so that more modest and selfishly motivated measures can be a great beneficence to the overall human condition.[23]

Lederberg's list of interventions closely followed those in the IOM report: global and domestic surveillance, entailing "the installation of sophisticated laboratory

[19] On discipline-building, see Susan Greenhalgh, "The Social Construction of Population Science: An Intellectual, Institutional, and Political History of 20th Century Demography," *Comparative Studies in Society and History* 38 (1996): 26–66.

[20] Centers for Disease Control and Prevention, *Addressing Emerging Disease Threats: A Prevention Strategy for the United States* (Atlanta, Ga., 1994); Committee on International Science, Engineering and Technology, *Global Microbial Threats in the 1990s: Report of the Committee on International Science, Engineering and Technology's Working Group on Emerging and Reemerging Infectious Diseases* (Washington, D.C., 1995).

[21] World Health Organization, *Emerging and Other Communicable Diseases: Strategic Plan 1996–2000* (New York, 1996), http://www.who.int/emc-documents/emc/whoemc961c.html.

[22] Ironically, attendance at this hearing was extremely low, due to a concurrent debate on Medicare. U.S. Senate Committee on Labor and Human Resources, *Hearing on Examining the Threat and Risk of Certain Old and New Infectious Diseases on the Nation's Health,* 104th Cong., 1st sess., Hrg. 104–298, 1996, 3, 21.

[23] Lederberg, "Infection Emergent" (cit. n. 12), 244.

capabilities"; vector management and monitoring of safe food and water; public and professional education; basic scientific research; and the promotion of private pharmaceutical development through the institution of proper regulatory and incentive structures.[24]

By the end of the decade, the basic premises of the emerging diseases campaign had gained acceptance in many American governmental agencies and international health organizations. In 1998, the CDC released its second comprehensive plan on the topic, *Preventing Emerging Infectious Diseases: A Strategy for the 21st Century.* Closely following the IOM's 1992 recommendations, the CDC—identifying the IOM report as the source of "a new consensus"—emphasized the need for better surveillance technologies to "detect, investigate, and monitor emerging pathogens"; applied research in laboratory science and epidemiology; and the ability to "ensure prompt implementation of prevention strategies and enhance communication of public health information about emerging diseases."[25] Three years later, another CDC plan argued that American biomedical and information technology were instrumental to the achievement of global health:

> Promoting international cooperation to address emerging infectious diseases is a natural role for the United States, whose scientists and business leaders are important members of the biomedical research and telecommunications communities that provide the technical and scientific underpinning for infectious disease surveillance and control. The United States can continue to lead from its strengths in medical science and technology to help protect American and global health.[26]

These reports used scale as a resource for transforming Morse's conceptual argument into a pragmatic political campaign, providing American policy makers with a rationale for funding international health. Ostensibly "global" causes produced "local" (American) consequences, as the proliferation of international transportation and trade networks threatened to spark epidemics within U.S. borders. In this respect, the emerging diseases campaign did not significantly deviate from previous international health efforts, which similarly equated national self-interest with global humanitarianism.

However, even as campaigners sought to convince Americans of the appeal of this equation, by mobilizing two universalizing scalar extremes—the "global" and the microbial—it reformulated other "local" interests in American terms. Identifying emerging diseases as the consequence of globalization, campaigners associated it with an ineffable and complex phenomenon. The solutions they offered had the advantage of immensely reducing the scale of intervention, from global political economy to laboratory investigation and information management. Whether the object was "global health" or national security, interventions would involve "passing through" American laboratories, biotechnology firms, pharmaceutical manufacturers, and information science experts.

The emerging diseases campaign also drew on previous campaigns to promote Big Science projects, endorsing basic research as both a value in its own right and a gen-

[24] Ibid., 245.

[25] Centers for Disease Control and Prevention, *Preventing Emerging Infectious Diseases: A Strategy for the 21st Century* (Atlanta, Ga., 1998), 14.

[26] Centers for Disease Control and Prevention, *Protecting the Nation's Health in an Era of Globalization: CDC's Global Infectious Disease Strategy* (Atlanta, Ga., 2001), 16.

erator of practical results and advocating greater collaboration between the private and public sectors. This vision of Big Science, however, differed from that of highly centralized projects shackled to huge machines and requiring industrial management of thousands of collaborators. Instead, it involved a decentralized, horizontally integrated network of small projects—academic researchers, private vaccine and drug developers, epidemiologic surveillance projects and field laboratories—whose coordination is made possible by information-processing and communications technology.[27]

MEDIATING EMERGENCE:
RICHARD PRESTON, LAURIE GARRETT, AND THE NEW "VIRAL PANIC"

Emerging diseases quickly became an object of interest for the mass media and culture industries. Discipline-building activities such as meetings and publications, often announced by accompanying press releases, provided mass media with "pseudo events" worthy of coverage in their own rights.[28] Morse's 1989 conference (to take one example) was covered by *Bioscience, Medical World News,* and *Science News* and leading science journalists, including *The New York Times'* Lawrence Altman and *Newsday's* Laurie Garrett.[29] Integral to the emerging diseases campaign was a program of "media advocacy" designed to publicize the release of the IOM report and promote changes in public policy.[30]

More generally, the concept of emerging diseases offered journalists a powerful scalar resource for characterizing individual outbreaks as incidents of global significance. The mass media incorporated this scalar narrative into their coverage of other events, acting as both outsiders reporting on a campaign and participants actively shaping content and meaning.[31] Two of the most important reporters were Richard Preston and Laurie Garrett.

Preston's 1992 *New Yorker* article, "Crisis in the Hot Zone," described a 1989 outbreak of Ebola hemorrhagic fever among laboratory monkeys at a primate quarantine unit in Reston, Virginia. Although eventually contained through the efforts of the CDC and the United States Army Medical Research Institute for Infectious Diseases, the outbreak led to the euthanization of several hundred monkeys and resulted in four subclinical infections among human workers.

Preston emphasized the global consequences of minute microbial changes and

---

[27] For historical antecedents of this version of Big Science, see Capshew and Rader, "Big Science" (cit. n. 6), 20–3.

[28] Sharon M. Friedman, "The Journalist's World," in *Scientists and Journalists: Reporting Science as News,* ed. Sharon M. Friedman, Sharon Dunwoody, and Carol L. Rogers (New York, 1986); and Dorothy Nelkin, "Managing Biomedical News," *Social Research* 52 (1985): 625–46.

[29] Lawrence K. Altman, "Fearful of Outbreaks, Doctors Pay Heed to Emerging Viruses," *New York Times,* 9 May 1989, C3; Laurie Garrett, "Emerging Viruses, Growing Concerns," *Newsday,* 30 May 1989, 1.

[30] Vicki Freimuth, Huan W. Linnan, and Polyxeni Potter, "Communicating the Threat of Emerging Infections to the Public," *Emerging Infectious Diseases* 6 (2000): 337–47; Robert J. Howard, "Getting It Right in Prime Time: Tools and Strategies for Media Interaction," *Emerging Infectious Diseases* 6 (2000): 426–7.

[31] Sharon Dunwoody, "The Scientist as Source," in Friedman, Dunwoody, and Rogers, *Scientists and Journalists;* Nelkin, "Managing Biomedical News"; Friedman, "Journalist's World" (all cit. n. 28); and Bruce V. Lewenstein, "Cold Fusion and Hot History," *Osiris* 7 (1992): 135–63. I thank Lynn Nyhart for her challenging comments on the role of the media and for bringing the Lewenstein essay to my attention.

concluded his account by noting that the IOM report (released earlier that year) considered the Reston episode to be a "classic example" of disease emergence.[32] After detailing the report's recommendations, Preston drew explicit connections between Ebola, HIV, and other emerging viruses. Interviewing Morse, Preston asked whether an emerging virus "could wipe out our species." Morse cautiously speculated on the possibility of an aerosolized form of HIV causing a pandemic of a hybrid "AIDS-flu": "The human population is genetically diverse, and I have a hard time imagining everyone getting wiped out by a virus. . . . But if one in three people on earth were killed—something like the Black Death in the Middle Ages—the breakdown of social organization could be just as deadly, almost a species-threatening event."[33] By casting the Ebola-Reston incident as an example of disease emergence, Preston shifted the scale of his account upward, transforming an anticlimactic account of a small, successfully contained outbreak in one primate facility into a narrowly averted disaster and a harbinger of pandemics to come.

Preston's article had considerable impact, receiving critical acclaim from other journalists and touching off a bidding war among producers eager to adapt it into a screenplay.[34] He immediately set to work on a book based on the story. *The Hot Zone*, published in September 1994, was a multiweek bestseller, garnering Preston several awards and a $3 million advance for his next book. The *American Scientist* would later name *The Hot Zone*—along with Morse's *Emerging Viruses* (1993)—as one of the "100 or so Books That Shaped a Century of Science."[35]

Preston's book expanded on his article's linkage of the microbial and global scales. Using a metaphoric association between global ecology and individual immunology, he concluded with a warning about the transgression of the borders between humans and nature:

> The emergence of AIDS, Ebola, and any number of other rainforest agents appears to be a natural consequence of the ruin of the tropical biosphere. The emerging viruses are surfacing from ecologically damaged parts of the earth. . . . In a sense, the earth is mounting an immune response against the human species. It is beginning to react to the human parasite, the flooding infection of people, the dead spots of concrete all over the planet. . . . Nature has interesting ways of balancing itself. The rain forest has its own defenses. The earth's immune system, so to speak, has recognized the presence of the human species and is starting to kick in.[36]

The Hot Zone was not the only major work on emerging diseases published in 1994. From 1992 to 1993, while a fellow at the Harvard School of Public Health, former *Newsday* correspondent Laurie Garrett was conducting basic research on the subject. Having previously covered the 1976 Swine Flu "epidemic that never was," as well as the HIV/AIDS pandemic in Africa and the United States, Garrett had long been interested in the science and international politics of infectious disease.[37] She was also

[32] Richard Preston, "Crisis in the Hot Zone," *New Yorker,* 26 Oct. 1992, 80.

[33] Ibid, 81.

[34] Thomas Kunkel, "A Friend Writes," *American Journalism Review* 20 (1998): 18–20; Eric Utne, "Tina's New Yorker," *Columbia Journalism Review* 31 (1993): 31–6; Marshall Fine, "A Contagious Fascination with Infections: An 'Outbreak' Sweeps through Hollywood," *USA Today,* 28 Feb. 1995, 4D.

[35] Philip Morrison and Phylis Morrison, "100 or so Books That Shaped a Century of Science," *American Scientist* 87 (1999): 543–53.

[36] Richard Preston, *The Hot Zone* (New York, 1994), 287–8.

[37] Background information on Garrett can be found in James Kinsella, *Covering the Plague: AIDS and the American Media* (New Brunswick, N.J., 1989), 225–41.

familiar with Morse's 1989 conference and the Ebola story, having covered possible bans on importation of research monkeys as a result of the Reston outbreak. Upon learning of Preston's book contract, she accelerated work on her own book so that it would be released simultaneously. The publication of *The Coming Plague: Newly Emerging Diseases in a World Out of Balance* contemporaneously with Preston's book gave her 750-page work an improbably large audience.

In contrast to Preston's romantic account of environmental transgression, Garrett argued that emerging diseases resulted from decades of declines in public health, increasing economic inequality, and widespread social injustice. She concluded *The Coming Plague* with a similar meditation on the causal interplay between the microbial and global scales:

> Ultimately, humanity will have to change its perspective on its place in Earth's ecology if the species hopes to stave off or survive the next plague. Rapid globalization of human niches requires that human beings everywhere on the planet go beyond viewing their neighborhoods, provinces, countries, or hemispheres as the sum total of their personal ecospheres. Microbes, and their vectors, recognize none of the artificial boundaries erected by human beings. . . . In this fluid complexity human beings stomp about with swagger, elbowing their way without concern into one ecosphere after another. The human race seems equally complacent about blazing a path into a rain forest with bulldozers and arson or using an antibiotic "scorched earth" policy to chase unwanted microbes across the duodenum.[38]

In its depiction of the proper scale of intervention, Garrett's scalar narrative departed from those produced by Preston, Morse, and the IOM. While recognizing the value of simplifying international health problems so that they might "pass through" laboratories, she also recommended that large-scale interventions in international environmental policy, political economy, and social justice be coupled with biomedical research and technological innovation. As she argued in the introduction to a later book, *Betrayal of Trust: The Collapse of Global Public Health:*

> Yes, scientific and medical tools invented in the twentieth century will form a vital basis to global public health efforts in the twenty-first century, as will bold innovations based on altering human and microbial genetics. But the basic factors essential to a population's health are ancient and nontechnological: clear water; plentiful, nutritious, uncontaminated food; decent housing; appropriate water and waste disposal; correct social and medical control of epidemics; widespread—or universal—access to maternal and child health care; clean air; knowledge of personal health needs administered to a population sufficiently educated to be able to comprehend and use the information in their daily lives; and, finally, a health care system that follows the primary maxim of medicine—do no harm.[39]

Following Garrett and Preston, authors along the continuum of science communication sought to capitalize on emerging diseases.[40] The culture industries responded in kind. After a competing production was canceled, Warner Brothers released

---

[38] Laurie Garrett, *The Coming Plague: Newly Emerging Diseases in a World Out of Balance* (New York, 1994), 618–9.

[39] Laurie Garrett, *Betrayal of Trust: The Collapse of Global Public Health* (New York, 2000), 13.

[40] I borrow the term "continuum of science communication" from Lewenstein, "Cold Fusion" (cit. n. 31), 137. See also Paula Treichler, "AIDS, Homophobia, and Biomedical Discourse: An Epidemic of Signification," *October* 43 (1987): 31–70.

Outbreak, which opened at number one in 1995. Though it wasn't a direct adaptation, the movie's central conceit—a rogue general threatens to use an African monkey virus called Motoba for biological warfare—bore strong resemblances to the main theme of Preston's original article. Two months later, NBC aired *Robin Cook's Virus,* a TV movie about a CDC researcher's encounters with Ebola and other viruses.

Though the latter film was derided by one reviewer as "so harebrained it commands a kind of perverse attention," two days later reports began filtering into the United States that seemed to justify its alarmist speculations.[41] On May 10, American media reported an outbreak of Ebola in the village of Kikwit, Zaire (now the Democratic Republic of Congo). Science journalists, including Altman and Garrett, rushed to the scene. For the next three weeks, they reported on the progress of the epidemic through the small central African village. Weekly magazines published cover stories on the outbreak, network news programs such as ABC's *Nightline* and PBS's *Nova* devoted special episodes to it, and CNN aired a special report, "The Apocalypse Bug."[42]

Many members of the media used their coverage of the Kikwit outbreak as a springboard to discussions of emerging diseases. *Newsweek*'s cover featured a pair of gloved hands gripping a test tube labeled "E.Bola" (sic) and the text "KILLER VIRUS. Beyond the Ebola Scare—What Else is Out There?" The accompanying article argued, "Ebola is a potent emblem of the microbial world's undiminished power over us. But it's not the only one. New viruses have emerged with terrifying regularity in recent decades."[43] This warning proved prescient, as each succeeding year seemed to bring with it the emergence of exotic new or newly drug-resistant diseases, assiduously tracked in the popular and medical literature: Hantavirus, Lyme disease, the Asian "bird flu," West Nile Virus, E. Coli, BSE/CJD, multidrug resistant tuberculosis, SARS, and the "flesh-eating bacteria," *streptococcus* A.

Some critics saw coverage of Ebola and emerging diseases as evidence of a media-driven "viral panic" or "viral paranoia." In its July 1995 issue, *The New Republic* devoted its cover—which read "Paranoia Strikes Deep. Ebola, *Outbreak, The Hot Zone* and the new panic about plagues"—to Malcolm Gladwell's review of Preston's and Garrett's books. Gladwell argued that the United States was "in the grip of paranoia about viruses and diseases" and blamed the entertainment industry for stimulating this "paranoia." He reserved some of his harshest criticism for Preston: "It is safe to say that it is because of the success of *The Hot Zone* that *Outbreak* was made, that the Ebola outbreak in Zaire was covered as feverishly as it was, that the idea of killer viruses has achieved such sudden prominence. In the epidemic of virus paranoia, *The Hot Zone* is patient zero."[44] Many critics agreed with Gladwell's assessment, condemning the media for exaggerating the threat of individual outbreaks or emerging diseases more generally.[45]

[41] John J. O'Connor, "Television Review: A 90's Kind of Predator, via Robin Cook," *New York Times,* 8 May 1995, C16.

[42] Several scholars have analyzed media coverage of the 1995 Kikwit outbreak. The most comprehensive review is Sheldon Ungar, "Hot Crises and Media Reassurance: A Comparison of Emerging Diseases and Ebola Zaire," *British Journal of Sociology* 49 (1998): 36–56. See also the critique presented in chap. 2 of Susan D. Moeller, *Compassion Fatigue: How the Media Sell Disease, Famine, War, and Death* (New York, 1999).

[43] Geoffrey Cowley, "Outbreak of Fear," *Newsweek,* 22 May 1995, 52.

[44] Ibid, 39.

[45] Among others, see Stephen Budiansky, "Plague Fiction," *New Scientist,* 2 Dec. 1995, 28–31; and John Schwartz, "Media's Portrayal of Ebola Virus Sparks Outbreak of Wild Scenarios," *Washington Post,* 14 May 1995, A3.

These critiques underscored the ambivalent role of the media in the emerging diseases campaign. The emerging diseases concept allowed otherwise "local" stories to take on a "global" significance, persuading an American public notoriously uninterested in disease outside its own borders to pay attention to infectious disease. While some critics derided coverage of individual outbreaks as overblown, that coverage ensured a steady stream of publicity that gave emerging diseases an ominous immediacy. Popular representations of emerging diseases also allowed scientists and public health officials to distance themselves from irrational "hype." While clearly benefiting from (and in many ways contributing to) this publicity, scientists portrayed themselves as more sober, judicious, and reasonable than their excitable counterparts in the media and entertainment industries. At issue was the relative cultural authority of the narrators; the scalar narratives presented by scientists and journalists were virtually identical.

BIOTERRORISM

The summer-fall 2001 issue of the *Georgetown Journal of International Affairs* featured a forum titled "Bioalert: Disease Knows No Borders." In his introduction to the forum, former WHO consultant James M. Wilson praised the essays, in which "leading infectious disease experts . . . challenge the reader to consider the problem of transnational movement of pathogens—either through 'natural' or 'intentional' mechanisms—and how this can affect U.S. national security."[46] The section included articles by experts from the WHO, CDC, and the World Bank who argued that emerging diseases and biological terrorism presented similar threats to American health and security.

The appearance of this issue of the journal coincided with the national panic over the October 2001 anthrax outbreak in several eastern states. However, the essays' mobilization of the familiar cliché and association of emerging diseases with biological terrorism were far from novel. Discussions of the threat of bioterrorism during the opening years of the twenty-first century were shaped by the emerging diseases campaign that closed the twentieth, and discursive similarities between the two were supported by strong personal and institutional linkages.

The use of the threat of an imminent biological weapons attack to justify funding for public health infrastructure had roots in the cold war. During the early 1950s, CDC (then the Communicable Disease Center) chief epidemiologist Alexander Langmuir capitalized on American anxieties about biological warfare to channel defense funds into laboratory investigation of infectious diseases, communicable disease control, and the creation of the Epidemic Intelligence Service (EIS). As Langmuir told Donald Henderson in a 1979 interview, "Part of the justification . . . was that biological warfare defense was a very hot issue, it was a major propaganda issue in the Korean War. . . . I argued that if there was anything to this, there was a need for epidemiologists . . . and by winning that, we had the clear charge to recruit and train epidemiologists for civilian and military needs."[47]

[46] James M. Wilson, "Prevention is Key," *Georgetown Journal of International Affairs* 2 (2001): 4.
[47] Alexander D. Langmuir, "The Alpha Omega Alpha Interview with Alexander D. Langmuir, MD, MPH," interview by D. A. Henderson, videotape, Bethesda, Md., National Library of Medicine, 1979. See also Elizabeth Fee and Theodore M. Brown, "Preemptive Biopreparedness: Can We Learn Anything from History?" *American Journal of Public Health* 91 (2001): 721–6; and Elizabeth W. Etheridge, *Sentinel for Health: A History of the Centers for Disease Control* (Berkeley, Calif., 1992), 36–42.

Henderson, one of Langmuir's recruits to the EIS and later his deputy at the CDC, went on to lead the WHO campaign to eradicate smallpox and later became dean of the Johns Hopkins School of Public Health. In the 1990s, he became a forceful advocate of bioterrorism preparedness, founding the Johns Hopkins Center for Civilian Biodefense Studies in 1998. In November 2001, he was appointed director of the newly created Office of Public Health Preparedness. Like his mentor Langmuir, Henderson argued that epidemiologic surveillance was the core of biodefense.[48]

Other figures in the emerging diseases campaign were active in public discussions of biodefense. Joshua Lederberg had argued since the 1970s that basic biomedical research was necessary to address the threats presented by both biological weapons and natural epidemics.[49] During the 1990s, he reiterated his argument in a number of articles and an edited collection on the subject and served as a member of President Bill Clinton's 1998 ad hoc committee to discuss bioterrorism.[50] Stephen Morse, who had advocated global epidemiologic surveillance as a means of monitoring adherence to international weapons conventions as far back as 1992, left his university position to join the Defense Advanced Research Projects Agency as manager of the Unconventional Pathogen Countermeasures program.[51]

Richard Preston also turned his attention to bioterrorism, reporting on the Russian biological weapons program and possible links between an outbreak of West Nile Virus in New York City and Iraqi biological weapons testing.[52] His 1997 book, *The Cobra Event,* a fictional account of bioterrorist attacks in New York City and Washington, D.C., was cited by Bill Clinton as one of the motivations for his interest in the topic. In addition, Preston testified in front of the Senate during hearings on biological weapons.[53] Laurie Garrett published several articles and devoted a chapter of *Betrayal of Trust* to the topic, warning that the close association between public health and national security might hurt the credibility of the former.[54]

The incorporation of bioterrorism into the emerging diseases campaign sharpened two aspects of the campaign's scalar narrative. It accentuated the view that "global" causes begat "local" consequences and that international transportation, trade, and information networks threatened the health and security of the nation-state.[55] Drawing on a centuries-old logic encapsulated in the slogan "disease knows no borders,"

[48] D. A. Henderson, "Bioterrorism as a Public Health Threat," *Emerging Infectious Diseases* 4 (1998): 488–92; and idem, "The Looming Threat of Bioterrorism," *Science* 238 (1999): 1279–82.

[49] Joshua Lederberg, "Our CBW Facilities Could Help against Pestilences," *Washington Post,* 7 March 1970, A15.

[50] Joshua Lederberg, ed., *Biological Weapons: Limiting the Threat* (Cambridge, Mass., 1999); and idem, "Infectious Disease and Biological Weapons: Prophylaxis and Mitigation," *J. Amer. Med. Ass.* 278 (1997): 435–7.

[51] Ali S. Khan, Stephen S. Morse, and Scott Lillibridge, "Public-Health Preparedness for Biological Terrorism in the USA," *Lancet* 356 (2000): 1179–82; and Stephen S. Morse, "Epidemiologic Surveillance for Investigating Chemical Biological Warfare and for Improving Human Health," *Politics and the Life Sciences* 11 (1992): 28–32.

[52] Richard Preston, "The Bioweaponeers," *New Yorker,* 9 March 1998, 51–65; idem, "The Demon in the Freezer," *New Yorker,* 12 July 1999, 44–61; and idem, "West Nile Virus Mystery," *New Yorker,* 18–25 Oct. 1999, 90–108.

[53] U.S. Senate Judiciary Subcommittee on Technology, Terrorism, and Government Information, *Chemical and Biological Weapons Threats to America: Are We Prepared?* 105th Congress, Hrg. 105–710, 1998.

[54] Laurie Garrett, "The Nightmare of Bioterrorism," *Foreign Affairs* (2001): 76–89; and idem, *Betrayal of Trust* (cit. n. 39), chap. 5.

[55] I have addressed these issues in more depth in Nicholas B. King, "Dangerous Fragments," *Grey Room* 7 (2002): 72–81.

emerging diseases campaigners argued that national security and international health were closely connected and that America had a "vital interest" in the health of other nations.[56]

At the same time, the campaigners sought to persuade American policy makers that intervention in "global" phenomena necessitated a particular set of large- and small-scale responses. Expanding on Langmuir's cold war logic, they translated parochial national defense concerns into a justification for large-scale international epidemiological surveillance projects, as well as smaller-scale interventions at the levels of state and local public health infrastructure and laboratory investigation. National security would thus have to "pass through" the laboratory and the epidemiologic database.

ABSTRACTION AND THE POLITICS OF SCALE

Detailing all of the reasons for the emergence and success of this campaign is beyond the scope of this essay.[57] Nevertheless, I hope I have demonstrated that its scale politics played a significant role in making it attractive to a variety of actors, including biomedical scientists, public health and national security experts, and the mass media. While these actors presented different scalar narratives, they all depended upon the principle of scalar equivalence first introduced by Morse: the universality of the laws of viral traffic allowed one to bypass the messiness of specific locations in jumping from vast transportation networks to individual microbes and back again. In contrast to Roemer's idealistic vision of global health, the emerging diseases campaign presented a set of scalar tools for reframing "international" problems in language palatable to American interests.

Understanding the emerging diseases campaign as an example of the political production of scale might also offer insight into the process of historical reconstruction. Historians and sociologists of science, no less than the actors I have described, construct narratives that often involve weaving together cause and consequence at different scales. As we turn our attention toward the interaction between environment and health, we might pay special attention to the multiple scales that both "the environment" and "health" can signify and to the hidden arguments contained within the seemingly neutral terms "global" and "local."

[56] Institute of Medicine Board on International Health, *America's Vital Interest in Global Health: Protecting Our People, Enhancing Our Economy, and Advancing Our International Interests* (Washington, D.C., 1997), 1.
[57] Nancy Tomes, "The Making of a Germ Panic, Then and Now," *American Journal of Public Health,* 90 (2000): 191–8.

ECONOMY AND PLACE

Gender and the Economy of Health on the Santa Fe Trail

By Conevery Bolton Valenčius[*]

ABSTRACT

Correspondence surrounding the death from consumption of a New England woman on the Santa Fe Trail in 1857 demonstrates how gender roles and economic networks influenced health travel and the search for healthy places in the nineteenth-century United States. Women did travel seeking healing or relief from sickness—sometimes, as here, in arduous, overland trips—but in ways subtly different from male health seekers: family attachments, as well as their own health concerns, impelled and justified women in their decisions to take journeys. Yet for women as for men, decisions about health travel were also bound up with the economic considerations that shaped their families' lives.

INTRODUCTION

Spring came dry in 1857. Not until late May–early June were the grasses of North America's southern plains tall enough to support the oxen and mules of a long caravan. That spring, almost 10,000 heavily-built freight wagons left Missouri and swayed out across the oceans of grass on a two-month journey to the New Mexican outpost of Santa Fe. Among the tightly packed bags and boxes of one caravan traveled a desperately ill New England woman. It was not Kate Kingsbury's first voyage across the plains, but it would be her last.[1]

[*] Department of History, Campus Box 1062, Washington University in St. Louis, One Brookings Drive, St. Louis, MO 63130-4899; cvalenci@wustl.edu.

This project was made possible by a 1994 fellowship from the Missouri Historical Society, St. Louis, and faculty research funding from Washington University in St. Louis. I would like to thank research assistants Joseph N. Bartels and Jeremy M. Mikecz; Lilla Vekerdy, of the Archives and Rare Book Room of the Becker Medical Library, Washington University School of Medicine; and staffers of the Missouri Historical Society and the Phillips Library of the Peabody-Essex Museum, Salem, Massachusetts. I greatly appreciate the comments of Emily Abel, Barbara Baumgartner, S. Charles Bolton, Gregg Mitman, Matthew G. Valenčius, David J. Weber, and two anonymous reviewers.

[1] The events of Kate Kingsbury's passing—as well as the affairs of the firm of Webb & Kingsbury—have been exquisitely researched by Jane Lenz Elder and David J. Weber. See Elder and Weber, *Trading in Santa Fe: John M. Kingsbury's Correspondence with James Josiah Webb, 1853–1861* (Dallas, 1996), especially 57–60; and idem, "'Without a Murmur': The Death of Kate Kingsbury on the Santa Fe Trail," in *The Mexican Road: Trade, Travel, and Confrontation on the Santa Fe Trail*, ed. Mark L. Gardner (Manhattan, Kans., 1989), 98–105; as well as Elder, "Homesick on the Road to Santa Fe: James J. Webb's Private Diary, 1856," *New Mexico Historical Review* 72 (1997): 141–57; and James Josiah Webb, *Adventures in the Santa Fé Trade 1844–1847*, ed. Ralph P. Bieber (Philadelphia, 1974).

Most archival sources on these protagonists are in the James Josiah Webb Papers (hereafter cited as Webb Papers), Missouri Historical Society, St. Louis. For a good introduction to the trail, see David Dary, *The Santa Fe Trail: Its History, Legends, and Lore* (New York, 2000) (on 1857 wagons, see 235).

Though her appetite had improved since the wagon train had left Westport, the Santa Fe Trail's western Missouri staging area, Kate was still wracked by the coughing, pain, and breathing difficulties that had long sickened her. As her older brother, William Messervy, put it, she was "in a deep consumption"; early-twenty-first-century practitioners encountering the same symptoms would likely conclude that she was in the throes of tuberculosis.[2] Her husband, John, and family had hoped that getting her out of Salem, Massachusetts, and on to the high plains would heal her or at least quiet the disease consuming her body and her life with ever-increasing fury. But on June 4, when the party stopped for the night before fording the Arkansas River, Kate Kingsbury was taken with shortness of breath, as she had been throughout the journey. This time, however, she could "get no relief." Her husband and sister, Eliza Ann Messervy, tended to her, but without much hope.[3]

As the night wore on, John and Eliza Ann told Kate they had done all they could. It was considered no kindness to let a Christian soul depart unprepared. Struggling to draw breath, Kate insisted, "[I]f I was in your place I would not give up, but would persevere in trying to do something to give relief until the last, and never give up." Yet as the narrative of a good death demanded, she was eventually convinced that this was her end. As family friend James Webb wrote, "She then commenced with perfect composure, and took leave of her sister and John." She asked them to forgive her for "every hasty expression, or unkind word that had passed her lips during her illness" and reassured them that "if my Heavenly Father has sent for me, I am ready to go." Toward dawn, she died.[4]

Kate Kingsbury and her companions traveled in one of the long caravans by which John Kingsbury and James Webb, his business partner and close friend, made themselves a major trading force in the New Mexico Territory. Every year in the spring train, they shipped the goods they would market. For this journey, John Kingsbury had also quietly packed a metal coffin—a gesture of respect in an era in which fear of un-Christian burial added to the weight of untimely death. "John what are you going to do with this body?" Kate asked him as her lungs failed. He reassured her that he would care for her in death as he had in life, and she was comforted.[5] After Kate's death, John, Eliza Ann, and two friends left the main wagon train and traveled swiftly with the coffin to Santa Fe. They buried Kate Kingsbury in the cemetery of the International Order of Odd Fellows, a fraternal club her husband had joined to find genteel companionship amid the rough and brawling life of Santa Fe.[6]

This narrative of frustrated healing and eventual death, like so much else about the mid-nineteenth century, can seem at once mawkishly sentimental and deeply moving. Yet this death along the Santa Fe Trail—one of many, during the forty years in which Mexican and American traders laboriously worked their wares the 800 miles between two outposts of the faded northern empire of Spain and the expanding western empire

---

[2] William S. Messervy to Webb & Kingsbury, Salem, Mass., 16 Aug. 1856, Webb Papers.
[3] James J. Webb to Lillie Webb, Santa Fe, 18 June 1857, in Elder and Weber, *Trading in Santa Fe,* (cit. n. 1), 57.
[4] Ibid., 57–8.
[5] Ibid., 58.
[6] On everyday violence, see John Kingsbury to James J. Webb, Santa Fe, 26 July 1853, Webb Papers.

of the United States—reveals how gender and economic networks influenced health travel and the search for healthy places in nineteenth-century America.

Kate Kingsbury's presence on the Santa Fe Trail is surprising. Most Americans of her era regarded Santa Fe and its trading routes as a fine place for rough-hewn men and Spanish-speaking women, but not for a middle-class white New England wife. Moreover, she was mortally ill. Yet she engaged in the kind of arduous, long-distance health travel historians of medicine have associated primarily with sick men of the time. Kingsbury's journey forces us to acknowledge the complex reasons nineteenth-century women seeking wellness traveled, often in particularly gendered ways. Because women's health was understood in a matrix of relation and affinity, women who traveled to be with closely connected family were in nineteenth-century terms taking measures appropriate, and even necessary, for their physical well-being.

Kate Kingsbury's final journey further demonstrates that decisions about health travel were embedded in the economic, as well as the personal, relations of nineteenth-century households. In her day, as in our own, decisions about health—particularly grievous illness—took place in a context not simply of emotional and familial relations but of economic pressures and financial decisions as well. Her travels were predicated on some of the same economics that drove the Santa Fe Trail itself. The extant letters and documents chronicling Kate Kingsbury's decision to embark on this journey are from the men in her family circle; unlike the letters written by the women involved, the men's letters were preserved as part of the trading firm's records. Yet in the business correspondence of this set of Santa Fe traders, we see the complex, emotional repercussions of ill health that historians have long identified with women's lives and concerns. Kingsbury's story thus opens for us moderns a world in which health discussions we normally associate with the female world of the family could take place in male correspondence and take center stage in commercial decision-making. Kate Kingsbury's arduous passage demonstrates the close connection between male and female worlds of health, the profoundly economic basis for decision-making about health travel, and the impact of gendered family roles in shaping women's interactions with "healthy" places.

KATE KINGSBURY'S TRAVEL AND THE SANTA FE TRAIL

In the latter part of 1853, Kate Messervy stood at an important crossroads. She was preparing to leave her family home in Salem, Massachusetts, and embark upon a partnership with a young man—significantly younger than she—with few means but promising prospects.[7] In 1851, John Kingsbury had gone to Santa Fe to work as a clerk for the flourishing establishment of Messervy & Webb, one of Santa Fe's foremost trading firms. Through undercapitalized and often risky trading ventures such as Messervy & Webb, American and Mexican firms used Santa Fe, perched at the tail end

---

[7] How much older she was is unclear. Elder and Weber note that her tombstone has her thirty years old in 1857 (thus born in 1827), but the 1850 Massachusetts census lists her as being 27—and therefore born in 1823. They conclude that her age on the tombstone may be a polite fiction, since the 1823 birth date would make her seven years older than her husband. Elder and Weber, *Trading in Santa Fe* (cit. n. 1), 23 n. 13. Reasoning backwards ahistorically from several older female relatives of mine, I might add that she could well have lied to her husband about her age.

[8] Mark L. Gardner, introduction to Gardner, *The Mexican Road* (cit. n. 1), 3–5; Daniel Drake, *A Systematic Treatise, Historical, Etiological, and Practical, on the Principal Diseases of the Interior Valley of North America . . .* (Cincinnati, 1850), 1:175; David A. Sandoval, "Gnats, Goods, and Greasers: Mexican Merchants on the Santa Fe Trail," in Gardner, *The Mexican Road,* 22–31.

of the long Rocky Mountain range, as a base to exchange American manufactured goods for Mexican commodities such as wool, corn, grains—and, especially, the precious metals so scarce in the States.[8] On his trips back East, Kingsbury apparently gained the notice of Kate Messervy, the sister of his firm's senior partner, William S. Messervy. When John returned to Santa Fe from a brief summer trip in 1853, he carried an image of Kate—more than a trivial token of courtship by midcentury middle-class codes. Prodded perhaps by his friend's new relationship, James Webb soon met and commenced courting Florilla Mansfield Slade in Cornwall Bridge, Connecticut. The two couples married within weeks of each other.[9]

John Kingsbury and Kate Messervy's marriage occurred at the midpoint of a series of related Santa Fe trading partnerships involving James Webb, William Messervy, and John Kingsbury. James Webb had started out in the Santa Fe trade in 1844. After five profitable years, his initial business partner withdrew, and in 1850, Webb joined with fellow Yankee William Messervy to form Messervy & Webb.[10] Thus the Messervys, who came from old sea captain stock in Salem and had long profited from the East India trade there, branched out into overland international trade with Santa Fe.[11] Messervy & Webb proved to be a successful firm. In 1851, the year they hired Kingsbury, the partners brought between sixty and seventy wagons to New Mexico in their spring train.[12] Three years later, after Messervy moved back East, they gradually dissolved their partnership, allowing Kingsbury to take over with Webb. When Webb and Kingsbury retired their firm in 1861, it proved an opportune time: they sold their remaining goods and property before sectional fighting within the United States made the long-distance overland trade unfeasible.

Kate Messervy, William's much younger sister, was involved with these overlapping Santa Fe partnerships in ways both direct and indirect. She had a stake in her brother's success—literally: with their older sister, Eliza, she had apparently provided a substantial share of Messervy & Webb's initial capital.[13] The historical record treats these successive mercantile partnerships as the product of their male principals, but women's money underlay them. By marrying John, Kate further tied herself to the economics of the overland trade. The family circle of her adult life revolved around the yearly rhythm of purchasing trips and spring caravans.

At all levels, financial ties among the Messervy, Webb, and Kingsbury families underscored personal ones. As Kate and John courted in 1853, William Messervy urged James Webb to take their clerk on as a junior partner.[14] Though he had little capital of his own, John Kingsbury did have experience in the business. Taking Kingsbury on as a partner would allow James Webb to move up from junior to senior partner

[9] Elder and Weber, *Trading in Santa Fe* (cit. n. 1), 10 n. 14.

[10] Ibid., xxiii.

[11] "Death of Ex-Mayor Messervy," *Salem Register,* 22 Feb. 1886.

[12] Elder and Weber, *Trading in Santa Fe* (cit. n. 1), xxiii.

[13] Notwithstanding note 7, Kate was significantly younger than William. (Elder and Weber, *Trading in Santa Fe* [cit. n. 1], xxv.) Negotiating over the terms of their partnership, James Webb wrote to John Kingsbury, "In the partnership between myself and Mr. Messervy there was unequal capital, but the *house* paid interest on $5,000 to the credit of his sisters on our books." (Webb, to Kingsbury, Cornwall Bridge, Conn., 5 Dec. 1853, Elder and Weber, *Trading in Santa Fe* [cit. n. 1], 10 n. 13.) The Messervy family included five brothers but only the two sisters. J. Alfred Messervy, *Genealogie de la famille Messervy* (Jersey, 1899), 48.

[14] Elder and Weber, *Trading in Santa Fe* (cit. n. 1), 9.

while maintaining the consistency of his business. From William Messervy's perspective, withdrawing from the trade would allow him to both enjoy the quieter pace and more refined environment of Salem, Massachusetts (where he soon became mayor) and ensure a good position for his younger sister's fiancé.[15] The Messervy-Kingsbury marriage and the Webb-Kingsbury business partnership thus created simultaneous bonds among the three families: James Webb was financially tied to both, while the Messervys and Kingsburys were related through marriage and a passed-on business partnership. Lillie Webb also became a close friend to Kate; Eliza Messervy, in turn, checked in on John Kingsbury's consumptive sister in Boston while he was in Santa Fe.[16]

Over the next several years, as John and Kate Kingsbury moved with James and Lillie Webb to a shared storefront and joint house in Santa Fe, and William Messervy and his wife raised their young children in Salem, the health and well-being of all their family members—Kate, in particular—formed a theme throughout the three men's voluminous business correspondence.[17] Such records offer us a chance to see how deeply related were the worlds of female domesticity and male economic partnership. For all those involved in this overlapping set of relationships, one important issue was whether or how Kate Kingsbury should travel in response to her husband's business needs and her own health.

HEALTH TRAVEL

James Webb met Kate Messervy in October 1853 on a visit to Salem. "I talked much of New Mexico," he wrote to John Kingsbury, "and much of her going out there, provided you desired her to do so." Acknowledging what all concerned probably knew very well, he noted, "Her health is very delicate," but added that "I can but believe that the trip and a residence of a few years in that country would establish her health upon a strong constitution."[18] Webb thus referenced two important ways in which people of his time tended to their health: by engaging in travel for its own sake, and by moving to another location.

Travel, held nineteenth-century wisdom, could be uncomfortable, dangerous, and unhealthy. Yet it could also be restorative, strengthening, and healing.[19] Not travel alone, however, but travel to particularly healthy places was a therapeutic re-

---

[15] "Death of Ex-Mayor Messervy" (cit. n. 11). Messervy also served briefly as acting governor of Santa Fe territory. Ralph Emerson Twitchell, *The Leading Facts of New Mexican History* (Albuquerque, 1963), 2:295–6.

[16] James Webb to John Kingsbury, Hamden, Conn., 18 June 1858, Webb Papers.

[17] See, e.g., the discussion about where Eliza Messervy should live: William Messervy to John M. Kingsbury, Santa Fe, 29 Jan. 1854, Webb Papers.

[18] James J. Webb to John Kingsbury, Cornwall Bridge, Conn., 5 Oct. 1853, Webb Papers. Nineteenth-century Americans often referred to areas outside their own regions, even places formally within the United States, as being part of another "country." Elder and Weber misrecord Webb's comment, noting that he thought a stay in Santa Fe would place Kate's health upon "a firm foundation." (Elder and Weber, "Without a Murmur" [cit. n. 1], 99.) Webb's comment implies an even deeper transformation of her "constitution," or intrinsic propensity to illness or good health.

[19] On the health effects of travel, see Sheila M. Rothman, *Living in the Shadow of Death: Tuberculosis and the Social Experience of Illness in American History* (New York, 1994), especially 19–20, 31–7, and 131–47; and Richard Wrigley and George Revill, eds., *Pathologies of Travel* (Amsterdam, 2000). Change of site itself—rather than "any special climatic agency" of a particular place—was strongly endorsed by some leading health authorities. See Austin Flint, *Phthisis* (Philadelphia, 1875), 395.

sponse. In the worldview of health and well-being, which would be largely re-placed by the "germ theory" of disease over the latter decades of the nineteenth century and the opening decades of the twentieth, places held influences that imparted themselves to human beings.[20] Living near a smelly, miasmatic, effluvi-ous swamp was likely to weaken and sicken the body, while spending time in re-gions of gentle winds, free from sources of putrefaction or decay, enjoying whole-some water and air, would keep a person in health—and potentially even restore someone who was ill. Though the exact characteristics of "healthy" places were subject to debate, most people agreed that elevated, warm places of fresh air, gen-tle climate, agreeable smells, and clear springs or ocean water were likely to bol-ster and cure the human form.[21]

Seasides and ocean resorts had long been the refuges of sick and weary health trav-elers.[22] In the middle decades of the century, American territorial expansion began to make available bright, sunny, dry places in the country's interior. In Kate Kingsbury's era, health travel burgeoned to what would ultimately become the American west and southwest.[23] Travelers could thus undertake voyages at sea—or on the oceans of grass in the country's interior.

The healing that would come from new places was predicated on a match between illness and clime: people beset by wet coughs, thick mucus, and heavy lungs might be sickened further if they stayed in cold and damp (or hot and humid) regions but could heal and feel stronger if they spent time breathing in lighter, drier air. High, dry places, with cool—not hot—summers and temperate winters were thus commonly held to be

[20] Indeed, concern for health and place was often incorporated into newer disease ideas: an 1880 source extolled the salubrity of seashores and mountains because the "abundance of oxygen . . . de-stroy[s] any poisonous germs which may float in [the atmosphere]." A. A. Hayes Jr., *New Colorado and the Santa Fe Trail* (New York, 1880), 180.

[21] On the perceived relationships between health and place, see Barton H. Barbour, "Westward to Health: Gentlemen Health-Seekers on the Santa Fe Trail," *Journal of the West* 28 (1989): 39–44; John E. Baur, "The Health Seeker in the Western Movement, 1830–1900," *Mississippi Valley Historical Re-view* 46 (1959): 91–110; James H. Cassedy, "Medical Men and the Ecology of the Old South," in *Sci-ence and Medicine in the Old South,* ed. Ronald L. Numbers and Todd L. Savitt (Baton Rouge, La., 1989), 166–78; Mary J. Dobson, *Contours of Death and Disease in Early Modern England* (Cam-bridge, 1997); Billy M. Jones, *Health-Seekers in the Southwest, 1817–1900* (Norman, Okla., 1967); Esmond R. Long, "Weak Lungs on the Santa Fé Trail," *Bulletin of the History of Medicine* 8 (1940): 1040–54; Linda Nash, "Finishing Nature: Harmonizing Bodies and Environments in Late-Nineteenth-Century California," *Environmental History* 8 (Jan. 2003): 25–52; idem, "Transforming the Central Valley: Body, Identity, and Environment in California, 1850–1970" (Ph.D. thesis, Univ. of Washing-ton, Seattle, 2000), chap. 3, "Disease"; Kenneth Thompson, "Climatotherapy in California," *Cali-fornia Historical Quarterly* 50 (1971): 111–30; idem, "Wilderness and Health in the Nineteenth Century," *Journal of Historical Geography* 2 (1976): 145–61; Conevery Bolton Valenčius, "The Geography of Health and the Making of the American West: Arkansas and Missouri, 1800–1860," in *Medical Geography in Historical Perspective,* ed. Nicolaas A. Rupke (London, 2000), 121–45; and idem, *The Health of the Country: How American Settlers Understood Themselves and Their Land* (New York, 2002), chap. 3, "Places."

[22] Rothman, *Living in the Shadow of Death* (cit. n. 19), 19–20, 31–7; Harriet Deacon, "The Politics of Medical Topography: Seeking Healthiness at the Cape during the Nineteenth Century," in Wrigley and Revill, *Pathologies of Travel* (cit n. 19), 279–98.

[23] Barbour, "Westward to Health"; Baur, "Health Seeker in the Western Movement"; M. H. Dunlop, *Sixty Miles from Contentment: Traveling the Nineteenth-Century American Interior* (New York, 1995), 181–5; John Mack Faragher, *Women and Men on the Overland Trail* (New Haven, Conn., 1979), 17; Jones, *Health-Seekers in the Southwest;* Long, "Weak Lungs"; Thompson, "Climatother-apy in California." (All previous cit. n. 21.)

[24] See Daniel Drake, *A Systematic Treatise, Historical, Etiological, and Practical, on the Principal Diseases of the Interior Valley of North America . . .* (1854; repr., Birmingham, Ala., 1990), 2:884–901.

good for bad lungs.[24] Beginning in the 1840s, travel writers and medical authorities publicized health effects already noted by early American explorers and traders, identifying the Great Plains and Rocky Mountains—in particular, Santa Fe—as being "healthy," especially for those with consumption.[25] Josiah Gregg's endorsement of the dry, healthy air of Santa Fe in his widely read *Commerce of the Prairies* (1844), an introduction to the Santa Fe Trail, proved particularly influential. (During their time in Santa Fe, James Webb and John Kingsbury were sent letters of introduction on behalf of young men headed west to Santa Fe for better health.)[26] The region typically featured temperate summers, sunny winters, dry air, high elevation, and healthful sources of stimulation—from a characteristically nineteenth-century combination of fresh air, travel, and often breathtakingly beautiful new sights.[27] Even the drearily slow pace of a wagon train could be beneficial to invalids trying to reach Santa Fe, allowing a weakened body plenty of time to acclimate to new and challenging airs and places.[28] In nineteenth-century terms, each of these factors helped produce a healthful and restored body.

Yet recommendations for health travel were not universal. Health travel was often implicitly understood in terms of male attributes and male bodies. The very difficulty of making one's way across rough terrain—a difficulty considered appropriate primarily to the male physique—was potentially vivifying: the bodily system would rise to the challenge, stimulated by fresh air, new experiences, and hearty living. (Many authorities added that the demands of overland trails would necessarily and usefully challenge the self-obsessive quality of invalidism.)[29] In the widely disseminated advice literature, camping on the ground was as essential as breathing fresh air—but women were not expected to participate in these rougher aspects of overland travel.

Travel on the American plains was particularly regarded as healthfully bracing in writings aimed either implicitly or explicitly at male invalids.[30] Daniel Drake, for instance, encouraged consumptive travelers to the Rocky Mountains in 1844 that in addition to salubriously rough living, they would benefit from "[t]he excitement connected with the danger of being lacerated by the bullets of the *Pawnees;* pierced to death by the arrows of the *Blackfeet,* or picked to death by the *Crows.*"[31] Such advice was clearly both tongue-in-cheek and addressed only to a male readership. Through-

[25] Barbour, "Westward to Health"; Jones, *Health-Seekers in the Southwest,* chap. 3 (both cit. 21). See, e.g.: W. W. H. Davis [he escorted Kate Kingsbury on her 1856 return East], *El Gringo, or, New Mexico and Her People* (New York, 1857), 297–8; Drake, *Systematic Treatise,* (cit. n. 8), 1:156, 175; J. J. Jones, "New Mexico as a Health Resort for Consumptives," *Medical and Surgical Reporter* (Philadelphia) 37 (15 Sept. 1877): 201–3; Daniel Millikin, "Notes on the Climate of Colorada [*sic*] and New Mexico," *The Cincinnati Lancet and Clinic,* n.s., 9 (1882): 577–9; W. D. Napton, *Over the Santa Fé Trail in 1857* (Kansas City, Mo., 1905), 3–4, 9; J. Hilgard Tyndale, "New Mexico: Its Climatic Advantages for Consumptives," *Boston Medical and Surgical Journal* 108 (1883): 265–316.

[26] Josiah Gregg, *Commerce of the Prairies,* ed. Max L. Moorhead (Norman, Okla., 1954), 105. In his endorsement of Santa Fe, Daniel Drake relied on Gregg's account (Drake, *Systematic Treatise* [cit. n. 8], 1:155–6). On health seekers' introductions, see, e.g., Glasgow & Brothers to Webb & Kingsbury, St. Louis, 17 Sept. 1856, Webb Papers.

[27] To this list of health benefits, later medical authorities added educated druggists and appropriate society. Well-being had many ingredients. Jones, "New Mexico as a Health Resort" (cit. n. 25), 202–3.

[28] For these reasons, late-nineteenth-century authorities warned that train travel was not as healthful as earlier mounted journeys. Millikin, "Notes" (cit. n. 25), 577.

[29] Drake, *Systematic Treatise* (cit. n. 8), 1:175.

[30] An 1882 article noted that western air offered not rest but healthy *arousal* of the lungs. Millikin, "Notes," (cit. n. 25), 577.

[31] Daniel Drake, "Traveling Letters from the Senior Editor," no. 4, *The Western Journal [of Medicine and Surgery],* n.s., 2 (Sept. 1844): 270–9, on 276.

out recommendations for overland health travel runs a similarly masculine bravado. In a series of *Harper's* articles from the early 1860s, Lieutenant George Douglas Brewerton advised

> any person who is suffering from dyspepsia or a tendency to consumption to pack up his traps, take leave of the doctor, . . . and then, with a good horse and one of Sharpe's patent rifles, a bowie-knife, and a Colt's six-shooter, let him "make a break" and go westward to the spurs of the rocky Mountains.[32]

Brewerton wrote of "any person," but he meant only free male people. Arduous conditions, fresh air, a simple diet heavy on wild meat, and ready guns all jumbled together as salutary factors in many midcentury health recommendations for the recovery of what were understood to be male bodies. Indeed, most overland health travelers were men, and men of some means.[33]

Women, too, traveled for health, but generally more locally and to the more refined locations of spas or health resorts. Women's health travel was often combined with customary long-term "visiting" with friends and relations.[34] Nonetheless, some of the same principles applied. Horseback riding, for instance, could be healthful for women precisely because of its physical challenge: for women with "obstructed" menses or rheumatism, the jouncing was thought to restore the body's healthy and natural rhythms.[35]

Yet precisely because of women's gendered natures, difficult travel could be considered healthy for women. Nineteenth-century American society expected women to embark on journeys because of family relationships, including family business demands.[36] A sick woman traveling because of her close attachment to family could therefore be understood to be making justifiable and comprehensible health decisions.[37] Leading physician Austin Flint, for instance, cited in his 1875 textbook *Phthisis* (consumption) the case of a middle-aged woman desperate to join her husband in California, despite her severe illness. When she insisted, against medical advice, Flint acceded; despite the arduous journey, being reunited with him helped bring her back to health. "Here was a powerful moral influence at work," concluded Flint, endorsing this cure.[38]

Discussions of Kate Kingsbury's potential health travel reflect these complex and

[32] George Douglas Brewerton, *Overland with Kit Carson: A Narrative of the Old Spanish Trail in '48,* ed. Stallo Vinton (New York, 1930), 265, quoted in Barbour, "Westward to Health" (cit. n. 21), 42.
[33] Barbour, "Westward to Health" (cit. n. 21), 42.
[34] Joan D. Hedrick, *Harriet Beecher Stowe, A Life* (New York, 1994), chap. 16, "The Water Cure: 1846–1848"; Valenčius, *Health of the Country* (cit. n. 21), 103. Sheila Rothman argues that women did not generally travel for health, but her evidence demonstrates extensive family visiting (and even western travel) by health-seeking women. Rothman, *Living in the Shadow of Death* (cit. n. 19), 77, 100, 102, 136–7.
[35] Susan E. Klepp, "Lost, Hidden, Obstructed, and Repressed: Contraceptive and Abortive Technology in the Early Delaware Valley," in *Early American Technology: Making and Doing Things from the Colonial Era to 1850,* ed. Judith A. McGaw (Chapel Hill, N. C., 1994), 68–113, on 79. Riding was widely recommended as healthful for both sexes. Drake, *Systematic Treatise* (cit. n. 8), 1:175.
[36] Women not infrequently insisted that they had the right—and duty—to follow their husbands west. Julie Roy Jeffrey, *Frontier Women: "Civilizing" the West? 1840–1880,* rev. ed. (New York, 1979), 43.
[37] Robert L. Munkres, "Wives, Mothers, Daughters: Women's Life on the Road West," *Annals of Wyoming* 42 (1970): 191–224, 219.
[38] Flint, *Phthisis* (cit. n. 19), 411.

often contrasting strands. Her female relations opposed her travels when she first contemplated going out to Santa Fe with her new husband. "I think she would go," wrote James Webb to John Kingsbury, "but her sister and Mrs Messervy have the greatest horror of that country, and the strongest attachments to home of any two persons I ever met."[39] Her emigration would fracture the bonds of a close-knit family circle. Kate's brother, by contrast, encouraged her travel, emphasizing her relationship with John and the potential benefit to her health. If he chose to accept James Webb's offer of partnership, William Messervy wrote to his new brother-in-law in January 1854, "you ought to bring Kate with you. In this matter you should not allow any one to influence you. Kate is your wife, it is her duty to consult your interests, welfare, & desires, more than that of any other person living." Messervy also believed the trip could save his sister: "If you think she is strong enough to stand the trip, I have no doubt it is the only thing which will prolong her life." Messervy, writing during the sunny winter of Santa Fe, had faith in the healing power of the town's climate and air. "[I]f there is hope," he concluded, "it is well to attempt it."[40]

The Messervy women's "horror of that country" underscores the *cultural* danger posed by travel to Santa Fe—and the rough West generally—for a traveler such as Kate. "In those days," wrote one memoirist of his 1858 journey to Santa Fe, "the women dreaded, worse than death, the perils of the Western trails."[41] American travelers' accounts of Santa Fe invariably stressed the raunchy and dangerous quality of the town's social interaction, as well as the colorful and often earthy allure of Mexican (and then New Mexican) women. As John Kingsbury and James Webb knew only too well, Santa Fe was a place of early prostitution and absconded husbands—not of respectable New England matrons.[42]

Nonetheless, Kate traveled with John to Santa Fe and spent the next several years there, sharing a household and business with Lillie and James Webb. Though not cured, she recovered enough to have a child. In at least a limited way, her experience demonstrated the benefits of change of latitude and climate. At the same time, Kate Kingsbury's journeys illustrate the provisional nature of health travel. Like other therapeutic responses, it could be overridden or rethought as circumstances warranted. In her case, she undertook a reverse trip, traveling back East in 1856 to her potentially insalubrious home territory because of the overweening imposition of another concern of well-being: sickness of the heart and soul.[43]

Many of Kate's difficulties may have stemmed from the isolation and challenges of

[39] James J. Webb to John Kingsbury, Cornwall Bridge, Conn., 5 Oct. 1853, Webb Papers.

[40] William Messervy to John M. Kingsbury, Santa Fe, 29 Jan. 1854, Webb Papers.

[41] "Narrative by Hezekiah Brake, 1858," in *On the Santa Fe Trail,* ed. Marc Simmons (Topeka, Kans., 1986), 39.

[42] James Webb paid a neighborhood girl's way into a convent to save her from prostitution; John Kingsbury was asked to help track down an Illinois woman's husband who had run off with her money and left her with a young child. (James Webb to Florilla Webb, Santa Fe, 10 July 1856; and George McKinney to John Kingsbury, Oquawka, Ill., 15 Nov. 1859, Webb Papers.) Few American women traveled to Santa Fe, especially with the trading caravans. (Barbour, "Westward to Health" [cit. n. 21], 42. For exceptions, see Dary, *Santa Fe Trail* [cit. n. 1], 136; and Susan Shelby Magoffin, *Down the Santa Fe Trail and into Mexico,* ed. Stella M. Drumm [Lincoln, Neb., 1982].) Josiah Gregg wrote of meeting two "respectable French ladies"; his insistence on their "respectability" is telling. (Gregg, *Commerce of the Prairies* [cit. n. 26], 33.) Most American sources simply ignored the Mexican women and children who regularly traveled the trail. Sandoval, "Gnats, Goods, and Greasers" (cit. n. 8), 23–4.

[43] Unlike travelers on the overland routes to the West Coast, many people on the Santa Fe Trail made multiple journeys. Barbour, "Westward to Health" (cit. n. 21), 42.

[44] Elder and Weber, *Trading in Santa Fe* (cit. n. 1), 22–3, 290.

raising a child with apparently heavy handicaps. The Kingsburys and the Webbs had children within a month of each other, in the winter of 1854–1855: Jimmie Webb grew to robust health (he eventually became a member of the Connecticut Superior Court), but George Kingsbury was born with some form of serious disability.[44] The existing male correspondence carefully elides the exact nature of his condition, but it was grave enough that after reading the letter John Kingsbury sent about his son, William Messervy burned it (probably at John's request). "I have not communicated one word of it to any one not even to my wife," William assured John; "none of the family suspect that Kate has yet been delivered." Presumably, if the baby boy died, as was expected and perhaps hoped, others in this closely bound family would not necessarily have to know that he had been born with problems. William added, "Should the Child live, there is no other way, but to give it that place in your Affections to which it would be entitled if it had been born perfect. We soon . . . lose sight of the imperfections of our Children, in our strong natural love for them." This was not bland reassurance; he continued, "I feel that I love my little Girl much more than if she had been perfect in her physical development. The extra claims her helplessness has upon me, causes me the more to love & protect her. Such will be the case, with you and yours" (his daughter had a condition involving fluid on her brain and did not live into adulthood).[45] Brothers-in-law separated by half a continent of distance and by almost two decades of age thus engaged in intimate communication as they tried to shield Kate from criticism or repulsion. But they also denied her some of the commiseration, sympathy, and practical help she might otherwise have received from family members.

George Kingsbury did live through infancy, but his condition was difficult for his mother. She faced an often suffocatingly limited circle of American women of her own social station. James Webb once wrote to Lillie with indignation of Kate's being shunned by a colonel's wife, perhaps because of Kate's illness or George's condition.[46] In physical, and apparently emotional, terms, Kate was increasingly challenged. By 1856, her husband was adamant that even the arduous journey should not prevent her from returning to her home of family and friends. On that trip, during the summer of 1856, George Kingsbury died at his uncle William's home in Salem.[47] Kate Kingsbury, too, began to fail physically. John left his post in Santa Fe to rejoin his wife in Massachusetts—which meant James Webb had to leave the East Coast, where he had recently moved his own family, to return to Santa Fe and take care of the New Mexico end of their business. In November, John wrote his partner that "Our Physician think[s] her in the consumption and past cure[;] still if she is able to return to Santa Fe she may last a long time yet. . . . [T]he thing is will she be able to stand the journey."[48]

Ultimately, Kate Kingsbury determined to travel back to Santa Fe with John with the firm's spring shipment. On this journey she was accompanied by Eliza Ann Mes-

[45] William Messervy to John Kingsbury, Salem, Mass., 13 March 1855, Webb Papers; Elder and Weber, *Trading in Santa Fe* (cit. n. 1), 53.

[46] Elder, "Homesick on the Road to Santa Fe" (cit. n. 1), 147–8.

[47] William Messervy to Webb & Kingsbury, Salem, Mass., 16 Aug. 1856, Webb Papers; Elder and Weber, *Trading in Santa Fe* (cit. n. 1), 38–9.

[48] John Kingsbury to James Webb, Salem, Mass., 12 Nov. 1856, Webb Papers.

[49] *Salem Register,* 20 Feb. 1893. William Messervy hoped the trip would help Eliza's health as well. (Messervy to John Kingsbury, Salem, 17 Aug. 1857, Webb Papers.) For similar familial travel, see Susan Armitage, "Another Lady's Life in the Rocky Mountains," in *Women and the Journey: The Female Travel Experience,* ed. Bonnie Frederick and Susan H. McLeod (Pullman, Wash., 1993), 25–38.

servy, who apparently overcame her aversion to leaving her home for the wilds of New Mexico because of her younger sister's dire need. (Her obituary in 1893 would note that Eliza Messervy "found her chiefest joy in making others happy.")[49] Both Messervy women thus engaged in travel for reasons of affinity well recognized as legitimate and appropriately feminine. This voyage—the third of Kate Kingsbury's cross-country travels in response to her changing states of well-being—would ultimately end in her death from consumption. Going to a healthy place was not a simple matter. Travel for health could be contingent, reversed, and retried—and it might in the end be unsuccessful.

THE ECONOMY OF HEALTH

This set of exchanges and travels thus reveals some characteristics of how health travel was discussed and undertaken. In this story we can also see some of the ways in which health travel, and indeed responses to health more generally, operated within the context of economic relations. Historians of medicine have identified ways in which nineteenth-century health, in particular women's roles in caring for those who were ill or needed care, took place in a deeply domestic context.[50] At the same time, these interactions also functioned as part of household economies.[51] Health travel occurred for some of the same reasons, and in some of the same ways, as decisions about economic interactions—here, the Santa Fe trading partnership in which the Kingsbury, Webb, and Messervy families participated.

Multiple and overlapping forms of personal relationship were crucial to the business of the Santa Fe trade and to decisions about health travel. For merchants in this trade, family connections reinforced commercial bonds.[52] Certainly, in the intertwined Messervy, Webb, and Kingsbury families, trust and mutual dependence had a financial, as well as social and familial, dimension. Kate Kingsbury's health was discussed and debated among a network of men joined by family and financial ties, men whose friendship was quite real and deep—as their sometimes exasperated, often affectionate business letters to one another attest—and who at the same time had to depend upon one another's judgment, probity, and hard work in the high-stakes venture of cross-continental trade. The forces shaping health decisions and those shaping economic networks were structured in parallel.

In the Santa Fe trade and in this story of health travel, the center of gravity was emphatically not the southwestern terrain about which conversation centered. In the conversations of these middle-class, East Coast Americans, decisions about Santa Fe, and the conditions determining Santa Fe as a possibility in financial and health-related terms, were all dependent on the East. In crude but fundamental terms, federal money was crucial to the economy of Santa Fe in that period.[53] Similarly, eastern relatives, eastern home, and an eastern social circle formed the center of Kate Kingsbury's ex-

[50] Judith Walzer Leavitt, "'A Worrying Profession': The Domestic Environment of Medical Practice in Mid-Nineteenth-Century America," *Bull. Hist. Med.* 69 (1995): 1–29; Emily K. Abel, *Hearts of Wisdom: American Women Caring for Kin, 1850–1940* (Cambridge, Mass., 2000), pt. 1.

[51] As indeed Abel (*Hearts of Wisdom* [cit. n. 50]) and others recognize. On the economics of women's health care, see, e.g., Laurel Thatcher Ulrich, *A Midwife's Tale: The Life of Martha Ballard, Based on Her Diary, 1785–1812* (New York, 1990).

[52] Mark L. Gardner, "The Glasgows: Missouri Merchants in the Mexican Trade," in Gardner, *The Mexican Road,* 13; Elder and Weber, *Trading in Santa Fe,* xxvii (both cit. n. 1); Sandoval, "Gnats, Guns, and Greasers" (cit. n. 8), 27.

[53] Elder and Weber, *Trading in Santa Fe* (cit. n. 1), xxix.

perience as a sick woman, the emotional environment to which Kate returned for care and emotional refreshment even though Santa Fe's physical environment helped check her consumption.

For these commercial migrants and middle-class Yankees, Santa Fe was a place to squeeze value from—and then leave. It was a place to travel to in hope of return (in both senses).[54] William Messervy, James Webb, and John Kingsbury all lived there while they built up their various partnerships. After each partnership became established, however, the senior partner—first Messervy and then Webb—retired in comparative comfort to raise a family back East. A somewhat parallel move marks Kate Kingsbury's interaction with Santa Fe. She went there to be with her husband and because it might help her get well, or at least keep her from growing weaker. Yet living for a while in the rough salubrity of Santa Fe did not alter the center of her interactions, which remained firmly in the Massachusetts trading port of her birth. This woman's decisions about health travel thus help us understand the physical underpinnings of the connections between western regions and eastern entrepreneurship: in their well-being as in their finances or social lives, those from the eastern seaboard could remain strongly tied to their home regions even as they engaged in long-term interaction with places such as Santa Fe.[55]

In practical terms, Kate Kingsbury went out West on her final trip largely because of her husband's business needs. Writing to James Webb in January 1857, John Kingsbury made clear that he knew he had to return to tend to their affairs. "It is true Kate is very feeble, & it is uncertain what change may occur," he wrote, "but if possible I wish to take her back & am making all calculations *now* to that effect." As this and other letters underscore, he was frantic with worry and grief at the loss of his son and the impending loss of his beloved wife. (His mother had also died not long before.) Yet he felt the pressing concerns of their business, as well as a responsibility not to keep James too much longer from his own family on the East Coast. John Kingsbury continued, "Should it prove impossible for her to go, then I feel that it will be my duty to go without her. I shall not give up untill the last moment, if she has nothing to put her back I think there is a chance for her yet."[56] Intertwined in John and Kate Kingsbury's decision to go back West along the Santa Fe Trail, like the multiple relationships linking the Messervy, Webb, and Kingsbury families, were concerns at once health-related and commercial.

In these many reasons for her travel, Kate Kingsbury was not alone. Many male as well as female "valetudinarians" sought to accomplish multiple goals in their journeys for health. In overland travel, proclaimed the influential physician Daniel Drake, invalids might discover "Excursions of health, science, and pleasure united": valuable geographic observation, lovely views, and improved health could all be found on the great grasslands of North America.[57] Similarly, military travelers might expect south-

---

[54] Ibid.; Gardner, "Missouri Merchants" (cit. n. 52).

[55] On the complex connections between eastern entrepreneurs and the regions they helped shape, see Jeffrey S. Adler, *Yankee Merchants and the Making of the Urban West: The Rise and Fall of Antebellum St. Louis* (Cambridge, 1991).

[56] John Kingsbury to James Webb, Salem, Mass., 12 Jan. 1857 and 12 Nov. 1856, Webb Papers.

[57] Drake, *Systematic Treatise* (cit. n. 8), 1:174.

[58] Henry Smith Turner, *The Original Journals of Henry Smith Turner with Stephen Watts Kearny to New Mexico and California, 1846–1847,* ed. Dwight L. Clarke (Norman, Okla., 1966), 139; Millikin, "Notes" (cit. n. 25), 578.

western service to improve their well-being; financial and medical imperatives were often explicitly linked in advice literature.[58]

Kate's eventual travel was not solely a business decision. William Messervy's letters, for instance, make clear that he, too, felt that his sister's only (slim) hope of survival was that she would "have strength enough to go as far as Santa Fe."[59] Indeed, medical reports reinforced the possibility of drastic cure, even for those deeply compromised by consumption, of a trip to the southwestern highlands. For Kate, going was also about being with her husband. During her last bout of illness she hated to be separated from John even for a night or two, as when the business of buying stock took him briefly away from Salem. She knew the needs of the spring merchandise run would take him away for a long time, and her only possibility to be with him—as well as perhaps to arrest her decline—was to accompany him on the wagon train. Kate Kingsbury traveled because she was sick, because of her affinity and social duty as a wife, and because she was intimately involved in an international trading network that sustained her family. This was how an elite American woman of the long-settled, urbane East Coast came to die in a canvas enclosure on a western pack train, far from home and material comfort.

Kate Kingsbury's travels for health on the Santa Fe Trail reveal much about nineteenth-century health choices, yet they could only have taken place during a certain period of time. By the late 1850s, the trail itself was becoming much less important than it had been in the tumultuous early decades of the Mexican Republic; simultaneously, the American sense of "healthy" places was undergoing change. By the 1870s, Colorado and areas further west had become the sought-after destinations of health pilgrims, who were lured out there by easy rail connections and the concomitant, aggressive promotion of western resort towns.[60] Santa Fe continued to promote itself as a health resort, especially for those with what was by then called tuberculosis, but it was the desperate gasp of a dying town.[61]

Yet even in its particularities, the structure of this story is revealing. Women participated in health travel much more than has been usually recognized—not as neutral bodies but in gender-laden ways. Even in her participation in the rough travel of the Santa Fe Trail, Kate Kingsbury ratified her era's truths of women's lives and women's well-being.[62] For both women and men, concern for health could be a concern for family economic connections as well. These letters demand that we recognize that a masculine business role could include fretting over the well-being of family members. The female world of health decision-making and the male world of economic interaction were much more tightly interwoven than we often acknowledge, especially in regard to health travel.

As this account demonstrates, looking at women's western experience can help reveal aspects of U.S. western history—and the history of the U.S. *east*. Kate Kingsbury's

[59] William S. Messervy to James Webb, Salem, Mass., 10 Dec. 1856, Webb Papers.

[60] Gregg Mitman, "Hay Fever Holiday: Health, Leisure, and Place in Gilded-Age America," *Bull. Hist. Med.* 77 (fall 2003): 600–35; Long, "Weak Lungs" (cit. n. 21), 1048.

[61] Dary, *Santa Fe Trail* (cit. n. 1), 301. See also *The Climate of New Mexico and Las Vegas Hot Springs* (Chicago, 1883); Edward Willcocks Meany, *Santa Fe, as a Health Resort* (Santa Fe, 1890); Bureau of Immigration of the Territory of New Mexico, *Santa Fe County New Mexico,* Bulletin no. 22 (Santa Fe, 1902), 13, 21–3.

[62] This was true of many overland trail experiences. Faragher, *Women and Men on the Overland Trail* (cit. n. 23), especially chap. 4, "Masculine Men and Feminine Women."

health travels help us understand how region and locale came to be defined in this period not only through commerce, military takeover, and cultural interaction but also through the lived realities of sick or healing bodies. The West was a place of danger, potential, and health; the East represented for this family a complicated combination of security, family connection, and threat of illness. Because these decisions about health travel were recorded in the correspondence of overland commerce, we learn intimate details of health travel that might otherwise have been lost in the robust and hearty encouragement of male-oriented advice literature. This narrative also demonstrates that for those going west as for those staying home, health was understood at every point in the context of gender and family relations. Moreover, even in areas that were mostly male—Santa Fe, the Santa Fe Trail, large regions of the American West in the nineteenth century—taking seriously women's experience tells us much about both women's and men's lives.[63] Here, Kate Kingsbury's health choices reveal concerns that could dictate the travels of ambitious long-distance traders as well as the economic realities that could determine the therapeutic decisions of a woman, wife, and mother ailing in spirit and in body.

Most of all, this story insists to us that going to a healthy place was a powerful, complex decision in the nineteenth century. Such decisions were informed and constrained by a host of factors involving trading networks and family ties, personal affinity and financial responsibility, calculations of climatic benefit as well as calculations of credit. Kate Kingsbury's illness and death show us moderns the many repercussions of the search for healthy places in the familial and economic worlds of the nineteenth-century United States.

---

[63] Katherine G. Morrissey, "Engendering the West," in *Under an Open Sky: Rethinking America's Western Past,* ed. William Cronon, George Miles, and Jay Gitlin (New York, 1992), 132–44, 139.

Geographies of Hope:
Mining the Frontiers of Health in Denver and Beyond, 1870–1965

By Gregg Mitman[*]

ABSTRACT

Across the western landscape of the United States, health was a natural resource, mined and sold by late-nineteenth and twentieth-century town boosters and physicians to those afflicted with chronic pulmonary illnesses such as tuberculosis and asthma. Regional economies of health were built upon climate and sunshine. After the Second World War, children, rather than nature, increasingly became a vital resource upon which institutions such as Denver's Jewish National Home for Asthmatic Children expanded the economic networks through which capital and drugs flowed. Despite these changes, the material traces of past landscapes lingered and resurfaced in the reconfigured places where hope dwelled.

INTRODUCTION

> What about this pulmonary cripple who has been swept aside by a fast moving world into the gutter of despair, gasping in the shadows of hopelessness, feebly holding on and with tired lusterless, fading eyes, scanning and searching the distant horizon for the sunshine of hope.
>
> —M. Murray Peshkin, 1955[1]

Where does hope dwell for the chronically ill? In a better place? In a child? In the curative promise of medicine? Sufferers seek hope—medicine trades upon it. In service of this relationship between wish and promise, lives have been relocated, cities have been built, and fortunes have been made.

Hope may spring eternal, but the location of the promised land for those pursuing health and those selling it has changed with the shifting contours of disease, environment, and place. To people suffering from pulmonary troubles in late-nineteenth-century

[*] Department of Medical History and Bioethics, University of Wisconsin–Madison, 1300 University Avenue, Madison, WI 53706; gmitman@med.wisc.edu.

This project was supported by National Science Foundation grant SES-0196204. I owe special thanks to Victoria Elenes for her invaluable research assistance and to the graduate students in my seminar "Geographies of Science, Technology, and Medicine" for the lively and thoughtful discussions. Thanks also to Jeanne Abrams at the University of Denver for her archival assistance and to my colleagues Michelle Murphy and Christopher Sellers, as well as the participants in the workshop "Environment, Health, and Place in Global Perspective," for their helpful comments and support.

[1] M. Murray Peshkin, "From a Home to a Hospital," 1 Aug. 1955, Box 10, Folder 7, National Asthma Center Archives, Special Collections, Univ. of Denver (hereafter cited as NAC).

America, the western frontier seemed to be that land, an image boomtowns such as Colorado Springs and Denver capitalized on. Climate, air, and sunshine were marketable health commodities, sold by railroads, civic boosters, and physicians to consumptives and asthmatics looking for relief, if not a cure. Those natural resources became as vital to the economies of these towns as their deposits of gold and silver. Across the western landscape, health itself was transformed into a natural resource, integral not only to local and regional economies but also to the identity and authority of physicians promoting the therapeutic benefits of the places in which they practiced and lived.

The regional physical environment proved an asset to the health promotion of place well into the twentieth century. By the 1950s, however, medical professionals had begun selling the biomedical, rather than the western, frontier as the promised land for relief from chronic asthma. Yet prominent institutional centers for the research and treatment of asthma were deeply embedded in, and profited from, the history of western cities as a last resort and hope for those afflicted with chronic pulmonary illnesses. Nowhere is this more evident than in the mile-high city of the plains, Denver. As the city's consumptive population declined in the 1930s, institutions such as the National Home for Jewish Children, which began as a sheltering home for orphans or dependents of Jewish tubercular parents, shifted care and treatment to asthmatic patients. Initially, the National Home had been dependent upon Denver's physical geography and regional economy. By the 1960s, however, this institution, which became the Children's Asthma Research Institute and Hospital in 1957 (CARIH), had positioned itself to be an important node in the national and transnational circulation of capital, resources, and products in the medicalization of asthma and the mining of new health frontiers and markets. CARIH was able to do so largely because of its history as a Denver orphanage, which enabled it to successfully forge links to John F. Kennedy's Alliance for Progress in Latin America and Lyndon B. Johnson's war on poverty. Johnson's presidential proclamation of National CARIH Asthma Week in 1967 looked to the promise of medical research, "valiantly seeking the cause and cure of asthma" for an estimated 3 million children living "as captives of asthma." Yet CARIH had been founded upon a landscape of hope initially embodied not in the universal prospects of medicine but in the localized therapeutic qualities of Colorado's foothills and mountains.

While forests, minerals, and game have been staple subjects of environmental historians in understanding the historical relations between economy and ecology in landscape change, the subject of health has been relatively absent from such scholarship. My aim in this essay is to suggest by way of modest example one possible avenue for creating a fusion of environmental and medical history. The changing ecology and economy of health in the Rocky Mountain region illustrates how central a role health as a resource and commodity has played in the material and social relations embedded in place. It is a story that could easily be extended to other cities, such as Tucson, Arizona, or Warm Springs, Georgia, that first capitalized on the health benefits of nature. Aspects of Denver's economic and urban geography, respiratory illnesses such as asthma and tuberculosis, and the bodies that contained them were made and remade. However, as this essay argues, the material traces of past landscapes lingered and resurfaced in the reconfigured places where hope dwelled.[2]

---

[2] In attending to how political economy shapes the production of space and place, I am most indebted to the tradition of historical-geographic materialism exemplified in works such as Henri Lefebvre, *The Production of Space* (New York, 1991); David Harvey, *Justice, Nature, and the Geog-*

SELLING PLACE

"sought gold here, but found a greater treasure—good health"
—George Ragan, *Colorado and Asthma,* 1873[3]

Denver was founded on the gold of prospectors, who came by the thousands in 1859 following strikes on the South Platte and the mountains to the west. By the 1870s, however, the arrival of the Denver Pacific, the Kansas Pacific, and the Denver & Rio Grande Railway had transformed this mining camp into the Queen City of the Plains, a place that traded not only upon its vast wealth of mineral resources but also upon a natural resource whose value, according to the Union Pacific Railway, was "above rubies": health.[4]

After the Civil War many who suffered from pulmonary troubles uprooted families, homes, and businesses and headed west in search of health. By 1890, an estimated 30,000 invalids—the majority consumptives—had come to Denver seeking a fortune in health. They made up almost a third of the city's population of 100,000. During the next three decades, the size of Denver's population of health seekers climbed. By 1920, an estimated 40 percent of the city's 250,000 residents—100,000 "lungers," as they were known—had come in search of health.[5]

Denver's town boosters in the late 1800s were aware that nature had bestowed upon the region not only rich veins of gold and silver but also health-restoring qualities that seemingly breathed new life into weary bodies. The *Rocky Mountain News,* founded above a saloon in 1859 and the city's leading newspaper through the 1890s, began advertising the health benefits of the area in the early 1870s. When Grace Greenwood arrived via the Denver Pacific in August of 1871, she reported to her *New York Times* readers that already "the town is crowded with tourists and invalids, and I sometimes wonder that the overtaxed hospitality of the people here does not give out."[6] In the wake of the Leadville silver boom in 1877, the *Rocky Mountain News* editors echoed this latter sentiment when they expressed concern that Denver citizens had forgotten the "very great importance" of the "invalid patronage of Denver." They reminded readers, in an 1880 editorial, that the "thousands of families afflicted with consumption or asthma" residing in the city composed an economic resource akin to "permanently improving gold mines within our limits, in which every citizen shares the profits." To attract wealthy health seekers from the East who would bolster local businesses, the newspaper proposed building an oasis within the heart of Denver with modern sewers, cement sidewalks, and a hydrant system that would keep the streets free from dust, the latter being Denver's greatest drawback in attracting such patronage. "Were Denver

raphy of Difference (Cambridge, 1996); as well as to the work of environmental historians such as William Cronon, *Nature's Metropolis: Chicago and the Great West* (New York, 1991).

[3] *Colorado and Asthma* (Denver, 1874), 2.

[4] *The Union Pacific Tourist: Illustrated Sketches of the Principal Health and Pleasure Resorts of the Great West and Northwest, Embracing Yellowstone Park, Shoshone Falls and Yosemite and the Chief Points of Interest in the Rocky Mountain Region, All Most Easily Reached Via the Union Pacific Railway,* 3d ed. (Buffalo, 1886), 56. On Denver's early history, see Stephen J. Leonard and Thomas J. Noel, *Denver: Mining Camp to Metropolis* (Niwot, Colo., 1990).

[5] Estimates are from Susan Jane Edwards, "Nature as Healer: Denver, Colorado's Social and Built Landscapes of Health, 1880–1930" (Ph.D. diss., Univ. of Colorado, 1994), 127; Billy M. Jones, *Health-Seekers in the Southwest, 1817–1900* (Norman, Okla., 1967), 97; Sheila M. Rothman, *Living in the Shadow of Death: Tuberculosis and the Social Experience of Illness in American History* (Baltimore, 1995), 132.

[6] Grace Greenwood, *New Life in New Lands: Notes of Travel* (New York, 1873), 46.

sweet and clean," wrote the newspaper's editor, "she would have enough invalid patronage to fill a city of her present size. 'All' does not 'depend on the mine.'"[7]

Mining and health were both prominent features in the economic and physical geography of the Rocky Mountain region after the Civil War. In newspaper copy, railroad promotional brochures, and the accounts of health seekers, the parallels drawn
between the mining and health frontier abound. Prospective travelers on the Union Pacific Railway first learned of the distinctive qualities of the Rocky Mountain region as
a sanitarium before turning the pages of *The Union Pacific Tourist* to learn about the
geography and economics of Colorado's mining industry. Mining and health were
seen as important natural resources upon which the economic, material base of the
region depended and the primary reasons why, beginning in the 1870s, hundreds of
railroad travelers arrived each day in Denver.

Hope propelled both miners and health seekers to Colorado, but the search for
health was often characterized as the nobler calling. This was, at least, the moral lesson to be learned from Helen Hunt Jackson's 1878 novel, *Nelly's Silver Mine*. Jackson had been a resident of Colorado Springs for four years when she composed her
novel for young women, drawing extensively upon her experiences as a health seeker
in the West. The daughter of Deborah and Nathan Welby Fiske, Jackson spent her
youth in Amherst, Massachusetts, in a household saturated with the presence of God
and death. Like her mother, Jackson exhibited literary talent—and a weakened constitution of the lungs.[8]

The search for health inspired Helen Hunt Jackson's westward journey, following
the deaths of her husband and two sons, as it does the March family's in *Nelly's Silver Mine*. Unable to preach in the summer because of his debilitating wheezing, Mr.
March makes the painful decision to leave his beloved New England congregation and
move his family west upon reports of an acquaintance and a physician that Colorado's
climate makes asthma unknown there. The Marches are accompanied on their trip by
a deacon and his wife, whose persistent coughing marks the advanced stages of consumption. In this novel of Colorado life, hope is linked to the material prospects of
both health and mining. The Marches' journey and new life are filled with numerous
adventures, especially when daughter Nelly finds an unworked silver mine that in-
spires the entire family to make future plans. When it is discovered that the mine's ore
has little value, the entire community, in turn, is inspired by Nelly's good-hearted
spirit and resolve in making the best of a disappointment that had broken "many a man
in this country." The Marches had reason enough to be thankful, for their father's
health was restored, and Nelly's brother, who had been a sickly, delicate child in New
England, was growing stronger by the day.[9] "The contagion of the haste to be rich is
as deadly as the contagion of disease," wrote Jackson after Colorado's economy fal-
tered when silver prices plummeted in the early 1890s. Keenly aware of the reckless
mining expansion and precariousness of Colorado's future, Jackson saw the state's
destiny not in the overexploitation of its mineral resources, but in its climate, in which
"asthma, throat diseases, and earlier stages of consumption" were "almost without exception cured by [the] dry and rarefied air."[10]

---

[7] Editorial, *Rocky Mountain News,* 2 March 1880; ibid., 11 Aug. 1880.
[8] For biographical details of Helen Hunt Jackson's family and life, see Ruth Odell, *Helen Hunt Jackson* (New York, 1939); Rothman, *Living in the Shadow of Death* (cit. n. 5), 83–127, 168–75.
[9] Helen Hunt Jackson, *Nelly's Silver Mine: A Story of Colorado Life* (Boston, 1878), 378.
[10] Helen Hunt Jackson, *Bits of Travel at Home* (Boston, 1898), 384, 226.

What were the qualities that set Denver and other Colorado towns apart in the health promotion of place? To many health seekers, "the pure dryness of the atmosphere, through which flow the constant currents of fresh air from the mountains," was "a wonderful and beneficent thing."[11] Among the many phrases invalids and railroad advertisements used to describe the air of the Rocky Mountain region were "buoyant and delicious," "bracing and exhilarating," "the ethereal properties of champagne," and "no subtle, malarious taint."[12] To those who found hope and a cure, "hallowed" could also be added to the list of descriptive adjectives. Mrs. W. H. Smith captured the feelings and thoughts of more than a hundred asthmatics gathered at Denver's Fifteenth Street Presbyterian church in 1873 when she spoke of how "thankful" she was every day "that there is a place in God's world, that I can breathe His air with comfort and ease."[13] Grace Greenwood similarly told "confirmed asthmatics,—unhappy men and women who, like shipwrecked mariners, perishing of thirst, with 'water, water everywhere,' gasp and fight for their scanty breath in a world of air,—they who find it not difficult to realize the sufferings of men suffocated in mines . . . "—that she did "not believe there is out of Heaven such a place as the mountain land of Colorado."[14]

God had bestowed upon Colorado a nature bountiful in health. "Scenic attractions and advantages of climate conspire[d] to make" the region "popular with tourists and healthseekers." Railroads such as the Denver & Rio Grande, however, viewed their capital investments as being built on "another nature," that of the "rich mineral deposits of Colorado." Nevertheless, as the region began to attract more and more invalids, enterprising physicians such as Charles Denison saw the "pecuniary advantages" offered by Colorado's beneficial climate to both the railroads and the medical profession.[15]

Denison migrated to Denver in 1873 after a pulmonary hemorrhage forced him to give up a newly established practice in Hartford, Connecticut.[16] During the Christmas season of 1874, he sent out a confidential letter to prominent physicians, wealthy businessmen such as Andrew Carnegie, and former territorial governors such as John Evans proposing the establishment of the The Climatic-Cure and Life-Prolonging Association. As envisioned by Denison, the association would operate along the lines of a life insurance company, offering reduced passenger rates to physicians holding

[11] Greenwood, *New Life in New Lands* (cit. n. 6), 47–8.

[12] Ibid., 48, 96; Denver and Rio Grande Railway Co., *Health, Wealth, and Pleasure in Colorado and New Mexico* (Chicago, 1881), 15.

[13] *Colorado and Asthma* (cit. n. 3), 15.

[14] Greenwood, *New Life in New Lands* (cit. n. 6), 95. For a sampling of other literature promoting the region's health benefits for asthmatics, see W. R. Whitehead, "Remarks on the Climatic Influence of Colorado in the Cure of Asthma, with a Review of a Large Number of Cases Reported by a Convention of Asthmatics, assembled at Denver, in December, 1873," *American Journal of Medical Science* 67 (1874): 388–95; H. A. Lemen, "Report on the Climate of Colorado in relation to Asthma," *First Annual Report of the Secretary of the State Board of Health, of the State of Colorado, for the Fiscal Year, Ending September 30, 1876* (Denver, 1877): 70–3; Frederick I. Knight, "The Climatic Treatment of Bronchial Asthma," *Transactions of the American Climatological Association* 6 (1889): 159–64.

[15] Denver and Rio Grande Railway, *Health, Wealth, and Pleasure* (cit. n. 12), 16. Charles Denison, Denver, 22 Dec. 1874, Charles Denison Reprint Collection, Denison Memorial Library, Univ. of Colorado Health Sciences Center.

[16] For biographical information on Denison, see A. McGehee Harvey, "The American Clinical and Climatological Association: 1884–1984," *Transactions of the American Clinical and Climatological Association* 95, suppl. (1983). "Charles Denison: [In Memoriam]," *Transactions of the American Climatological Association* 25 (1908): xxiii.

stock in the association and to their patients. Given the increased invalid traffic to Colorado, Denison was convinced that "railroads can carry no more remunerative freight . . . than that selected class of invalids *who get well* here, and vice versa, no less remunerative freight than that other class *who only come here to* die."[17] If a passenger boarded a train only after paying a fee and consulting an association physician about the climate and locale best suited to his or her health needs, disappointment and distrust could, Denison argued, be avoided.

Denison was trying to address a serious issue. Too many people were arriving back East in coffins after an unsuccessful trip west in search of health. Lieutenant Benjamin MacIntire complained to the editor of the *Boston Herald* in 1875 of the "injustice" being done to "those suffering with lung troubles" in the health promotion of Colorado as the "Switzerland of America." The truth, wrote MacIntire, was that Denver "has almost as many undertakers as New York, and of course all have to make a living."[18] Such bad publicity, harmful to the business of both western railroads and cities, could easily be averted, argued Denison, if health seekers first sought the advice of physicians, who would know better than to prescribe the climate cure for individuals with advanced stages of consumption or those suffering from cardiac diseases. Profits from the association would be invested in "building up and advertising the health-resorts of the country," which were "located in the sparcely [sic] settled regions at the extremities of the railroads," thereby strengthening the region's health economy.[19] Denison's company never materialized, but a decade later, he became one of the early members of the American Climatological Association, devoted to "the study of the climatology and diseases of the respiratory organs." Many of its officers, like Denison, had financial investments or lucrative practices in the healthful places they espoused.[20]

Railroads were not the only nineteenth-century technology vital to physicians and town boosters in mining the West's therapeutic landscape after the Civil War. So, too, was the telegraph. In both medical practice and civic promotion, western physicians such as Denison became increasingly reliant on precise meteorological data and statistics in matching the particularities of patients and their respiratory ailments to specific geographic locales. The list of meteorological and physical geographic factors doctors attended to was extensive: temperature, relative humidity, barometric pressure, atmospheric electricity, wind direction, rainfall, diathermancy, cloudiness, and soil types all became important in distinguishing the therapeutic benefits of one location from those of another. None of these observations would have been possible without the U.S. Signal Service.

Established by Congress in 1870, the Signal Service operated a network of weather stations linked to commercial telegraph lines, which gradually moved westward with the commercial expansion of the frontier. In 1873, the service established the highest meteorological station in the country, and the first in Colorado, atop Pikes Peak. Financed by General William Jackson Palmer, a Civil War hero and Union Pacific surveyor, the seventeen-mile telegraph line further tied health to the region's economic

---

[17] Denison, 22 Dec. 1874 (cit. n. 15).

[18] Benjamin F. McIntire, "The Climate of Colorado," *Boston Herald,* 15 Feb. 1875. Reprinted in *Rocky Mountain News,* 24 Feb. 1875.

[19] Denison, 22 Dec. 1874 (cit. n. 15).

[20] Harvey, "American Clinical and Climatological Association" (cit. n. 16), 8. See also Jones, *Health-Seekers in the Southwest,* 123–49; Rothman, *Living in the Shadow of Death,* 150–60. (Both cit. n. 5)

prosperity.[21] Two years earlier, Palmer and his partner, Dr. William A. Bell, had amassed enough capital to build a narrow-gauge branch of the Denver & Rio Grande from Denver to a newly created resort for wealthy invalids—Colorado Springs. Meteorological data gathered from the Pikes Peak and surrounding stations was worth its weight in gold in promoting the climatic and therapeutic virtues of Colorado Springs. Samuel Edwin Solly, an English physician recruited by Palmer and Bell in 1874, claimed himself living proof of the region's health benefits and backed local empirical knowledge with quantitative weather data in his popular 1883 book, *The Health Resorts of Colorado Spring and Manitou.*[22] To medical climatologists, the sheer mass of data offered important corollary facts to the personal testimony of asthmatics, consumptives, and others proselytizing the climate cure. Denison, too, utilized the Signal Service data to produce a highly detailed health resource map of the eastern mountain slope of Colorado. It included colored elevation and isothermal lines of mean annual temperature and rainfall as well as average seasonal temperatures, relative humidity, wind speed and direction, and cloud cover from Cheyenne, Wyoming, south to Las Vegas, New Mexico. (See Figure 1.) Included in an 1880 book by Denison, *Rocky Mountain Health Resorts,* the map fit conveniently in a pocket at the end of an advertisement section for western railroads and Denver hotels and businesses in which Denison had vested interests.[23]

By the end of the nineteenth century, railroad financiers, enterprising physicians, and town boosters had built a regional economy in which health as a natural resource played a significant role in western towns such as Denver and Colorado Springs. To those in search of wealth or health, hope dwelled in the physical landscape. Those searching for wealth might find it in Colorado's mineral deposits and deep sandy loams of the earth; those searching for health might find it in the air, climate, and sunshine that enveloped the region. To many of the latter, the "clear, brilliant, intoxicating air" was as tangible as the dry, loose soil of the plains or the cold, shiny gray ore of the mine. With the aid of the railroad and telegraph, physicians surveyed, extracted, and marketed health as a natural resource, but throughout the late nineteenth century, hope rested not in the medical profession but in the salubrious environment. The promotion of Denver and the Rocky Mountain region as a sanitarium, a word René Dubos suggests remained popular "as long as the faith in the healing power of nature prevailed," signified that hope resided first and foremost not in medicine but in nature.[24]

---

[21] On the establishment of the Signal Service, see James Rodger Fleming, "Storms, Strikes, and Surveillance: The U.S. Army Signal Office, 1861–1891," *Historical Studies in the Physical and Biological Sciences* 29 (1999): 315–32; idem, *Meteorology in America, 1800–1870* (Baltimore, 1990); Patrick Hughes, *A Century of Weather Service: A History of the Birth and Growth of the National Weather Service, 1870–1970* (New York, 1970); Phyllis Smith, *Weather Pioneers: The Signal Corps Station at Pikes Peak* (Athens, Ohio, 1993).

[22] S. Edwin Solly, *The Health Resorts of Colorado Springs and Manitou* (Colorado Springs, 1883). See also idem, "The Comparative Merits of Resorts in New Mexico, Colorado, and Arizona," *Transactions of the American Climatological Association* 13 (1897): 171–85.

[23] Charles Denison, *Rocky Mountain Health Resorts: An Analytical Study of High Altitudes in Relation to the Arrest of Chronic Pulmonary Disease* (Boston, 1880). On the early use of isothermal lines in medical cartography, see Nicolaas Rupke and Karen E. Wonders, "Humboldtian Representations in Medical Cartography," in *Medical Cartography in Historical Perspective,* ed. Nicolaas Rupke (London, 2000), 163–75. On Denison's financial investments in Denver, see Edwards, "Nature as Healer" (cit. n. 5), 110.

[24] Woods Hutchinson, "Climate and Health," *The Outing Magazine* 53 (1909): 751. René J. Dubos, "The Philosopher's Reach for Health," *Transactions of the Association of American Physicians* 66 (1953): 36.

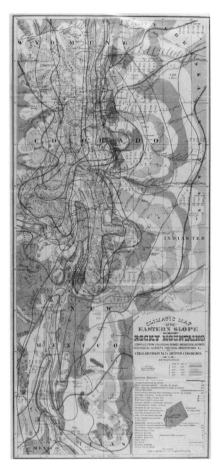

Figure 1. *This map of the eastern mountain slope of Colorado featured in Charles Denison's 1880 book,* Rocky Mountain Health Resorts, *served as a valuable resource map for health prospectors. (Courtesy of Middleton Health Sciences Library, University of Wisconsin–Madison, Madison, Wisconsin.)*

SHIFTING LANDSCAPES

Throughout the late nineteenth century, Denver appealed as a last resort to health seekers afflicted with a variety of pulmonary diseases. Although outward signs marked differences among those suffering from lung troubles, the curative benefits of the Rocky Mountain region as a sanitarium made little distinction. By the early 1900s, however, the growth of sanatoria, the clinical mark of bacteriology, and the identification of tuberculosis as a disease of the poor had physically and socially isolated tuberculosis sufferers from others afflicted with breathing difficulties in the city. To towns built on health as a natural resource, contagion and poverty were certainly less attractive than wealthy invalids. Medical climatotherapy suddenly appeared a less promising medical and business prospect.

In the rise of the tuberculosis sanatorium in the early 1900s, the healing powers of the physician gradually replaced those of nature. Sanatorium, Dubos notes, "from *sanare* (to treat), replaced sanitarium," from *sanitas* (health), as the word of choice "when active forms of treatment such as collapse therapy, surgery, and chemotherapy

became the vogue" for the care of consumptives.[25] The American sanatorium movement, which began in 1885 with the establishment of Edward L. Trudeau's famous
Adirondack Cottage Sanatorium at Saranac Lake, shattered the consumptive's belief
in the restorative powers of nature. Certainly the establishment of sanatoria in western cities such as Denver, which in 1925 had 11 of the nation's roughly 500 sanatoria, owed much to the history of the Rocky Mountain region as a health resort for
those suffering with lung illnesses. However, the closed institutional model that developed in America undermined the importance of regional geography in tuberculosis treatment, investing power in the physician instead. Colorado doctors such as
Samuel Fisk raised objections to Dr. Paul Kretzshmar's claims that the open-air resorts favored by Americans violated the principles and practice of germ theory. But
Kretzshmar's appeal to Robert Koch in introducing the German sanatorium model,
based upon close medical supervision and confinement, had by the early 1900s
gained a considerable foothold in the medical care and treatment of consumption in
the United States.[26]

Although by the early 1920s physicians had extricated their identity and authority
from place in the treatment of tuberculosis, such was not the case for asthma. If the
science of bacteriology offered a powerful fulcrum in shifting tuberculosis from the
resources of nature to those of the laboratory, the science of immunology strengthened the bond between asthma and the health benefits of nature. In 1910, Rockefeller
Institute researcher S. J. Meltzer suggested that the constriction of the bronchi observed in an asthmatic attack resulted from an anaphylactic reaction. Meltzer's observation helped shift medical attention away from asthma as a nervous disorder (the
nineteenth-century view) to that of an immunological disease, though the two concepts would continue to intersect in medical and popular thought.[27] Understanding
the local environment of the asthmatic sufferer became critical to diagnosis and treatment. In the United States, the clinical specialty of allergy, which fell under various
headings (such as "protein sensitization," "human hypersensitiveness," and "clinical
anaphylaxis"), first developed around a set of practices that included preparation of
pollen extracts, skin testing, and hyposensitive injection treatments. Pollen extracts
quickly became an important diagnostic tool for physicians since only allergy sufferers would exhibit skin reactions to small doses of particular protein extracts applied through scarification of the skin or intradermal injections. By 1920, there were
at least a half dozen clinics—located in Boston, Chicago, New York City, New Orleans, and San Francisco—testing allergy sufferers and administering desensitiza-
tion treatments, which often included a combination of pollen and vaccine therapies.

[25] Dubos, "Philosopher's Reach for Health" (cit. n. 24), 36.
[26] See, e.g., Barbara Bates, *Bargaining for Life: A Social History of Tuberculosis, 1876–1938*
(Philadelphia, 1992); Mark Caldwell, *The Last Crusade: The War on Consumption, 1862–1954* (New
York, 1988); Rothman, *Living in the Shadow of Death* (cit. n. 5). Denver sanatoria statistics are from
Edwards, "Nature as Healer" (cit. n. 5).
[27] S. J. Meltzer, "Bronchial Asthma as a Phenomenon of Anaphylaxis," *Journal of the American
Medical Association* 55 (1910): 1021–4. See also M. B. Emanuel and P. H. Howarth, "Asthma and
Anaphylaxis: A Relevant Model for Chronic Disease? An Historical Analysis of Directions in Asthma
Research," *Clinical and Experimental Allergy* 25 (1995): 15–26; Kenton Kroker, "Immunity and Its
Other: The Anaphylactic Selves of Charles Richet," *Studies in the History and Philosophy of Biolog-
ical and Biomedical Sciences* 30 (1999): 273–96; Mark Jackson, "Allergy: The Making of a Modern
Plague," *Clinical and Experimental Allergy* 31 (2001): 1665–71.

The unreliability of pollen extracts, however, and limited samples of pollen-producing plants from different regions in the United States, created a great demand for botanical knowledge. Almost all U.S. allergy clinics during the thirties and forties worked closely with botanists and meteorologists in developing standardized pollen extracts and in conducting regional surveys of climate and vegetation important to their therapeutic practices. Climatology and regional natural history, which fit readily into a Hippocratic model of disease to which medical climatotherapy ascribed, were widely employed in the research, treatment, and practice of allergy well into the 1950s.[28]

The shift in the interior geography of asthma from a constitutional nervous disorder to an allergic reaction marked an important ecological shift in the geography of hope. Nature, beneficial in some places, could be harmful in others. Only the physician, through intimate regional knowledge and through skin welts left on the body, could detect which parts of nature were corrosive. For the intractable asthmatic, who failed to respond to immunotherapy and for whom relocation to another environment offered the only promise for relief, the task of the physician was to know "what the patient [is] being taken away from, as well as what he is moving into."[29] This was the aim of the 1955 volume *Regional Allergy,* edited by Max Samter, chief allergist at the research and educational hospitals of the University of Illinois. Each of the chapters written by a resident physician covered the geography, climate, social structure, and allergenic factors for one of thirty-nine different regions across the United States, Canada, Mexico, and Cuba "to acquaint physicians and patients, at least in broad terms, with the territory in which they are to live." Not unlike the authority of medical climatologists of the late nineteenth century, that of the physician treating asthma before the advent of antihistamines and corticosteroids rested largely in the power of place.[30]

The beneficial climates of Denver and Tucson, among other locales, continued to serve as havens for intractable asthmatic sufferers into the 1950s. In his contribution on Colorado, eastern Wyoming, and western Nebraska for *Regional Allergy,* physician Frank Joyce noted that the "family histories of the present generation of allergic patients frequently reveal that an ancestor came here, and 'never had any more asthma.'" Joyce even speculated that the lower net incomes, salaries, and wages of Colorado compared with those of other states might be attributed to the fact that people came "to Colorado either because of their health or the climate and [were] willing to accept a lower income in order to remain here." The high ratio of physicians to other professional groups in Denver also reflected the importance of health to the

---

[28] On the history of allergy clinics in America, see Sheldon B. Cohen, "The American Academy of Allergy: An Historical Review," *Journal of Allergy and Clinical Immunology* 64 (1979): 332–466; Gregg Mitman, "When Pollen Became Poison: A Cultural Geography of Ragweed in America," in *The Moral Authority of Nature,* ed. Lorraine Daston and Fernando Vidal (Chicago, 2003), 438–65; idem, "Natural History and the Clinic: The Regional Ecology of Allergy in America," *Stud. Hist. Phil. Biol. Biomed. Sci.* 34 (2003): 491–510. For a sampling of primary literature on climatotherapy in the practice of allergy, see, e.g., Editorial, "Institutions for the Care of Asthma," *Journal of Allergy* 25 (1954): 379; S. D. Klotz, "Environmental Climatologic Therapy in Bronchial Asthma," *Annals of Allergy* 14 (1956): 502–5; Meyer B. Marks, "Climate as an Influencing Factor in Childhood Allergy," *Annals of Allergy* 12 (1954): 403–8; Ian P. Stevenson, "The Therapeutic Effect of the Climate of Arizona," *Archives of Physical Medicine* 28 (1947): 644–52.

[29] Klotz, "Environmental Climatologic Therapy" (cit. n. 28), 504.

[30] Max Samter and Oren C. Durham, eds., *Regional Allergy of the United States, Canada, Mexico, and Cuba* (Springfield, Ill., 1955), xii.

city's early economic development.[31] The decline of tuberculosis during the 1930s and 1940s left a void in the beds and institutional marketplace that had catered to Denver's large tuberculosis population and upon which the city had partially been built. Intractable asthmatic children would partially fill that void. Separated in the early part of the century, the geographies of asthma and tuberculosis once again converged in the 1940s as residential asthma treatment centers, many of which had institutional affiliations to sanatoria, were established in Denver, Tucson, and other cities throughout the United States. Such homes proved critical in shaping a new landscape of hope for the asthmatic sufferer.

FROM HOME TO HOSPITAL

> "On that small 17 acres of land in Denver, *hope* was born for us very lucky kids!"
> —CARIH alumnus, 12 Nov. 1980[32]

Of the approximately dozen convalescent homes for asthmatic children established between the late 1930s and early 1960s, at least one-third were institutions founded in places where climate and the physical environment had long attracted those suffering from pulmonary illnesses. Two achieved national attention. In Tucson, the National Foundation for Asthmatic Children established the Sahuaro School in 1954, a sixty-bed residential facility built upon a rehabilitation program for low-income children with intractable asthma. The program was based on "climate, individualized medical care, change of environment," educational programs, and recreational activities. In Denver, the National Home for Jewish Children began admitting children with intractable asthma in 1939. It became one of the leading institutional centers for the research and treatment of asthma with the creation of the Children's Asthma Research Institute and Hospital in 1957.[33]

Both facilities, but particularly the Denver National Home for Jewish Children, were based upon a therapeutic treatment for intractable asthmatics proposed in 1930 by Dr. Murray Peshkin, chief of Mt. Sinai Hospital's Children's Allergy Clinic in New York. Peshkin argued that "sensitizing substances" were "merely exciting factors, not the basic cause of the symptoms" found in asthmatic children. Believing allergy the result of a "physiochemical disturbance" that "manifests itself through the nervous system," he attempted to restore the "physiochemical balance" by separating children who did not respond to conventional medical therapy from their home environments and placing them, for six months or longer, in convalescent homes prepared to accommodate the needs of allergic patients. He found that children with intractable asthma responded favorably to this environmental change, and in many cases their clinical symptoms disappeared without medication. A decade passed, however, before anyone acted upon Peshkin's plea for a home that could administer his parentectomy treatment.[34]

---

[31] Frank T. Joyce, "Colorado, Eastern Wyoming, and Western Nebraska," in ibid., 312.

[32] CARIH alumnus, 12 Nov. 1980, Box 14, Alumni Correspondence, 1980 Folder, NAC.

[33] W. B. Steen, "Rehabilitation of Children with Intractable Asthma," *Annals of Allergy* 17 (1959): 871. See also M. Murray Peshkin, "Survey of Convalescent Institutions for Asthmatic Children in United States and Canada," *Journal of Asthma Research* 2 (1965): 181–94.

[34] M. Murray Peshkin, "Asthma in Children: IX. Role of Environment in the Treatment of a Selected Group of Cases: A Plea for a 'Home' as a Restorative Measure," *American Journal of Diseases of Children* 39 (1930): 776.

Founded in 1907, the National Home for Jewish Children was organized by Fanny Lorber and Bessie Willens as a sheltering home for Denver's underprivileged Jewish children. Most of its residents were orphans or dependents of poor eastern European immigrants who had come to Denver in search of a cure for tuberculosis and found hope instead in the free sanatorium care offered by the Jewish Consumptive Relief Society. By the early 1920s, the home had established a separate hospital and began accepting Jewish children on a national basis. The majority, however, came from referrals of charitable organizations, social work agencies, and hospitals servicing the needs of New York City's Jewish population. Peshkin probably became familiar with the National Home, and its possibilities as a care facility for asthmatic children, through his active role in Jewish causes within New York City. His interest in catering to a new population of children came at an opportune time. The number of applications for children with tubercular parents was declining. By 1947, the admission of thirty-eight allergic or asthmatic children compared with six children of tubercular parents marked a turning point in the institution's focus. Six years later it changed its name to the Jewish National Home for Asthmatic Children.[35]

Lorber and Peshkin initially promoted the beneficial climate afforded by the mile-high dry air of Denver as an important environmental factor in the rehabilitation of the asthmatic child. The city's long history as a therapeutic landscape thus occupied an important place in the promise of hope offered by the National Home. Staff soon found, however, that change of climatic environment was not in itself a sufficient treatment for a portion of children admitted. As Peshkin noted:

> I told the parents of one child that their son needed a radical change of treatment and suggested, among other procedures, the removal of the boy to a high and dry altitude. With little concern for my plan of treatment, the father with his entire family went West and to deprivation. The boy remained asthmatic. That was a tragic blunder because the boy required not only a change of climate and environment but also medical supervision and separation from his parents, in other words "parentectomy!"[36]

The embrace of a medical institutional model hinted at in Peshkin's anecdote occurred slowly and was first centered in psychiatry. In the late forties, for example, the number of psychiatric and social work staff in the National Home's Social Service Division far exceeded the number of attending physicians in its Medical Service Division (two). The large Social Service Division also indicated the increased attention paid by Peshkin and other clinical allergists to the emotional and psychological environment of the asthmatic child. By the 1950s, one could pick up any popular health magazine and read that "deep seated emotional conflicts between the child and his parents" or an "over-protective mother" played a primary or secondary role in the development of asthma.[37] Consequently, although the home still traded upon the promise offered by Denver's climate and physical geography, it increasingly focused at-

[35] On the early history of the Denver National Home, see Marjorie Hornbein, "National Jewish Hospital and National Asthma Center Merge: A Historical Perspective," *Rocky Mountain Jewish Historical Notes* 1 (1977): 3–6; Fannie Lorber, "The Story of the National Home for Jewish Children at Denver," Box 11, Histories, 1928–83 Folder, NAC. On the Jewish Consumptive Relief Society, see Jeanne Abrams, *Blazing the Tuberculosis Trail: The Religio-Ethnic Role of Four Sanatoria in Early Denver* (Denver, 1990).

[36] M. Murray Peshkin, "The Child Rescue Work of the National Home for Jewish Children," Box 11, Histories, 1929–83 Folder, NAC.

[37] John E. Allen, "How We Help Asthmatic Children," *Saturday Evening Post*, 3 Sept. 1960, 54.

tention upon the places of hope being built inside its walls through therapeutic treatment and research.

Peshkin's emphasis on the emotional environment as a contributive factor in asthma reflected the prevalence of psychosomatic explanations given for illnesses throughout the 1950s. This explanation was also nourished by the home's institutional history as an orphanage and refuge for other children, one that ministered to the health, education, religious training, physical well-being, vocational opportunities, and recreational activities of the underprivileged child. Each of these continued to occupy a place in the home's rehabilitation program for asthmatic children. Traces of the past also appeared in how asthma's social environment was understood. Poverty became incorporated into the new architecture of the National Home and asthma as a disease. When asked whether the "ravages of asthma are more prevalent among the poor," Peshkin answered that the disease was "more severe, more continuous, and more prevalent in the under-privileged class than the so-called privileged class." He cited inadequate food, overcrowding, and "undesirable, unprepared and untidy homes" as a few of the reasons. The ghosts of tuberculosis continued to haunt the halls of the National Home, even as the institution sought to place itself on a new foundation.[38]

In the early fifties, the National Home for Jewish Children struggled to establish credibility at a national level. At times, the staff members felt it "was neither a hospital nor an orphanage" but simply "a 'glorified home' for parentectomy."[39] In 1951, the entire social service staff quit, the resident allergist resigned, a measles epidemic swept the dormitories, and a child died, but when the storm had passed, it left a brighter Denver sky. The major reorganization that followed saw a significant expansion of medical staff, including the hiring of a medical director, the creation of an Allergy Section, the reestablishment of a Counseling Department run by an attending psychiatrist, and the hiring of four caseworkers trained in psychology.[40] By 1954, the Jewish National Home for Asthmatic Children had become a 112-bed facility with sixty-two staff members, including doctors, nurses, house parents, social workers, and psychiatrists. "Hope," Fannie Lorber told the audience at the National Home's forty-ninth Annual Convention of Auxiliaries in 1956, now lay in the "laborious work of research."[41]

The National Home did indeed present an "ideal laboratory" for biomedical research on children's asthma.[42] Few other U.S. hospitals or convalescent homes housed such a large number of asthmatic children under daily medical supervision who stayed an average of two years. To medical technicians, the children were model research subjects. One staff member remarked, for example, that she took a research position at the home reluctantly, based upon her past experience of screaming child patients "being dragged into the Lab." In Denver, she found her experience the complete opposite. Not only were the children not afraid, having been poked and prodded by medical experts all their lives, but also the social service staff had done a remarkable

---

[38] M. Murray Peshkin, "Questions and Answers," n.d., Box 9, Folder 15, NAC; Peshkin to Fannie Lorber, 22 April 1951, Box 5, Folder 5, NAC.

[39] Evelyn Walsh to Peshkin, 11 May 1951, Box 5, Folder 5, NAC.

[40] See Peshkin to Fannie Lorber, 25 Jan. 1951, Box 5, Folder 5, NAC; Walsh to Peshkin, 11 May 1951 (cit. n. 39); "Annual Report of the Medical Director," 1 Jan. 1952, Box 2, Meetings, 1952 Folder, NAC; "Agenda: Special Board Meeting," 19 Aug. 1956, Box 2, Minutes, 1955 & 1956 Folder, NAC.

[41] Lorber, "Welcome Address," c. 1956, Box 5, Folder 3, NAC.

[42] "Jewish National Home for Asthmatic Children at Denver," Box 11, Description of Programs, 1937– Folder, NAC.

job of "conditioning these children for the slight, but very real, pain which all patients have to experience in being scratch tested, having blood taken, [and receiving] the continuous injections which have to be given over a long period of time for inhalants to do any good."[43] Pharmaceutical companies, too, quickly saw the research opportunities presented at the Home as they sought outlets to test the first generation of antihistamines, such as Chlor-Trimeton, and corticosteroids, such as hydrocortisone, being developed for allergy sufferers. In December 1954, the home entered into agreement with major pharmaceutical firms—including Charles Pfizer & Co; Smith, Kline and French; National Drug Company; and Schieffelin & Co.—to test drugs in exchange for free pharmaceuticals as well as research grants, which resulted in considerable savings in medical supply costs.[44] In 1956, the home's board of trustees resolved to incorporate the affiliated Children's Asthma Research Institute and Hospital (CARIH), a nonprofit corporation aimed at establishing a "large-scale research program" and facilities in "the field of asthma and related diseases."[45] Aided by a $100,000 grant from the U.S. Public Health Service, CARIH opened its doors in 1959 and within a decade had grown into a world-renowned residential treatment and research center, with an annual net income in excess of $1 million in federal grants and charitable donations.[46]

Founded upon the economic and physical geography of the Rocky Mountain region as a health resort, the Jewish National Home for Asthmatic Children secured its economic future in the extraction of resources beyond the local environs of Denver. The changeover became all the more critical by the mid-sixties as many of the pollution problems confronting other American cities sullied Denver's salubrious environment. Rapid growth and the heavy reliance upon automobile transportation by the city's residents, coupled with high-altitude factors and restrictive meteorology, had resulted in severe air pollution problems, equal to, or greater than, those found in larger metropolitan areas.[47] Climate and sunshine offered the raw materials upon which a previous landscape of hope had been built. But children increasingly became a critical resource in mining the frontier of biomedical research and opening the Queen City of the Plains to a transnational market through which capital and drugs flowed.

THE PROMISE OF CHILDREN

In the rapidly expanding postwar economy of biomedical research, administrators and physicians at the National Home knew they were sitting on ore that far exceeded Denver's rich natural health resources: children. If the landscape of tuberculosis shaped the home's regional past, that occupied by polio on a national scale shaped the home's future. No other charitable, disease-based organization had so successfully

[43] Evelyn Walsh, "The Aims of a Medical Laboratory," 9 March 1949, Box 1, Minutes of Board, 1948 & 1949 Folder, NAC.

[44] See, e.g., "Annual Report of the Medical Director," 1954–1955, Box 2, Minutes, 1955 & 1956 Folder, NAC; "News from the Home Front," Sept. 1952, Box 9, Folder 3, NAC.

[45] "Agenda: Special Board Meeting," 19 Aug. 1956" (cit. n. 40).

[46] On the U.S. Public Health Service grant, see "President's Annual Report," 21 Jan. 1959, Box 2, Minutes, 1957–58–59 Folder, NAC. For income statistics, see Executive Director's Report, 1972, Box 11, Jack Gershterson Report, 1972 Folder, NAC.

[47] Colorado Department of Health, *Colorado Air Pollution Control Transportation and Land Use Plan* (Denver, 14 May 1973); Air Pollution Control Division, *Assessment of Air Quality: Metro Denver Region* (Denver, Aug. 1975).

marketed children for medical research as the National Foundation for Infantile Paralysis did during and after World War II. Its March of Dimes campaign, launched in 1937 by President Franklin Delano Roosevelt, had, in less than eight years, raised $25 million in funds for the National Foundation. Through its poster children, the foundation, aided by its organizational base of women, tugged at the emotions and opened the pocketbooks of Americans by portraying polio as the childhood crippler. Asthma's disfiguring effects were not as visible as polio's, but the National Home, in its efforts to get itself and asthma placed on the federal agenda for biomedical research, followed the path laid by the National Foundation.[48] (See Figure 2.)

Before its success on the federal research front, the National Home was funded almost entirely through national mail campaigns and local chapters run by female volunteers, who met once a year at the national auxiliary convention, where purses were collected. Geography and ethnicity, however, severely restricted the home's economic base. During the forties and early fifties, Lorber and other board members maintained a rigid admission policy that excluded non-Jewish children, even the severest cases. Peshkin thought it an embarrassment and believed that unless the policies were changed, the National Home would never be able to capitalize on its greatest asset. "From a standpoint of publicity," Peshkin told Lorber sharply, "there is no boundary we could not reach. . . . [T]he child is the best medium for making appeals for funds and attracting the public's attention. . . . We, who have a 100 percent child-caring program and have something constructive to offer, hide in the shade of prejudice, ignorance and, for shame, bigotry."[49] Closed regional networks also hampered the institution's ability to catapult itself into the national arena. In 1951, a major scandal broke out behind the scenes when the Chicago chapter complained to the board of trustees that it "saw no reason why the good people of Chicago should spend their money and their efforts in collecting money for the home when the number of children from Chicago was so very few." Chapter members could not understand why more kids from Chicago, a city with the second largest Jewish population in the United States, could not get into the home. Statistics confirmed perceptions. Of the 158 children residing in the National Home in 1950, only 1 was from Illinois; more than 100 were from New York. To move beyond the local arena, the home would have to overcome deep-seated policies that had entrenched it in limited regional and ethnic networks. Children's bodies would have to be made universal.[50]

Economic motives and medical recognition forced the board to publicly announce the National Home as a nonsectarian institution in 1953. Similarly, the board incorporated CARIH in 1956 largely "for the purpose of expanding the raising of funds from nation-wide, non-sectarian sources."[51] Although administrative records suggest that a preferential admission policy for Jewish children continued throughout the 1950s, the home's public open-door policy enabled it to launch a widespread national publicity campaign with children as the centerpiece.[52] Features on the home appeared

[48] See, e.g., Naomi Rogers, *Dirt and Disease: Polio before FDR* (New Brunswick, N.J., 1996), 165–90.

[49] Peshkin to Lorber, 12 Sept. 1949, Box 5, Folder 5, NAC.

[50] Leon Unger to Peshkin, 29 July 1951, Box 5, Folder 5, NAC. See also "Board of Trustees Meeting," 20 July 1951, Box 1, Minutes: House Committee, 1937–1950 Folder, NAC; "Administrator's Annual Report, 1950," Box 2, Minutes, 1951 Folder, NAC.

[51] "Special Committee Report," n.d., Box 2, Minutes, 1957–58–59 Folder, NAC.

[52] Evidence of continued discriminatory admissions can be found in "Annual Report—Mr. Gershtenson," 16 Jan. 1956, Box 2, Minutes, 1955 & 1956 Folder, NAC.

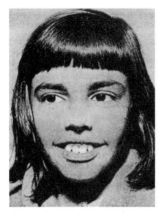

This is the face of what certainly appears to be a healthy child. But pictures can be deceiving. For example, you can't look into this girl's lungs. You can't see the severe asthma which grips and strangles her body in periodic attacks. It's difficult to imagine that before she was admitted to the National Home she was doomed to be a "pulmonary cripple."

It is for girls and boys like this youngster that the National Home is dedicated. Regardless of race, creed, or color, the Home's doors are open without cost to all children suffering from chronic intractable asthma.

Yes, when you give to the Jewish National Home for Asthmatic Children you return desperately ill children to a new life!

Figure 2. This advertising brochure, detailing the hidden effects of the "childhood crippler" asthma, illustrates how influential the March of Dimes' disease campaign for polio was in the marketing of asthma by the Jewish National Home for Asthmatic Children. (Box 9, Folder 15, NAC. Courtesy of National Jewish Hospital Collection, Back Archives, Penrose Library, University of Denver.)

in national magazines such as *Collier's, Newsweek,* and *Parade,* thanks to the "calculated planning" of the public relations director.[53] Some parents with intractable asthmatic children may have "given up hope." But photos of children running, boxing, and actively engaged in other recreational sports demonstrated that prayers for a "miracle" were being answered on a daily basis.[54] Films, shown annually at the national auxiliary conventions, also played a major role in the home's publicity efforts. They pulled on emotions and visually portrayed asthma as a "childhood crippler" in ways magazine articles could not. In *Heritage of Hope,* made by the National Home in the mid-fifties, shots of the feet and pigeon-chest of an emaciated boy are accompanied by the following narration: "This is not a concentration camp victim. This is a victim of chronic intractable asthma. . . . No concentration camp did this. This is the childhood crippler: asthma." Carefully composed scenes featuring children of different races playing together also instilled the belief that "even though their families have no money, without regard to race or religion, wherever they live, here is hope, free and

[53] Jack Frank to Sam Robinson, n.d., Box 2, Meetings, 1952 Folder, NAC; Albert Rosenfeld, "They've Got Asthma on the Run," *Collier's,* 16 April 1954, 25–7; "For the Forgotten," *Newsweek,* 28 Nov. 1955, 70; "Karen Walczak's Wonderful Second Life," *Parade* (1955): ?, 18, 20.
[54] Rosenfeld, "Asthma on the Run" (cit. n. 53), 25.

unfettered. Here in the shadow of the Rocky Mountains is the one possibility of break-ing the vicious circle of" asthma.[55]

Through the bodies of children, the National Home gradually expanded from its regional and ethnic network to a national, nonsectarian, fundraising organization and research facility, with annual charitable contributions in the early sixties that exceeded a half-million dollars. Small change compared with the March of Dimes campaign. However, five years later, in 1967, when Lyndon B. Johnson proclaimed the first week of May National CARIH Asthma week, charitable contributions from the National Home's 140 local chapters had nearly doubled. Not only had the home achieved na-tional prominence, but it had also helped place asthma on the national map of biomed-ical research and hope in the curative promise of medicine.[56]

CONCLUSION

In July 1961, a twelve-year old, forty-six pound malnourished boy from a *favela* in Rio de Janeiro came to Denver under the national media spotlight. "One of a num-berless multitude, a symbol, a medium through which one could understand the tragedy of poverty," Flavio da Silva became CARIH's most famous poster child. One month earlier, Gordon Parks's exposé of the violence, hardship, and despair that per-meated life in the Rio slums appeared in *Life*. The second of a five-part series, Parks's photographs and article "Freedom's Fearful Foe: Poverty" focused upon the life of the da Silvas and their eight children. The article was a way to portray the threat and chal-lenges faced by John F. Kennedy's Alliance for Progress program, which sought to curtail the spread of social unrest and communism in Latin America through massive economic aid. If Flavio represented "the spark of hope and warmth and care which keep life going," the *Life* article ended by suggesting that Flavio, "this labor of love and courage," would not triumph. "Wasted by bronchial asthma and malnutrition he is fighting another losing battle—against death." (See Figures 3a and 3b.)[57] Letters and donations poured into *Life*'s New York offices as eager Americans sought to ease their consciences and help Flavio and his family. CARIH, motivated either by hu-manitarianism or self-interest, offered to take "Flavio as a free emergency case and try to cure him." Four issues later, a bright-eyed, smiling boy in bed with clean white sheets and cuddling a stuffed animal graced the cover of *Life*. The headline read, "Flavio's Rescue: Americans bring him from Rio slum to be cured."[58]

Flavio was more than a symbol, however. He was also a boy, made of living flesh, whose body bore the marks of the impoverished economic and social conditions in which he lived. When he arrived at CARIH, that body connected him materially to Denver's past and present landscapes. Weekend camping trips during the summer to his host family's cabin in the Rocky foothills linked him to a history of the Rocky Mountain region as a sanitarium. His isolation from the boy's dormitory—until the results of a tuberculin test assured staff members that his body did not carry the

[55] *Heritage of Hope,* n.d., NAC.
[56] Executive Director's Report, 1972 (cit. n. 46); Proclamation 3777, *Federal Register* 32 (13 April 1967).
[57] Gordon Parks, *Flavio* (New York, 1978), 90; idem, "Freedom's Fearful Foe: Poverty," *Life,* 16 June 1961, 93. On the Alliance for Progress, see Michael E. Latham, *Modernization as Ideology: American Social Science and National Building in the Kennedy Era* (Chapel Hill, N.C., 2000).
[58] "The Compassion of Americans Brings a New Life for Flavio," *Life,* 21 July 1961, 25.

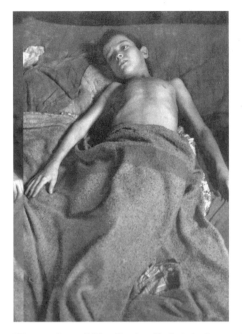 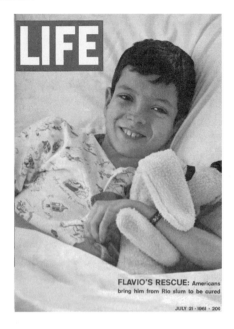

Figures 3a and 3b. Gordon Parks's before-and-after photographs of Flavio captured the hearts of Americans and symbolized the promise of hope in children, through which the Jewish National Home for Asthmatic Children expanded beyond its regional geographic and economic base. (Courtesy of Gordon Parks.)

tubercle bacillus—revealed the historical geography of consumption upon which the National Home and Denver were partially built. The frequent and painful scratch tests, the pulmonary function studies, the battery of psychological and intelligence tests, and the drugs that coursed through his veins and lined his bronchial membranes all situated him within a place of hope built through the bodies of children and biomedical research.[59]

The scars of poverty upon Flavio's body linked him not only to past landscapes but to future ones as well. When a *Life en Espanol* version of the Parks article appeared, the Brazilian magazine *O Cruzeiro* ran a counterpoint story of a Puerto Rican family in the slums of New York City, with a sleeping child covered in cockroaches and another child in anguish from hunger. A *Time* magazine correspondent found the *O Cruzeiro* article to be staged, a length to which the Brazilian magazine need not have gone.[60] Conditions of poverty, hunger, and asthma were readily present in American cities. For five consecutive years, a poor African American neighborhood in a heavy industrial sector of New Orleans had experienced an epidemic outbreak of asthma each fall. Similarly, in New York City and Chicago the year before the article was published, an unexplainable rise in emergency clinic visits for the treatment of asthma in neighborhoods serviced by Harlem Hospital and Cook County Hospital, respectively,

[59] For a provocative analysis of the material relationships between bodies and landscapes, see Christopher Seller, "Thoreau's Body: Towards an Embodied Environmental History," *Environmental History* 4 (1999): 486–514.
[60] See Parks, *Flavio* (cit. n. 57), 90–2.

made visible another environment of asthma: the degraded physical environment in which Americans of color and lower socioeconomic status lived. In this newly emerging landscape, hope for the asthmatic sufferer resided not in biomedical research but in political action, as citizens sought to combat the history of environmental and social injustice in places where they lived, worked, and played.[61]

The medical profession had once been intimately tied to the regional economic and physical geography of Denver in trading upon the hopes of chronic asthmatics. The rapid postwar economic growth of the city and federal expansion of biomedical research presented new opportunities and challenges for institutions such as the National Home for Jewish Children, which had built itself upon Denver's history as a health resort. Rescued from their homes and the places of poverty in which asthma allegedly dwelled, children would be given new hope in a space built through the latest advances in biomedicine, psychiatry, and social work. When hope failed in the miracle of modern medicine—and for many it did—there was still hope in a promised land. "Last fall," wrote a former CARIH patient in 1980, "I moved down to Tucson Arizona in hopes of clear sky, warm temperatures and NO ASTHMA!" The search for a better place, for a landscape free of disease, resurfaces again and again in the history of asthma in the postwar years—in Flavio da Silva's flight to America, in the fight for environmental justice, and in the continued migration of allergic citizens to America's southwest. It is upon hope that past historical landscapes and today's $20 billion allergy industry were built.[62]

[61] "A City Chokes Up," *Newsweek,* 14 Nov. 1960, 104; Hans Weill, M. M. Ziskind, R. C. Dicherson et al., "Epidemic Asthma in New Orleans," *JAMA* 190 (1964): 811–4; Stella Booth, Robert Markush, and Robert J. M. Horton, "Detection of Asthma Epidemics in Seven Cities," *Archives of Environmental Health* 10 (1965): 152–5; Leonard Greenberg, Carl Erhardt, Franklyn Field et al., "Air Pollution Incidents and Morbidity Studies," *Archives of Environmental Health* 10 (1965): 351–6. For historical analyses of environmental justice issues prior to the rise of the environmental justice movement, see, e.g., Dolores Greenberg, "Reconstructing Race and Protest: Environmental Justice in New York City," *Environmental History* 5 (2000): 223–50; Andrew Hurley, *Environmental Inequalities: Class, Race, and Industrial Pollution in Gary, Indiana, 1945–1980* (Chapel Hill, N.C., 1995); Robert Gottlieb, *Forcing the Spring: The Transformation of the American Environmental Movement* (Washington, D.C., 1993).

[62] CARIH alumnus to Mr. G., n.d., Box 14, Alumni Correspondence, 1980 Folder, NAC.

"Living Is for Everyone":
Border Crossings for Community, Environment, and Health

By Giovanna Di Chiro*

ABSTRACT

This article examines the transnational networking practices of Teresa Leal, an environmental justice activist living and working on the U.S.-Mexico border. It shows, through the method of engaged ethnography, how she and other community activists respond to the effects of global economic restructuring policies such as NAFTA. Grounded in an ecological epistemology, Leal blends "local" and "scientific" knowledges about the deteriorating health, economic, and environmental conditions at the border and constructs a "global sense of place" that brings into focus the everyday realities of neoliberal globalization. The article documents a daylong "toxic tour" of the Ambos Nogales region and highlights the multiple border crossings (epistemic, geographic, political, cultural) undertaken by Leal, other activists, and the author, a visitor to the region, to narrate a history of community health and environmental action in a transnational context.

INTRODUCTION

We can't alienate ourselves. We can't be hermits and continue on in our own ways because that means we're not evolving. To me, this is not about revolution, but evolution. We're not stopping at boundaries or borders. The biggest challenge is that we've got to get it all over the world, and that's not easy if we stay within our traditional parameters.
                                    —Teresa Leal, June 12, 2001[1]

It's a sad thing to have to say, but as long as we have faith—we have no hope. To hope, we have to *break* the faith.
                                    —Arundhati Roy, *The Cost of Living*[2]

* Earth and Environment Department and Women's Studies Program, Mount Holyoke College, 50 College St., South Hadley, MA 01075; gdichiro@mtholyoke.edu.
    I am grateful for the engaged readings and helpful comments of Anne Wibiralske, Gregg Mitman, Michelle Murphy, Chris Sellars, Marita Sturken, Dana Polan, and Enid Milagros Santiago. The successful completion of the Ambos Nogales toxic tour depended heavily on the agile driving of Joni Adamson and her intrepid car, which met with only one (major) mishap while faithfully transporting us to and fro. Teresa Leal's "good *mitote*" continues to inform my research and teaching in environmental studies. I thank her for her participation in these extensive interviews and for her feedback on, and endorsement of, the accuracy of her words in the final publication.
    [1] Teresa Leal, interview by author, Nogales, Sonora, Mexico, 12 June 2001.
    [2] Arundhati Roy, *The Cost of Living* (New York, 1999), 29.

It is by now a familiar trope in global environmental discourse that the current era of globalization is marked by unrestrained flows of capital, commodities, ideas, technologies, people, *and* pollution across international borders. While the worldwide expansion of global political economic restructuring orchestrated by the well-known ensemble of powerful states, corporate elites, and international trade and financial institutions is well documented in the literature of environmental studies, accounts of the transnational activities and movements of *community*-based actors and organizations are much less visible.[3] With the aim of tracking and documenting the cross-border "flows" of environmental thinking and acting undertaken by presupposed *local* activists, I set out on a research trip in June 2001 to the twin cities of Nogales, Arizona, and Nogales, Sonora, (Ambos Nogales, or "both Nogales") to study the post-NAFTA (North American Free Trade Agreement) transnational networking practices of women environmental justice activists along the U.S.-Mexico border. One segment of the trip brought me into contact with Teresa Leal, a Mexican Opata Indian environmental justice activist who works with marginalized communities living within the "biological and social corridor" of the Santa Cruz River basin. On a daily basis, Leal traverses multiple borders—political, geographic, cultural, epistemic—and produces an *engaged* environmental expertise that articulates diverse knowledge systems, both professional and lay, to build a transnational advocacy network devoted to environmental justice on the border.

Leal's transnational advocacy network-building practice brings into critical conversation many different ideas, voices, and perspectives. In this essay, I sketch an account of this interactive process in the context of a daylong "toxic tour" hosted by Leal for a small group of women academics interested in broadening the scope and sharpening the interdisciplinarity of our environmental studies. While this account focuses on the environmental justice theories and practices of Leal and her colleagues, it reflects an ethnographic-historiographic stance in which the ethnographer-historian is understood as being involved in the intercultural exchange and, therefore, part of the socio-ecosystem under study. The ethnographic–oral historiographic approach I use is situated in the field of "action" or "public anthropology" (what Roy A. Rappaport termed "engaged" anthropology).[4] This branch of qualitative research contests the conventional objectivist model, which, many argue, assumes the spectator's (researcher's) neutrality and distance from the object of study (being "above the fray").[5] In contrast, "engaged" ethnographers acknowledge that they enter the field as "witness, active and morally committed, [those] who take sides and make judgements."[6]

The method of engaged anthropology-history suggests a role for environmental ethnographies, environmental justice histories, or both that moves beyond simply

---

[3] There has been some recent attention to the transnational political stakes of community groups that join "transnational advocacy networks" to bring attention to the local impacts of global processes and to gain leverage on the international stage. See, e.g., Margaret Keck and Kathryn Sikkink, *Activists Beyond Borders: Advocacy Networks in International Politics* (Ithaca, N.Y., 1998).

[4] Roy A. Rappaport, "Disorders of Our Own: A Conclusion," in *Diagnosing America: Anthropology and Public Engagement,* ed. Shepard Forman (Ann Arbor, Mich., 1994), 235–94.

[5] See, e.g., Nancy Scheper-Hughes, "The Primacy of the Ethical," *Current Anthropology* 36 (1995): 409–20; Gavin Smith, *Confronting the Present: Towards a Politically Engaged Anthropology* (Oxford, 1999); C. P. Kottak, "The New Ecological Anthropology," *American Anthropologist* 101 (1999): 23–5; Amanda Coffey, "Ethnography and Self: Reflections and Representations," in *Qualitative Research in Action,* ed. Tim May (London, 2002), 313–31.

[6] David Gow, "Anthropology and Development: Evil Twin or Moral Narrative?" *Human Organization* 61 (2002): 299–313, on 306.

documenting a seemingly endless succession of ecological misadventures and then exposing and critically analyzing the inherent tensions, contradictions, and failures of environmentalists' efforts to rectify the problems. While such tales of defeat abound, and provide crucial insight into the intransigence of political, economic, and cultural systems, they can gloss over coexisting and equally important narratives of endurance. It becomes an environmental justice strategy to trace the persistence of alternative worldviews and lifeways envisioned by communities that suffer the negative consequences, but receive few of the benefits, of modern industrial society. What does it mean to persevere, holding on to a sense of hope in one's actions to create a better world, while facing apparently overwhelming odds?[7]

The following portrayal of the ever-expanding and peripatetic networking practices of women activists such as Leal highlights their rejection of a hypercritical cynicism in the face of the excesses of neoliberal globalization. This article, therefore, strives to present an "ethnography of hope." Dismissing cynicism as a political "luxury" that tends to breed alienation and ultimately, inaction, Leal chooses instead to "hit the road."[8] In so doing, she *breaks the faith* in the presupposition of a democratic system that purports to secure the welfare of all peoples and all environments; her *hope* for a different world is based not on passive yearning or faith but on active, steadfast engagement. The perpetual *movement* across many borders that constitutes Leal's work embodies the hope-filled activism practiced by many environmental justice activists materializing the conditions of possibility for social and environmental change. This story, an unabashed ethnography of hope, begins with one such border crossing.

NOT THE USUAL ROAD TRIP: TOXIC TOURING ON THE BORDER

"U.S. citizens?" The stern immigration agent eyeballs the car full of women, his gaze, gliding past the Anglo travelers to pause on the one café con leche–complected passenger, who, smiling discreetly, had already produced her identification papers for inspection. Teresa Leal, a resident of Ambos Nogales, was our tour guide for the day's journey in June 2001 across the U.S.-Mexico border to see the social and environmental impacts of NAFTA and its manufacturing and trade policies spawned by the rise of economic globalization. Satisfied with the validity of our tour group's credentials, the federal agent waved us through *la frontera*, and we left the sleepy border town of Nogales, Arizona, and entered the bustling, crowded streets of Nogales, Sonora.

A veteran social activist, cultural historian, women's rights advocate, environmentalist, mother of eight, and grandmother of ten, Teresa Leal has worked diligently as part of an ever-expanding, transnational network of women activists fighting for environmental justice along the border. For many years, her attention and energies have focused on improving the environmental conditions endangering the health of her family, community, and native land—the biocultural region comprising southern Arizona and northern Sonora, a region bifurcated by the official, national boundary dividing the twin cities of Ambos Nogales. Yet Leal's ecological vision for achieving

---

[7] The sociologist John Brown Childs has written on this subject and has developed the idea of "transcommunality" to describe the cross-cultural alliances that marginalized peoples forge to create more sustainable communities. See Childs, *Transcommunality: From the Politics of Conversion to the Ethics of Respect* (Philadelphia, 2003).

[8] Leal interview, 12 June (cit. n. 1).

environmental justice extends beyond her local environment in the Sonoran Desert, encompassing a broader reach and assembling a larger "community." In the following pages, I examine how the movements, actions, strategies, epistemic systems, and passions of women activists working on the border reveal the systemic interconnections between humans and the environments in which they live and on which they depend. By focusing in on the transnational actions of one activist, Teresa Leal, I explore how her sense of community and "place" is shaped by a collision at the border between the circuits of capital and its "global assembly line" and the toxic flows of industrial pollution through the environments and bodies of the Sonoran Desert inhabitants.

As I will present in this essay, this community-environment concept defines a new *ecosystem*—one that includes *people* and their interrelationships with the social and physical landscapes and with the circulating air, water, and industrial poisons that travel across international, geographic, watershed, neighborhood, and corporeal borders. I examine how women activists such as Leal construct a "global sense of place"[9] and a new concept of a healthy environment by articulating scientific knowledge about ecological stability with local knowledge about the impacts of global environmental change in their communities and the bodies of their children, their families, and their neighbors. Resisting an oversimplified partitioning of "lay" and "professional" environmental health knowledges, these activists argue for a strengthened eco-epistemological standpoint, one blending diverse understandings and visions for a sustainable world.[10]

TWIN CITIES, LINKED DESTINIES

Although crossing the international border in our car that June day went smoothly enough, the fourteen-foot-high, two-mile-long steel walls flanking the city belied this perception of a permeable boundary. At one time, Ambos Nogales had enjoyed a relaxed border ambience, in which Mexican and Indian day-crossers casually passed through the flimsy, chain-link fence to take a job, to visit family, or to do some shopping. In the early 1990s, after the Immigration and Naturalization Service (INS) instituted a border crackdown in California and Texas, immigrants flooded Ambos Nogales, which soon became a people-smuggling hot spot crawling with the ubiquitous green and white Ford Explorers favored by the INS, or *la migra*.[11] The Border Patrol's deployment in the 1990s of high-tech surveillance systems and military maneuvers enhancing the policing of the 2,000-mile-long U.S.-Mexico border transformed Ambos Nogales into a "low-intensity conflict" zone and one of the more reinforced entry points along the international frontier.[12]

Ambos Nogales, part of the Santa Cruz River watershed region, has suffered other

---

[9] See Doreen Massey, *Space, Place, and Gender* (Minneapolis, 1994), 146–56.

[10] A large body of scholarship examines the confrontation between local, "anecdotal" knowledge and professional, "scientific" knowledge. See, e.g., Steve Kroll-Smith, Phil Brown, and Valerie Gunter, eds., *Illness and the Environment: A Reader in Contested Medicine* (New York, 2000); Stephen Couch and J. Stephen Kroll-Smith, *Communities at Risk: Collective Responses to Technological Hazards* (New York, 1991); Steven Epstein, *Impure Science: AIDS, Activism, and the Politics of Knowledge* (Berkeley, Calif., 1996); Carolyn Raffensperger and Joel Tickner, eds., *Protecting Public Health and the Environment* (Washington, D.C., 1999).

[11] Miriam Davidson, *Lives on the Line: Dispatches from the U.S.-Mexico Border* (Tucson, Ariz., 2000).

[12] See Timothy Dunn, *The Militarization of the U.S.-Mexico Border, 1978–1992: Low Intensity Conflict Doctrine Comes Home* (Austin, Texas, 1996).

changes as well. Once a perennially flowing river lined with majestic cottonwood trees, the Santa Cruz River has sustained decades of reckless overpumping of its surface and groundwaters and years of unregulated dumping of toxic chemicals, copper mine tailings, and untreated sewage, which have contaminated its already-depleted water resources.[13] As a result, the river has become what the Tohono O'odham people refer to as *hik-dan,* a parched "cut in the earth."[14]

I first met Teresa Leal on a toxic tour she had organized for a group of academics attending the 1999 meeting of the American Society for Environmental History. On that expedition, Teresa and her *comadres,* Rose Marie Augustine and Ann Montaño, wove a collaborative narrative tracing the history of the environmental destruction of the Sonoran Desert and its inhabitants. Speaking clearly into the tour bus's P.A. system, the women recounted grim stories and appalling statistics of unusually high incidences of cancer, neurological disease, miscarriage, and birth defects suffered by the low-income, Latino and Indian communities living within the biological and social corridor drained by the Santa Cruz River, a regional watershed extending from Nogales, Sonora, to Phoenix, Arizona.

Two years later, I returned to the Santa Cruz basin for my research trip and again crossed into Mexico's free-trade zone, motoring alongside the dry streambed of the Nogales Wash, one of the primary tributaries emptying into the north-flowing Santa Cruz River. Perennially dry since the mid-1950s due to years of groundwater overdraft accompanying the development boom on both sides of the border, the Nogales Wash has become a convenient dumpsite for garbage, sewage, and all manner of industrial waste.[15] As I would soon learn from Leal and other activists, the protection of the ephemeral and "sacred" water resources of this desert ecosystem is at the center of the story of environmental injustice in Ambos Nogales and motivates the struggle for environmental health along the border.

Our party drove along the Carretera Internacional to the city's industrial park district, navigating Nogales, Sonora's busy streets teeming with midday shoppers, schoolchildren lugging heavy backpacks, and North American tourists seeking bargains. Nogales, a city of 350,000, is home to 100 industrial facilities, or *maquiladoras,* foreign-owned "export-processing" plants erected on the Mexican side of the border. The facilities began appearing in the mid-1960s as an economic development strategy to encourage foreign investment and create jobs to bolster Mexico's flagging economy.[16] With minimal tax and tariff liabilities and the open-door atmosphere afforded by NAFTA, the maquiladoras have come to play a pivotal role in Nogales's economic "health." Moreover, profiting from the free-trade zone's loose environmental restrictions, the industrial parks have also, according to many environmental,

[13] For a more detailed study of the environmental justice activism in the Santa Cruz River Basin, see Giovanna Di Chiro, "Steps to an Ecology of Justice: Women's Environmental Networks across the Santa Cruz River Watershed," in *Seeing Nature through Gender,* ed. Virginia Scharff (Lawrence, Kans., 2003). For an exhaustive history of the Santa Cruz River, see Michael Logan, *The Lessening Stream: An Environmental History of the Santa Cruz River* (Tucson, Ariz., 2002).

[14] Di Chiro, "Steps to an Ecology of Justice" (cit. n. 13), 282.

[15] See Helen Ingram, Nancy Laney, and David Gillilan, *Divided Waters: Bridging the U.S.-Mexico Border* (Tucson, Ariz., 1995); and Udall Center for Studies in Public Policy, *Water in Nogales: Survey of Use, Issues, and Concerns* (Tucson, Ariz., 1993).

[16] For a history of the maquiladora system, see Milo Kearny and Anthony Knopp, *Border Cuates: A History of U.S.-Mexican Twin Cities* (Austin, Texas, 1995); and Kathryn Kopinak, *Desert Capitalism: Maquiladoras in North America's Western Industrial Corridor* (Tucson, Ariz., 1996).

public health, and labor activists, come to contribute to the deteriorating health of the workers, local residents, and the surrounding desert environment.[17]

Thus activists seeking to improve the health of Nogales's residents while also looking out for their livelihoods cannot simply take to the streets to shut down the factories. "We know they'll just go elsewhere and do the same things to other people," says Teresa, "and then the people here will be left without jobs."[18] Supporting a strategy that brings the community and the corporations to the table as equal partners, she argues for a middle ground:

> We need to bring the maquiladoras into compliance. A *new* science is urgently needed to come up with environmentally friendly alternatives, substitutes that will continue to provide a profit for these corporations. That's the bottom line for these corporations. My son has educated himself as an environmental engineer for General Instruments–Motorola. You don't see me saying he's "copped out" because he's working for a maquiladora. There are many ways to get things done, and this might mean building nontraditional partnerships and thinking outside our own little sectarian corridors.[19]

SPREADING THE "GOOD GOSSIP": SYNCHRETIC SCIENCE AND WOMEN'S WORK

En route to our first "tour" stop, the Samson-Samsonite maquiladora, Teresa crossed her fingers in hopes that we would meet up with members of Comadres, a binational, grassroots organization of women seeking to improve the working and living conditions of women factory workers and their families. Teresa established Comadres in the early 1970s after being approached by women workers who knew of her community work and had listened to her weekly radio program, *Hablando de Mujeres*. The program provided information about women's and children's health, domestic violence, and the importance of ensuring water quality in the *colonias,* the squatter settlements encircling the city in which many maquiladora workers live.

One Samson-Samsonite luggage seamstress, Panchita, had sought Teresa's help after contracting mysterious skin rashes that resulted in bleeding sores on her legs and arms. Fearing the materials she was working with were harmful, Panchita asked Teresa to translate the instructions label stamped on a bolt of the textile she used every day for sewing suitcase liners. The doctor Panchita had previously consulted, rather than investigating the serious skin problem, had admonished her, saying, "You don't clean your house; you live in the dump and that's why you have mange. Because your hygiene is so bad."[20] Although she was poor and lived in humble conditions in the colonias, Panchita knew she and her children were not dirty. Reading and translating the label at Panchita's request, Teresa discovered that the material was made of fiberglass. She informed Panchita that "sewing through that material created microscopic little shards of glass that fly through the air and into her clothes and body. That's why she had all these rashes."[21] Emboldened by Teresa's workplace environmental expla-

---

[17] For discussions of the health problems along the border resulting from the maquiladoras, see Tom Berry and Beth Simms, *The Challenge of Cross-Border Environmentalism* (Albuquerque, 1994).

[18] Teresa Leal, interview by author, Nogales, Sonora, Mexico, 11 June 2001.

[19] Ibid.

[20] Joni Adamson, "Throwing Rocks at the Sun: An Interview with Teresa Leal," in *The Environmental Justice Reader: Politics, Poetics, and Pedagogy,* ed. Joni Adamson, Mei Mei Evans, and Rachel Stein (Tucson, Ariz., 2002), 44–57, on 49.

[21] Ibid., 48.

nation for the workers' health problems, Panchita began to attend meetings of Co-madres and talked with other women with similar stories.

Translated as "comothers," Comadres aims to empower women to stand up for their rights for a clean and healthy environment and to exchange valuable information about preventing chemical toxicity in the workplace, purifying the water used in their homes, accessing public health and environmental services, and developing income-generating strategies, such as weaving and sewing co-ops, to supplement their poverty-level wages. Through their associations with the Comadres network, the women share their experiences of environmental illnesses and blend this knowledge with the technical expertise of medical doctors and environmental scientists. In this way, they produce a hybrid ecological standpoint epistemology that emerges from their everyday lives and their encounters with professional knowledge.[22] This aggre-gate environmental health perspective—the product of anecdotal *and* scientific knowledge—comprises "women's talk," argues Teresa, but it is much more than triv-ial "gossip."

> As Latina women [Comadres have] always been . . . put down by the machos; [we] are considered *mitoteras*. Mitoteras comes from *mitote,* which means "gossip." Mitoteras are gossipy. But we are changing the meaning of "gossip" from negative gossip to positive gossip. We put out the good mitote, or good gossip. We get out the news about what's go-ing on in the villages, or *colonias,* and in the factories. We share information.[23]

As they learn about the widespread environmental impacts of the maquiladora op-erations and become aware of the epidemiological data documenting a dizzying array of health problems suffered by the people living adjacent to the polluting facilities, the Comadres are transformed into eco-political agents who shift their attention from individual self-help efforts to organizing strategies that address wider systemic prob-lems. These "macro-Comadres," Teresa explains, are

> [those] that help not only individuals, but also share the results of their efforts with their community. They're scavengers; they find materials that can be used to help build shelters for workers who have just come to the community. They scavenge food, clothes. The people who are seeking aid see what the Comadres are doing and go to them for help . . . . They're the ones challenging the system, challenging the government, and trying to stop the railroad tanker cars filled with toxic materials that roll through our communities. They're yelling about the fact that there's no water, no electricity, no police, no safeguards.[24]

Clean, potable water is a limited resource in the colonias, and it is essential to maintain the purity of the water delivered weekly by *pipas,* private vendors who pro-cure clean water from distant wells and, for a small fee, transport it by truck to fami-lies living in the colonias.[25] After discovering that the fifty-five-gallon drums women workers were scavenging from the maquiladoras or the local dump to store water had

---

[22] For a similar analysis focusing on white working-class women in the United States, see Celene Krauss, "Challenging Power: Toxic Waste Protests and the Politicization of White, Working Class Women," in *Community Activism and Feminist Politics: Organizing across Race, Class, and Gender,* ed. Nancy Naples (New York, 1998), 129–50.

[23] Adamson, "Throwing Rocks at the Sun" (cit. n. 20), 47.

[24] Ibid., 47–8.

[25] Ingram et al., *Divided Waters* (cit. n. 15), 76–8. The price of delivered water ranges from 4.5 to 5.0 pesos ($1.32 to $1.47) per 200 liters of water.

previously contained hazardous chemicals, the Comadres organized training sessions to teach people who had no choice but to use the barrels to line them not with concrete, tar, or lead-based paint, but with heavy polyethylene plastic.[26]

Evidence of the success of the Comadres' popular education efforts to protect the water supply dots the hilly landscape of Nogales' poorest barrios, where virtually every dwelling—small houses constructed of discarded plastic sheets, wooden shipping pallets, and the occasional cinder-block wall—is equipped with one or two water barrels stationed next to the road. Teresa describes how the Comadres designed and distributed through word-of-mouth networking a water-labeling system to differentiate between the barrels filled with salvaged gray water, which could be used for gardening or washing the house, and barrels reserved for drinking, bathing, and cooking water. "It's a simple system, but it works," she says. "The women cover the top of the 55-gallon drums that are meant for clean water with metal or wooden boards," a sign to the pipas that only those should be replenished with the precious commodity.[27] "So, you see the beauty of word-of-mouth strategies, of mitote, of gossip," Teresa exclaims, "Once it starts going, you can't stop it. That's how Comadres uses the power of 'gossip.'"[28]

Although these are small-scale transformations, Comadres is committed to broader and more institutional change in the living and working conditions of maquiladora workers. Organizing from the ground up, women's organizations such as Comadres seek to empower maquiladora workers to understand their environmental and labor rights. One woman worker commented: "I used to think that the factory was more important than I was, because that's how they make you feel. . . . [Now], I realize that I have rights and that I have to make them worth something at work, at home, and in the streets."[29]

SACRED WATERS AND TOXIC FLOWS

Arriving at the entryway to the Samsonite factory, we waited in the car as Teresa inquired about a meeting with Panchita and other Comadres who worked at the plant. A lovely, peach-tinted stucco façade decorated the external walls of the Samsonite maquila, which was elegantly landscaped with fountains, ornamental palms, and sweet-smelling flowers. The loading zone at the rear of the factory painted a very different portrait of the maquila's operations: a small, attached shed labeled *Químicos Peligrosos* (dangerous chemicals) stores the powerful glues and solvents used to assemble the pricey suitcases. The women who work all day on the assembly line hovering over vats of glue are known as *las loquitas* (the crazy ones), a double meaning referring to both their willingness to take the job and the glue's deleterious effect on their brain cells.

The security guard informed us that no workers were allowed on breaks that day due to an unscheduled "inventory." Annoyed at the dubiousness of this explanation, we drove instead to the home of María Luisa, another Comadre who had sought Teresa's advice in dealing with the Nogales Health Department after the death of her

[26] Teresa Leal, interview by author, Nogales, Arizona, 13 June 2001.
[27] Leal interview, 12 June (cit. n. 1).
[28] Adamson, "Throwing Rocks at the Sun" (cit. n. 20), 49.
[29] Julie Light, "Engendering Change: The Long, Slow Road to Organizing Women Maquiladora Workers," *CorpWatch,* 26 June 1999, http://www.corpwatch.org/issues/PID.isp_articleid=691.

two young daughters—the four-year-old from pancreatic cancer and the sixteen-year-old from leukemia. The family, suspecting that the rebar used to fortify the house's foundation had been salvaged from a shipment tainted with depleted uranium, asked Teresa to help them demand that health department officials inspect the house for radiation. With her aid, the family was able to schedule a home visit by the health department. The health inspector found no "unusual levels of radiation" and offered no other environmental explanation for the children's cancers.

María Luisa invited us into her home, a small stucco bungalow perched alongside the Nogales Wash. We sat in her tidy front room and listened to a painful story of the illnesses and deaths of the two girls, and the public health professionals' medical exegesis of their untimely demise: "It was probably a genetic defect transmitted through the females."[30] Still grieving six years later, María Luisa resigned herself to "the will of God" and held tightly to Teresa's hands. Although her three surviving children were healthy, they had grown up in the same house and so she continued to worry about their fates. We shared our condolences and exited our host's front door to an unobstructed view of the Nogales Wash, the desiccated channel that serves as an industrial dumpsite, an open sewer, and the local water source for the city of Nogales, Sonora. Teresa sighed, obviously frustrated at the explanatory preeminence of either genetic reductionist theories of disease or religious ones. "The family has been drinking this polluted water for years, but you have to start where people are, you can't impose your ideas on them and tell them they're crazy. In this overpumped, overused desert wash, people are more obsessed with the *quantity* of water available to them, and this can sometimes make them overlook the quality."[31] As for officials, Teresa said that despite the unassailable physical evidence of a potential health hazard provided by the proximity of the polluted streambed to people's homes and to the city's wells, "the health department is unwilling to accept a causal relationship."[32]

The preservation and safeguarding of water as the "lifeblood of Mother Earth" in an arid ecosystem has swirled at the center of Teresa's environmental justice organizing for many years.[33] Leaving María Luisa's house and the crowded urban center, we drove past the Nogales Wash and the makeshift shanties people had erected along its banks and proceeded eighteen kilometers west to the mountainous terrain that is home to the new Nogales Sonora Sanitary Landfill. We parked along a hillside road just north of the landfill, grateful for the peace and quiet and the panoramic view of the Sierra Madre mountains, but this bucolic scene would be short-lived. Barreling up the steep road appeared one fully loaded dump truck after another, conveying undifferentiated domestic refuse and hazardous waste from the city to this out-of-the-way sacred grove, now an out-of-sight disposal site. After years of binational grassroots pressure succeeded in shutting down the former landfill, a toxic stew notorious for bouts of spontaneous combustion, located just a few kilometers south of the border, city officials selected this distant locale to solve its waste disposal problems.[34] To escape the clouds of dust and grit kicked up by the constant trail of trucks as well as the

[30] María Luisa Garcia, interview by author, Nogales, Sonora, Mexico, 12 June 2001.
[31] Leal interview, 12 June (cit. n. 1).
[32] Ibid.
[33] For an extensive study of the status of groundwater resources in Arizona and New Mexico, see The Southwest Network for Environmental and Economic Justice and the Campaign for Responsible Technology, ed., *Sacred Waters: The Life-Blood of Mother Earth* (Albuquerque, 1997).
[34] See, Davidson, *Lives on the Line* (cit. n. 11); and Ingram et al., *Divided Waters* (cit. n. 15).

stench of garbage, our group set out into the sage brush and scrub oak, heading down a precipitous canyon in search of the Venero San Antonio, an artesian spring flowing in a southerly direction toward the Mambutu River.

As a young mother, Teresa had brought her children to the San Antonio Spring, where she told them stories, passed on to her by grandparents, of indigenous ancestors who had lived in this magnificent territory long before the Spanish *conquistadores* claimed the land. The miracle of a steady source of underground fresh water in the Sonoran Desert had always awed Teresa, and she had hoped to transmit this respect for the natural environment to her offspring. In recent years, however, rather than embarking on leisurely excursions to these beloved mountains, she has found herself struggling hard to protect their precious watercourses. After learning that her ancestral canyon lands would become the dwelling place for the new "sanitary" landfill in 1995, Teresa embarked upon a water-testing program to ascertain the *pre*-landfill quality of the surface and underground water. Since the landfill's installment, Teresa has worked with scientists and community activists from the Grupo Ecologista Independiente to determine if the hazardous chemicals and liquid wastes from the dump have seeped into the groundwater and contaminated the spring. Not owning a car, Teresa regularly hitchhiked to the site, where, wearing her favorite huaraches, she trekked the ten-mile route from the springs' headwaters near the landfill to its confluence with the Mambutu River, collecting water samples along the way. Through her itinerant water-sampling methods, she hopes to build a water-quality record that can provide early warnings and evidence of contaminants escaping the landfill in violation of Mexican environmental law and polluting the Mambutu, a river that has often been touted as pristine and therefore a candidate to replace the dwindling and contaminated water resources of the Santa Cruz.

Although she and the *ecologistas* with whom she collaborates have been unable to afford the more sensitive $100-per-assay water-quality tests (those that test for specific chemicals and calculate precise concentrations), they have sent samples to the Sonoran Water Laboratory for "general" assessments. "The pre-landfill samples showed that the springwater was very pure," Teresa says. "But, after the landfill went in, the samples have come back with descriptions like 'it's not potable' or there's residue of oils, gasoline, or industrial acetone."[35] When she presented this evidence to the Sonoran ministry for environmental quality, she was told "but the Mambutu is a south-flowing river, so it will not impact Nogales at all."[36] In spite of this geomorphology, immense aboveground water pipes pump water from the Mambutu to Nogales's water distribution system to supplement the subsiding water table around the border, thus belying this simplistic assessment of the region's hydrological processes. Ignoring the ecosystem comes at a risk, Teresa argues. "They think the landfill is so far away and won't affect Nogales, Sonora." Through human disregard and manipulation of the desert's subtle yet inexorable water cycles the industrial wastes from the maquiladoras are returned to the city, circulate through the environment into the rivers and washes, and eventually find their way into the bodies of the people. "And we know the Santa Cruz River flows from south to north, so the chemicals go over the border into Arizona."[37]

[35] Teresa Leal, telephone interview by author, 12 Feb. 2002.
[36] Ibid.
[37] Ibid.

MIXED EPISTEMOLOGIES, CROSS-BORDER ACTIONS

Returning to Nogales, Sonora, we once again crossed the Nogales Wash as we made our way to the site of the old landfill, a massive scar in the desert landscape located within striking distance of the international border. In the early 1980s, Teresa had targeted the old landfill as a site of action because of the local waste management authority's negligent regulatory practices and the increasing health and environmental dangers associated with the dump. Like that of many of her collaborators, Teresa's social activism is multifaceted yet informed by a unifying principle—people need to survive, to live a decent life, while sustaining the air, water, and land around them. Knowing that the so-called tunnel children scavenged in the dump and faced a myriad of hazards, including street violence, starvation, AIDS, toxic contamination, and *desamor,*[38] Teresa worked to set up Mi Nueva Casa (my new house), a drop-in center, only yards from the international fence, that offered food, condoms, drug treatment, and literacy classes to the kids. [39]

In addition to seeking out and assisting the various landfill scavengers (which included mothers and children living in nearby colonia *Bella Vista*), Teresa had a water monitoring agenda. She and other activists had long suspected that high levels of rare cancers—multiple myeloma, pancreatic cancer, childhood leukemia—and autoimmune diseases, such as lupus and scleroderma, were caused by the seepage of poisons from the landfill into the groundwater and eventually into the city wells. Meanwhile, the landfill was growing beyond its fenced borders, and semitrucks brimming with noxious cargo were filling up the lagoons, gullies, and washes surrounding the engorged site, which happened to lie a few hundred meters from an elementary school. Working with the schoolchildren, parents, and teachers, Teresa organized a grassroots campaign citing Mexican child safety laws that prohibit the siting of a potentially harmful entity within 2,500 meters of institutions such as schools, daycare centers, and churches. Aware that this law was originally intended to protect children from exposure to bars, adult bookstores, liquor stores, and pharmacies, Teresa asserted that going to school next to a hazardous landfill was an even greater danger to the welfare of children. As Teresa learned, teachers and students at the nearby school were suffering a long list of maladies. "They were dealing with diarrhea, conjunctivitis, skin diseases, and the roaches in the schoolyard were actually eating the 'carnage' [the disposed body parts from the local hospital], and the stench was incredible."[40]

Gaining confidence and knowledge about the environmental and human costs of the landfill, a mixed binational community of street children, students, parents, teachers, environmentalists, and workers mobilized forces in what Teresa called "urban anarchy at its finest":

> We got masks and buckets, and we brought water from the filled-up lagoon and dumped it on the steps of the mayor's office—we washed his steps with this water—and it was just awful. The fathers got pieces of old, junk cars and scrap materials at the dump, and

[38] Teresa describes *desamor,* inadequately translated as "dis-love," as encompassing the institutional "hazards" of alienation, disinterest, disinvestment, and lack of care, all faced by homeless children along the U.S.-Mexican border.

[39] The "tunnel children" are homeless kids who inhabit the binational network of street-size, underground storm channels built in the 1930s by the U.S. Civilian Conservation Corps for flood control of the Nogales Wash. See Davidson, *Lives on the Line* (cit. n. 11), chap. 4.

[40] Leal interview, 13 June (cit. n. 26).

at night they would drag those pieces and put them across the road where the dump trucks would come in the morning. They would stand on top of the scraps during the day, in very macho style, just challenging the drivers to come and get them out of the way. The truck drivers would say, "Well, we couldn't deliver the garbage because of these people." So it started escalating. It was a very citizen movement, and finally, we got [the landfill] closed down. Two weeks later the mayor had to hire dump trucks from Hermosillo to remove the landfill; it took them two weeks day and night![41]

We drove alongside the decommissioned landfill's rusty Cyclone fence, the solitary barrier separating the dump from the city limits, and stared out at the sacrificial scene before us: a vast, dirt-capped landmass littered with a field of methane off-gassing vents. Perched just outside the fence sat the *Torreon* water tank, an old well that was still being used, Teresa bemoaned, by several barrios in the area. The shallow wells (fifty feet or less) sunk along the Nogales Wash, she explained, had long been contaminated through seepage from the landfill and dumping from the factories, with high levels of TCE (trichloroethylene), PCE (tetrachloroethylene), trichloroethane, chloroform, and various toxic metals such as lead, mercury, chromium, and copper.[42]

In the years following the heterogeneous grassroots mobilizations to shut down the landfill, Teresa had joined forces with a binational community-based organization called LIFE (Living Is for Everyone). LIFE was founded in the early 1990s by residents of Carrillo Street, a pleasant, Latino subdivision in Nogales, Arizona, built on the grounds of Camp Little, a U.S. Army garrison established in 1910 to defend the town from possible invasions spilling over from the Mexican Revolution. Located a mile north of the border and a few blocks from the Nogales Wash, the Carrillo Street community had for several years shown signs of an emergent disease cluster of uncommon illnesses, including multiple myeloma, pancreatic cancer, anencephaly, and lupus. To collect evidence in support of this hypothesis, a group of neighbors led by Susan Ramirez, Anna Acuña, and Jimmy Teyechea launched their own "popular epidemiological" health studies of their neighborhoods.[43] The three leaders had each been collecting informal data on the rising numbers of illnesses in their community and "talked about what could be causing the diseases. Could it be, as Susan believed, the water? Or, was it, as Anna believed, the air?"[44] Gathering at Acuña's home, neighbors used colored pins signifying particular diseases (red for blood cancers, black for lupus, orange for ovarian cancer) to make a cluster map, which graphically revealed that most of the victims' homes were grouped on streets straddling the wash. LIFE contested the conventional argument that the illnesses were caused by either "genes or lifestyle or God or fate."[45] Group members became increasingly convinced that they and their families had become sick due to polluted air drifting northward from the constant plastic and refuse fires at the Nogales, Sonora, landfill and poisoned groundwater from the landfill, the maquiladoras, and the dilapidated Nogales International Wastewater Treatment Plant (NIWTP).

---

[41] Ibid.

[42] Berry and Simms, *Challenge of Cross-Border Environmentalism* (cit. n. 17); Davidson, *Lives on the Line* (cit. n. 11), chap. 2.

[43] See Phil Brown, "Popular Epidemiology and Toxic Waste Contamination: Lay and Professional Ways of Knowing," *Journal of Health and Social Behavior* 33 (Sept. 1992): 267–81.

[44] Davidson, *Lives on the Line* (cit. n. 11), 58.

[45] Ibid., 59.

Committed to pursuing their epidemiological research and encouraged by the support of two University of Arizona researchers, Doctors Joel Meister and Larry Clark, the LIFE organizers recruited Teresa Leal to help generate a more comprehensive health survey along the Ambos Nogales border.[46] Teresa recalls the origin of this binational health study:

> [LIFE] thought the natural conclusion was that if there were high levels of cancer, neurological and respiratory diseases on the U.S. side, there should be an equal amount, if not worse, on the Mexican side. And they were running into stone walls everywhere in the medical community, so they invited me to shadow [their study] . . . so we could do coverage on both sides.[47]

Explaining that her interest in conducting community-based epidemiological research was "in the name of searching for better quality information," Teresa elaborated on the virtue of blending the often-discordant epistemic systems of the scientific method with local, experiential observation:

> It's our human nature to be opinioned, and we have our views, we have faith in our instincts, which makes us suspicious, or inquisitive, and so we need to do more than just feel thoughtful. . . . [W]e also have to be open to the possibility that there are broader parameters than what we in our own mind have determined. I think that's the virtue of science, that it can take an objective view. And sometimes that objective view corrects things that have just been handed down as half-truths. I think that science does have the power to center us, and that's not a bad idea. But it's not always so powerful that some other information or view cannot share. So I think we have to be humble about the fact that we don't know everything and that we get more powerful the day that we admit that we don't know everything. It makes us think more clearly. In the name of searching for better quality information, we come up with synergisms that are much better, much more powerful, and it's nobody's possession, nobody's property.[48]

Publication of the results of the binational community-sponsored health studies in Ambos Nogales brought some long-awaited attention to the region. *USA Today* did a story titled "Nightmare on the Border," the news program *A Current Affair* aired a show about the rising death toll on the border, and in 1993, Arizona governor Fife Symington and Senator John McCain visited the homes of LIFE members and promised to allocate funding for a $100,000 health study in Nogales, Arizona.[49] After NAFTA went into effect in 1994, both the U.S. and Mexican governments began to pay attention to the environmental and health problems escalating along the border by tracking hazardous waste shipments from the maquiladoras, licensing waste haulers, and providing limited funding to support the factories' institution of pollution prevention systems in their manufacturing and waste management practices.[50] Although these are signs of improvement, Teresa argues, "globalization is just making things worse, and there are no provisions in NAFTA for environmental remedial clean-up along the border."[51]

[46] Ibid., 60.
[47] Leal interview, 13 June (cit. n. 26).
[48] Ibid.
[49] Davidson, *Lives on the Line* (cit. n. 11), 70.
[50] Ibid., 72; Frank Clifford and Mary Beth Sheridan, "Borderline Efforts on Pollution," *Los Angeles Times,* 30 June 1997, A1.
[51] Leal interview, 13 June (cit. n. 26).

CLEAN WATER, CONTINGENT ALLIANCES

Anxious for some more agreeable scenery, our tour party left the dump and recrossed the international border into Nogales, Arizona, on our way to Teresa's home, a small apartment just blocks from the line, which served as a home base for children and grandchildren, a busy office, and a safehouse for battered women.[52] Entering her sun-lit apartment, its walls painted a dazzling yellow, we witnessed Teresa move into action. Her answering machine was crammed with a remarkable array of messages, among them an interview request from NPR, a message from an environmental justice colleague wanting to firm up their plans to attend the United Nations conference on Racism, Xenophobia, and Related Forms of Discrimination in Durban, South Africa, and a jubilant reminder from her granddaughter of the school play to be performed that afternoon in Nogales, Sonora. Another message was from a member of the Sierra Club's Grand Canyon Chapter, calling to update Teresa on the status of their joint lawsuit against the International Boundary Water Commission (IBWC), the City of Nogales, Arizona, and the U.S. Environmental Protection Agency (EPA).

Binational strategies had proved successful in pushing the Mexican government to shut down the overfull, uncontrolled landfill in Nogales, Sonora, and in raising awareness and generating scientific and government attention to the environmental justice issues at the border. So when she learned of the decades-long negligence of various binational regulatory authorities charged with protecting the water quality of Ambos Nogales, Teresa decided to get involved in yet another cross-border action. The IBWC, the City of Nogales, Arizona, and the EPA had apparently been shirking their duties to monitor and upgrade the Nogales International Wastewater Treatment Plant (NI-WTP) and thus were listed as defendants in the suit charging violations of the Clean Water Act of 1972.[53]

The NIWTP is located twelve miles north of the U.S.-Mexico border, just downstream of the convergence of the Nogales Wash and the Santa Cruz River.[54] Built in the 1950s, the plant treats wastewater at a capacity of 17.2 million gallons per day and discharges 16,800 acre-feet of effluent per year into the Santa Cruz River, with approximately 65 percent of the wastewater flowing from the Mexican side of the border.[55] Teresa explained that the "three-way partnership" charged with overseeing the treatment plant had consistently engaged in "shoddy" management and had failed to install necessary technological improvements, even though "testing from EPA, from the Arizona Department of Environmental Quality (ADEQ), and the Friends of the Santa Cruz River has constantly flagged the high levels of contaminants that are still present in the water when it's recharged back into the river."[56]

When the district court decided that the EPA has "absolute discretion to ignore

[52] Presently, due to lack of funding, Teresa has been forced to relinquish this apartment, which, she laments, has left the women who need shelter "houseless." Teresa Leal, telephone interview by author, 16 June 2003.

[53] The lawsuit was heard in two federal courts: The United States District Court, *Sierra Club and Teresa Leal vs. Robert Ortega, Commissioner, IBWC et al.* (CIV 00–184-TUC-RCC); and the United States Court of Appeals for the Ninth Circuit, *Sierra Club and Teresa Leal vs. Christine T. Whitman, EPA, Robert Ortega, IBWC, and Marco A. Lopez, Mayor of Nogales, Arizona* (00–16895).

[54] Jana Fry and Luis Ernesto Cervera, *Water Resources in the Upper Santa Cruz Basin* (Tempe, Ariz., and Nogales, Sonora, 1995.)

[55] Hugh Holub, "The Santa Cruz River: A Resource Shared by Two Cities" (paper presented to the Border XXI EPA Regional Water Sub Work Group Meeting, Nogales Sonora, 6 March 2001).

[56] Leal interview, 13 June (cit. n. 26).

Clean Water Act violations at the Nogales Plant,"[57] the plaintiffs (Teresa Leal and the Sierra Club), with representation from the Arizona Center for Law in the Public Interest, challenged its decision in the U.S. Court of Appeals for the Ninth Circuit. They argued that, in accordance with the legislative history of the Clean Water Act, the EPA has "a *non*-discretionary mandate to act upon consistent violations at the International Wastewater Treatment Plant in order to avoid contaminating the river and to efficiently clean the residual waters that come into the waste water treatment plant as best they can."[58] In late 2001, the defendants signed a consent decree in which "the IBWC and the city of Nogales agreed that our complaints are real and that between now and 2005 they will work on them. We have the power to intervene, and if we feel it's not being done right, we have the power to seek explanation."[59]

Allying with the Sierra Club—an organization that "in working to preserve natural landscapes, often ignores or denies that some people need to *live* off the land, and depend for their survival on natural resources . . . a blind spot [that] does not set us up to be 'good neighbors'"[60]—was a decision Teresa mulled over at length. "But," she argued, "in a world increasingly affected by globalism, we cannot afford to work against each other. [In] this lawsuit . . . the Sierra Club and I are working together for a common goal. . . . We're coming from different perspectives and yet we are working towards a common goal."[61] Teresa believes that "if the decay of our river basin is out of control, then we have to do all that is in our power to help."[62]

According to Teresa, and many other local activists, the ecological and economic health of the region, as well as its social and cultural vitality, requires that the border's hegemonic authority be minimized and that binational democratic processes be put in place. Transnational organizing has become a staple of the environmental justice movement, and Teresa's activist history is no exception. As the cochair of the Southwest Network for Environmental and Economic Justice (SNEEJ), a network of eighty-five organizations focusing on environmental health, workplace safety, sustainable jobs, and cultural survival, Teresa works with her colleagues to build alliances with organizations and indigenous tribal councils that span many national borders. As an organization supporting leadership development in low-income communities of color, SNEEJ organizes training workshops that address environment, health, and work in a global framework. As Teresa explains:

> We are working on what we call a "just transition," which is not about "free trade," but "just trade." We try to prepare workers for globalization so they will not be victims of globalization. The real problem with globalization is that it threatens people's cultures and identity, and an identity is necessary for people to consider that they have something to fight for . . . and care for. Our workshops teach people about . . . sustainable economic development in their communities.[63]

[57] Vera Kornylak, staff attorney for Arizona Center for Law in the Public Interest, to Teresa Leal, 6 June 2001, Tucson, Ariz.
[58] Leal interview, 13 June (cit. n. 26).
[59] Leal interview, 12 Feb. (cit. n. 35).
[60] Adamson, "Throwing Rocks at the Sun" (cit. n. 20), 56.
[61] Ibid.
[62] Leal interview, 13 June (cit. n. 26). In June 2003, Leal reported that "the IBWC, the EPA and Nogales, Arizona continue to defer on next steps while the river and ecology continue to erode." Leal interview, 16 June (cit. n. 52).
[63] Adamson, "Throwing Rocks at the Sun" (cit. n. 20), 51.

After traveling to, and participating in, the World Trade Organization demonstrations in Seattle, Washington, Teresa was convinced that SNEEJ members needed to "come out of our little trenches" and "join forces with other environmentalist groups, with the turtle people, the whale people, with the monks from Tibet, with the Raging Grannies."[64] For her, eroding the borders that have separated environmental groups from women's groups from human rights groups and from labor organizations generates a grassroots-based globalization "from the ground up" and broadens the capacity to fight the destructive forces of globalism.[65] She argues that the long history of sectarian divisions, whether in the form of international boundaries, environmental and medical expertise, or grassroots organizing based on identity politics, corrodes the potential for cross-border community alliances. Furthermore, she contends:

> The air is for all of us; the water is for all of us. Shit and pollution, toxic substances, do not ask for permission to come into your house; they do not need a passport to cross the border. Without permission, these substances come into our lives. We can't say, "Oh, that person has cancer because they're poor." No. Cancer hits everybody. . . . [T]oxins can find their way into everyone's house, whether they're rich or poor. . . . There is plenty of research to suggest that POPs are released into the environment by industrial processes and by the spraying of pesticides, etc. These toxins are flowing through intercontinental airways. Do you know what that means? They're spreading all over the planet! Why don't we stop it? Because corporate profiteering is paramount; corporate heads don't seem to be able to live without exorbitant profits. So different groups need to come together to fight the corporations. No one group can do it alone.[66]

BEYOND GEOGRAPHIC AND EPISTEMOLOGICAL BORDER ZONES: IN SEARCH OF "COALESCENCE"

After returning several phone calls and setting up numerous appointments, Teresa prepared fresh tortillas for lunch, and we sat around her dining room table discussing the prospects for "coalescence" in the Ambos Nogales bioregion. In recent years, Teresa had joined with other activists to form the Cuenca (watershed) Network, a coalition of different groups interested in the preservation of the Santa Cruz River basin. As she explained:

> My concern is the holistic approach to saving the river, that it's both for ample water supply and quality of water. We're all dependent on the river in many ways. On the Mexican side of the river, the users like the *ejidatarios* [collective farmers and ranchers] need to be participating; they need to be included, and they're not being included as we speak. . . . We could get a biological corridor strategy going that transcends the border, that takes in cultural concerns and makes a constant effort to be bilingual and bicultural. . . . We have to be civilized enough—or maybe *primitive* enough—to do that. We have to stop creating limits to what we can do.[67]

The Cuenca Network has brought her into contact with the Friends of the Santa Cruz River, the Sustainable Border Group, and Amig@Naturales, as well as university researchers in Arizona and Sonora. They have discussed protocol design for an

[64] Ibid., 53.

[65] Jeremy Brecher, Tim Costello, and Brendan Smith, *Globalization from Below: The Power of Solidarity* (Cambridge, Mass., 2000).

[66] Adamson, "Throwing Rocks at the Sun" (cit. n. 20), 54–5.

[67] Leal interview, 13 June (cit. n. 26).

inventory of "the natural wildlife of the river to create a holistic plan . . . so that we can better understand what the river is about."[68] Two graduate students working at the Biosphere II station outside Tucson helped develop a Web site on the diverse, cross-border "coalescences" to protect the health of the Santa Cruz River and its human and nonhuman desert inhabitants.[69] Working with Amig@Naturales, Teresa is developing environmental education curricula for the schools and is conducting river walks with children on the Sonoran side of the border. "The focus is on children because they're the future," she asserts. "If they're exposed to these ideas, they'll have a backdrop that their parents didn't have."[70]

Cuenca networking creates a "global sense of place" and relies upon many forms of border crossing by its members to understand and articulate the many different affinities and antagonisms that flow throughout the Santa Cruz watershed and bubble up in the Ambos Nogales region. For Teresa, the border is a "meeting place" of a tangled network of uneven social and economic relations, environmental circumstances, and ecosystemic limitations, all constructed on a far larger scale than the *place* itself. In this sense, Teresa's ecological imaginary retains the idea of specificity and uniqueness of place without appealing to a static, nativist formulation of "community," because it "allows a sense of place which is *extroverted,* which includes a consciousness of its links with the wider world, which integrates in a positive way the global and the local."[71]

Teresa's most recent campaign (with the binational women's organization Las Sinfronteras) to buy a small house on the Sonoran side of the border that would serve both as an environmental education center and a safehouse for women maquiladora workers exposed to violence on the job and in the home,[72] illustrates this articulation of the global and the local. "I remember [at the Earth Summit] in Rio, the incipient message was that women's rights were environmental rights," Teresa recalled. "Now, we're getting ready for Rio-plus-10 in Johannesburg in September [2002]. . . . OK, let's see how we fared."[73] The terrible environmental conditions at the border and the miserable living and working conditions of the maquiladoras' predominantly female labor force cannot be disarticulated, she argues, and neither is faring very well. The violence against young women in the global assembly line, whose livelihoods and survival needs are tied to the factories, parallels the wholesale poisoning of the water and air of the Sonoran Desert by transnational corporations supplying low-priced products for upscale markets around the globe. The struggle for environmental justice on the border necessitates paying attention to these interlocking layers of injustice and devising locally rooted and globally informed strategies to confront them.

Like all who dwell in arid landscapes, Teresa and her Comadres express a heightened awareness of the sacredness of water, the lifeblood of daily existence. It would not be overstating the issue to say that for those people living in the Sonoran Desert, survival depends on the safekeeping of water. Protecting the *local* watershed, however, requires a larger-than-local water politics that spans borders of all kinds— national, racial, gendered, economic, linguistic, ecological, technological, spiritual,

---

[68] Adamson, "Throwing Rocks at the Sun" (cit. n. 20), 56.
[69] See http://www.geocities.com/woborders/2nogales1.
[70] Leal interview, 12 Feb. (cit. n. 35).
[71] Massey, *Space, Place, and Gender* (cit. n. 9), 155 (my italics).
[72] See, Davidson, *Lives on the Line* (cit. n. 11), chap. 1
[73] Leal interview, 12 Feb. (cit. n. 35).

and epistemic. The "popular" knowledge that grounds struggles for environmental justice on the border is *more* than local, more than anecdotal, more than just personal experience. It is the outcome of shared observation, careful research, and the forging of synchretic assemblages of "experts" of all stripes. As Teresa argues, our partial and limited worldviews cannot hope to address environmental problems of such magnitude; it is the "coalescence of people with differing views that is really inspiring."[74]

Quoting a favorite Mexican saying, "en rio revuelto ganancia de pescadores" (fishermen thrive on a wild river), Teresa maintains that "chaos" is a good place to begin change because it might, like a roiling river, sweep us out of our "sectarian trenches." There are signs of progress, she asserts, but there is still a long way to go. The diverse, occasionally contentious communities nestled along the dry streambeds of Ambos Nogales are crossing many borders to *coalesce*, inspired by the uniqueness of the people and landscapes that compose the desert ecosystem in which they live. Teresa smiles. "That's how I would calibrate what is progress and what is not—if you're moved by a beautiful inspiration, rather than a negative inspiration, then I think that's when we're making progress."[75]

[74] Ibid.
[75] Leal interview, 13 June (cit. n. 26).

MATERIAL FLOWS AND PUBLIC HEALTH

Mapping a Zoonotic Disease:

Anglo-American Efforts to Control Bovine Tuberculosis Before World War I

By Susan D. Jones[*]

ABSTRACT

Before World War I, British and American public health officials correlated tuberculosis in dairy cattle with severe infections in milk-drinking children. They traced
bacteria in municipal milk supplies, mapped the locations of infected animals, and
sought regulatory power to destroy them. Consumers, milk producers, municipal officials, veterinarians, and physicians all influenced the shape of antituberculosis regulations. Many condemned pasteurization as too costly and as masking tubercular
contamination and poor sanitation. They saw milk-borne tuberculosis as an environmental as well as a bacteriological problem. Similar to other zoonotic diseases
such as BSE, bovine tuberculosis blurred the boundaries between urban and rural,
production and consumption, and human and animal bodies.

INTRODUCTION

In July 1901, eminent German bacteriologist Robert Koch shocked delegates to the
International Tuberculosis Congress in London when he asserted that tuberculosis in
cattle represented no public health threat to human beings. His declaration made
headlines around the world because it threatened longstanding beliefs and practices
surrounding the prevention of tuberculosis (the nineteenth century's greatest killer) in
cattle and humans. Believing meat and milk from tubercular cattle to be infectious,
citizens and sanitary officials had long reviled such food products. Koch's address
ignited scientific and public debates over how often the bovine form of the disease
infected human beings and whether costly efforts to eradicate it or pasteurize milk
should be undertaken. British and American scientists and public health officials were
particularly interested in these questions. They were keen to refute Koch and to prove
the connection between tuberculosis in cattle (bovine tuberculosis, or BTB) and nonpulmonary tubercular disease in children.[1]

[*] Department of History, 234 UCB, University of Colorado, Boulder, CO 80309-0234; jonessu@
colorado.edu.

I thank the guest editors and participants in the "Environment, Health, and Place in Global Perspective" conference, Kevin R. Reitz, and two anonymous reviewers for valuable comments; and historian
Keir Waddington and the staff at the Wellcome Library for the History and Understanding of Medicine
(London) for research guidance. A Distinguished Scholar Award, U.S.-U.K. Fulbright Commission
(2002), supported my British research.

[1] Koch's address was published in *Transactions of the British Congress on Tuberculosis* (London,
1902), 1:23–37. "The Congress on Tuberculosis," *Nature* 64 (1 Aug. 1901): 327–8; "The Congress on

To do so they developed procedures to identify, map, and control the course of BTB, the first time this combination of techniques would be applied to a zoonotic disease (one transmissible from animals to humans). This required associating with groups and considering ideas usually excluded from public health efforts (and most historians' accounts): veterinarians and livestock owners, and concerns about the power of the state to regulate free-market activities. Public health officials interested in BTB conceived it as an "environmental" disease, grounded in local living and climatic conditions. Its natural history unfolded in rural stables as well as urban tenements, and its incidence and the policies to control it depended on local conditions. Nowhere was this environmental conception more developed than in the United States and the United Kingdom. Both nations had also been long accustomed to localized jurisdiction over public health issues. Both based anti-BTB regulations on voluntary participation and market incentives rather than state control and compulsory measures. As this article demonstrates, Anglo-American officials and scientists concerned with BTB in children began their campaign against the disease by creating tools that would allow them to identify and control it at its source—food-producing animals.

Mapping infected animals' locations proved more straightforward than developing anti-BTB policies and regulations. BTB caused tensions among relevant actors because it evaded easy definition and assignment of responsibility. Was it a rural or urban problem? Were livestock producers and the agricultural community responsible for controlling it, or was it primarily a public health concern for (mainly nonrural) consumers and their physicians? As the case of BTB demonstrates, animal diseases could challenge locational boundaries, disrupt traditional linkages, and forge unlikely unions. BTB blurred the boundaries between urban and rural, and production and consumption. Individual animals' bodies regularly crossed these barriers, and the disease traveled with them. By instituting and maintaining surveillance over individual animals, British and American municipal regulations redefined traditional social linkages and environmental perceptions and recognized the permeability of the organic boundary between animal and human bodies.

LINKING BOVINE TUBERCULOSIS WITH DISEASE IN CHILDREN

Tuberculosis cruelly altered the bodies of sufferers, and public health officials used the corporeal imprint of the disease as evidence of its transmission from bovines to humans. Children were often affected by nonpulmonary manifestations such as tabes mesenterica (infection of the intestines and mesenteric glands) or scrofula (large swellings in the neck, armpits, and elsewhere). Children also suffered from "the horrible disease"—meaning miliary skin tuberculosis (also known as lupus), which riddled the skin with lesions; or tuberculosis of the bones and joints, which deformed the skeleton and could cause a hunchback.[2] These manifestations, particularly of the intestines and glands, were widely suspected to have resulted from the ingestion of BTB-infected foods. The bodies of cattle, too, could display the "corruption" charac-

Tuberculosis," *Saturday Review* 92 (27 July 1901): 102–3; "The British Congress on Tuberculosis," *Popular Science Monthly* 59 (Sept. 1901): 508; Barbara Gutmann Rosenkrantz, "The Trouble with Tuberculosis," *Bulletin of the History of Medicine* 59 (1985): 155–75.

[2] "Congress on Tuberculosis," *Saturday Review* (cit. n. 1), 102; Linda Bryder, *Below the Magic Mountain: A Social History of Tuberculosis in Twentieth-Century Britain* (Oxford, 1988), 20.

teristic of tubercular infection,[3] and the milk produced by cows with infected udders was considered to be "the chief source of danger" in transmitting tuberculosis to children.[4]

In the bodies and secretions of infected cattle and humans, tuberculosis bacilli lived and multiplied, even if the infected host appeared healthy. Scientists exposed and displayed these bacilli under the microscope, creating a record of cellular as well as somatic tubercular corruption. By 1898, the bacterium responsible for BTB, *Mycobacterium tuberculosis bovis,* had been distinguished from that which caused pulmonary tuberculosis (*Mycobacterium tuberculosis*). Robert Koch's 1890 development of tuberculin, a preparation of attenuated bacilli, created an instrument that could identify externally healthy-looking bodies in which either species of tubercle bacilli lurked; positives were labeled "reactors."[5] Those humans and animals marked by tuberculosis in any of these visible ways assumed the identity of tuberculars and reactors: as they supplied useful data points for scientists, they became defined by their disease.[6]

American and British scientists also saw bodies as contained environments, labeling them the "soil" in which the "seeds" of infection could take hold.[7] The relationships of the bovine tubercle "seeds" to the "soil" of human and animal bodies underlay much of the debate over how easily BTB could be transmitted to people. In 1901, British bacteriologist Edgar Crookshank asserted that infection of humans with BTB was complicated because "man is not the natural soil of bovine tuberculosis. . . . [A]n animal may be markedly susceptible to infection with the virus from a foreign soil" only under certain conditions. To eradicate BTB, which Crookshank believed "desirable," the cattle with "tissues prone" to tubercle bacilli and their unhealthy environment had to be eliminated as the "fitting soil" for the infection. Influential American veterinarian Alexandre Liautard quoted Crookshank's words in the next editorial of *The American Veterinary Review* (the journal read by most U.S. veterinarians).[8] Crookshank, Liautard, and others focused on the environment in which BTB throve rather than only advocating eradication of the bacilli. These were complementary goals. British and American scientists understood the bodies that hosted tuberculosis as interactive locations, as complex milieux with which the tubercle bacilli had to interact to produce and display signs of the disease.

Because of its effects upon the body, the external environment of tubercular humans and animals also came under great scrutiny. By tracking the bodies of human and animal sufferers as well as the bacilli, disease and its social and environmental locations

---

[3] James Arthur Gibson, "The Eradication of Bovine Tuberculosis," *The Westminster Review* 156 (July 1901): 1–13, on 13.

[4] Ibid., 3.

[5] Ibid.; German Sims Woodhead, "Koch's Recent Researches on Tuberculin," *Nature* 55 (15 April 1897): 567; Herbert Maxwell, "Tuberculosis in Man and Beast," *Nineteenth Century* 44 (Oct. 1898): 673–87, especially 677–8.

[6] Several studies have made this point for human sufferers. See Barbara Bates, *Bargaining for Life: A Social History of Tuberculosis, 1876–1938* (Philadelphia, 1992); Bryder, *Below the Magic Mountain* (cit. n. 2); Sheila Rothman, *Living in the Shadow of Death: Tuberculosis and the Social Experience of Illness in American History* (Baltimore, 1995).

[7] Michael Worboys, *Spreading Germs: Disease Theories and Medical Practice in Britain, 1865–1900* (Cambridge, 2000), 6–7, 161–4.

[8] [Alexandre Liautard], "Tuberculosis," *American Veterinary Review* 25 (1901): 785–9, on 786. Crookshank's original use of "soil" may be found in *British Congress on Tuberculosis* (cit. n. 1), 4:71; he further developed this theme in his 1 October 1901 speech given to open the Royal Veterinary College term (reported in *The Field* 98 [5 Oct. 1901]: 566).

could be mapped more precisely.[9] Tuberculosis was an "environmental" disease, and its remedy involved the alteration of an environment that was "unhealthy" into one conducive to health.[10] An unhealthy environment contributed to the transmission of the disease in several ways: it weakened the normal defenses of the host body, harbored the infective bacilli, and prevented treatment from succeeding. These three principles were generally understood among not only scientists but also producers and consumers of milk. As Harold R. White reminded British readers of *The Westminster Review,* the best prevention against tuberculosis was "to strengthen the body in order to ward off, or withstand, the attacks of consumptive bacilli."[11]

Human bodies were weakened by what physicians and public health authorities classified as poor living and working conditions: cramped and unventilated houses and factories, poor food, and overwork or lack of rest. Unsanitary conditions and human behaviors introduced the bacillus into ill-ventilated workplaces and homes. These conditions, as historians have noted, were most often found in a particular physical and moral landscape: the city.[12] Tuberculosis throve on conditions in the cities, which included not only what the *Saturday Review* described as "the crowding of huddled families in closed alleys and sunless courts" but also alcoholism and other personal vices readily identified with urban life.[13] The body and its environment were necessary partners in purging the bacilli and in conquering the disease, and Anglo-American physicians believed a wholesome nonurban environment gave the patient the best chance at recovery.[14]

Scientists and some livestock owners on both sides of the Atlantic asserted that tuberculosis infection in cattle was likewise encouraged by "unhealthy" living conditions. This included stabling cows in cities, where they could not go outdoors or exercise and received poor quality food. (The by-products of distilleries, fed to city cows, were singled out as particularly dangerous, an assessment corresponding with the theme of human alcoholism.)[15] In the countryside, profit-conscious livestock owners believed cows kept in dark and unventilated stables gave more milk. Re-

[9] David Armstrong, *Political Anatomy of the Body: Medical Knowledge in Britain in the Twentieth Century* (Cambridge, 1983), 7, 11–2, 16.

[10] See, e.g., Charles Denison, "A Suggested Law to Regulate House Ventilation for the Prevention of Tuberculosis," *Transactions of the Sixth International Congress on Tuberculosis,* vol. 4, pt. 1 (Philadelphia, 1908), 323–30, on 326; Rothman, *Living in the Shadow of Death* (cit. n. 6), 3, 133–40; F. B. Smith, *The Retreat of Tuberculosis, 1850–1950* (London, 1988), 175–94.

[11] Harold R. White, "The Problem of Tuberculosis," *The Westminster Review* 156 (Nov. 1901): 545–52, on 547.

[12] For a general introduction to this concept in Britain and the United States, see Asa Briggs, *Cities and Countrysides: British and American Experience, 1860–1914* (Leicester, 1982); James L. Machor, *Pastoral Cities: Urban Ideals and the Symbolic Landscape of America* (Madison, Wis., 1987); Daniel J. Monti, *The American City: A Social and Cultural History* (Malden, Mass., 1999); and Raymond Williams, *The Country and the City* (New York, 1973).

[13] "Congress on Tuberculosis," *Saturday Review* (cit. n. 1), 102. David Barnes has thoroughly explored these themes in *The Making of a Social Disease: Tuberculosis in Nineteenth-Century France* (Berkeley, Calif., 1995), chaps. 4, 5.

[14] Charles Denison, *Rocky Mountain Health Resorts: An Analytical Study of High Altitudes in Relation to the Arrest of Chronic Pulmonary Disease* (Boston, 1880); Rothman, *Living in the Shadow of Death* (cit. n. 6); T[heophilus] N. Kelynack, ed., *Tuberculosis in Infancy and Childhood: Its Pathology, Prevention, and Treatment* (London, 1908); White, "Problem of Tuberculosis" (cit. n. 11), 547. For a theoretical overview, see Wilbert M. Gesler, "Therapeutic Landscapes: Medical Issues in Light of the New Cultural Geography," *Social Science and Medicine* 34 (1992): 735–46.

[15] Gibson, "Eradication of Bovine Tuberculosis" (cit. n. 3), 3–4; M[ilton] J. Rosenau, *The Milk Question* (Boston, 1912), 260.

formers in the United States and Britain sought to convince these owners that allowing cows to be outdoors in decent weather was a sound financial investment and a stimulus to good health. Summing up this position in a *Westminster Review* article aimed at land and livestock owners, James Gibson specifically linked reformers' housing recommendations for cows to similar ones for people seeking to avoid tuberculosis, asserting that the "laws which govern health in man govern it equally in beast."[16]

In this way, public health officials sought to create analogies linking tuberculosis in cattle and in humans through its embodied and environmental characteristics, refuting Koch's 1901 assertion that bovine tuberculosis was only rarely transmitted to humans. If tuberculosis was a shared "environmental disease," then environmental analogies were strong evidence that the disease in animals and humans should be correlated. In the years leading up to World War I, American and British medical officers of health (MOsH), physicians, and veterinarians collected data on the incidence of BTB in humans and animals not only as a step toward refuting Koch and strengthening Anglo-American scientific ties but also as support for their visions of appropriate public health policies against BTB.

MAPPING BOVINE TUBERCULOSIS

As a cornerstone of urban regulatory control, public health officials routinely ferreted out human tuberculosis sufferers and their families.[17] Urban animals, too, became targets of scrutiny. Using a combination of bacteriological and epidemiological techniques, reformers mapped BTB and the urban populations vulnerable to it. As sources of milk for children, large numbers of dairy cows were kept in English and American cities at the century's turn. San Francisco, for example, boasted one cow for every five human inhabitants; thousands of animals lived in London's dairies. Urban families were generally allowed one or two backyard cows.[18] In their initial efforts to link infected milk to children ill with nonpulmonary tuberculosis, American reformers in cities such as Minneapolis and New York, and their British counterparts in places such as Edinburgh and Manchester, targeted urban animals.

In part due to its well-established medical and veterinary medical schools, Edinburgh boasted an active urban dairy inspection program carried out by its tuberculosis officer and later by its Veterinary Department (staffed by graduates of local schools). At the beginning of the century, more than 100 urban byres housed about 5,000 cows, of which 15 to 20 percent were found to be tubercular and consequently

[16] Gibson, "Eradication of Bovine Tuberculosis" (cit. n. 3), 6–7; A. M. Trotter, *Tuberculosis: A Preventable Disease* (Edinburgh, 1899), 16–23 (Pamphlet Collection, 58 [16], Royal Highland and Agricultural Society, Edinburgh, Scotland); Harold Sessions, *Cattle Tuberculosis: A Practical Guide to the Agriculturalist and Inspector* (London, 1905), 14. For the United States, see Denison, "A Suggested Law" (cit. n. 10), 326–7; Henry E. Alvord and R[aymond] A. Pearson, "The Milk Supply of Two Hundred Cities and Towns," *Bureau of Animal Industry Bulletin* no. 46 (Washington, D.C., 1903), 25, 123; Leonard Pearson and M[azyck] P. Ravenel, "Tuberculosis of Cattle and the Pennsylvania Plan for Its Repression," *Pennsylvania Department of Agriculture Bulletin* no. 75 (Harrisburg, 1901), 43–4, 168–71; and Henry L. Shumway, *A Hand-Book on Tuberculosis among Cattle* (Boston, 1895), 170–1.
[17] Barnes, *Making of a Social Disease* (cit. n. 13), 112, 113, 137, 213; Bryder, *Below the Magic Mountain* (cit. n. 2), 41–3, 107–9, 131–2.
[18] Alvord and Pearson, "Milk Supply" (cit. n. 16), 9–26; Harold Swithinbank and George Newman, *Bacteriology of Milk, With Special Chapters Also by Dr Newman on the Spread of Disease by Milk and the Control of the Milk Supply* (New York and London, 1903), 121–3.

condemned.[19] The results of milk production by these infected animals were only too obvious to Edinburgh physicians, including Robert W. Philip, elder statesman of the city's antituberculosis campaign. Philip complained in 1905 that although childhood deaths due to pulmonary tuberculosis had declined by 23 percent since 1891, mortality due to nonpulmonary tuberculosis had actually increased by about 1 percent. Thus even children living in Edinburgh, with its relatively high degree of regulatory control, suffered the urban liability of BTB. Belfast physician John Byers asserted that the situation for urban children living in Ireland, with its lack of regulation, was considerably worse. He linked the fact that about 40 percent of the tubercular children in Belfast Children's Hospital suffered from bovine-origin nonpulmonary tuberculosis to the prevalence of BTB in Irish cattle (at least 30 percent). He concluded that, in terms of childhood tuberculosis, urban conditions in Ireland were "most deplorable."[20]

Officials in American cities likewise saw BTB as one of the liabilities of urban life for children. Minneapolis provided an instructive example. Within the city limits in 1902, about 1,500 cows lived in dairies and another 1,200 were kept by private families.[21] As early as 1894, the eminent Minneapolis pediatrician Thomas S. Roberts warned of the high prevalence of tuberculosis in milk-drinking children in the city. He found an eager colleague in veterinarian Charles E. Cotton, who had already begun administering tuberculin tests to Minneapolis's cattle. Employed as the city's veterinary official, Cotton used his tuberculin tests, with the support of Roberts, to persuade the Minneapolis licensing board to withhold a dairy license until an applicant's cattle had all been tested for BTB. Physician Arthur Myers, who came to Minneapolis in 1914 and became medical director of Lymanhurst School for Tuberculous Children, later described Cotton's tenacity in finding and removing tubercular animals from the city's milk producers. Due to these efforts, Myers believed, in 1895 Minneapolis became the first American city effectively to require tuberculin tests of cattle being licensed to supply urban milk.[22]

As Cotton and other municipal reformers recognized, however, the problem of bovine tubercle bacilli in the urban milk supply arose not only from within but also from without the city's boundaries. Infected milk sold in the city served as a marker of disease in extra-urban cows, but how could these animals be found? Health officials and physicians in the United States and MOsH in the United Kingdom usually began their searches by sampling the milk supply of the cities and towns under their jurisdictions. Their strategies were based on previous campaigns concerned with diphthe-

[19] "Tuberculosis in Cows," Annual Reports of the Public Health Department, SL 27/1/2, City Archives, Edinburgh, Scotland; for retrospective figures from the Veterinary Department, see Annual Report, 1913, 72–6.

[20] John Byers, "Tuberculosis among Children in Ireland," in Kelynack, Tuberculosis in Infancy and Childhood (cit. n. 14), 200–7, on 200, 203. Historians have also called this urban liability a "penalty" (Michael Worboys and Flurin Condrau, "Urban Penalty to Urban Advantage: TB Mortality in Britain and Germany, 1880–1920" [paper presented at the conference "From Urban Penalty to Global Emergency: Current Issues in the History of Tuberculosis," Society for the Social History of Medicine, Sheffield Hallam University, Sheffield, United Kingdom, 23–5 March 2002]. For a contemporary summary correlating bovine bacilli in tubercular lesions in children with the prevalence of infected cows, see Louis Cobbett, The Causes of Tuberculosis: Together with Some Account of the Prevalence and Distribution of the Disease (Cambridge, 1917), 570–611.

[21] Alvord and Pearson, "Milk Supply" (cit. n. 16), 26, table 1.

[22] J. Arthur Myers, Tuberculosis: A Half Century of Study and Conquest (St. Louis, [1970]), 6–8, 13–4, 222–4.

ria and other milk-borne hazards. BTB, however, quickly proved itself a somewhat different problem, one requiring the integration of bacteriological and epidemiological techniques. Unlike diphtheria and most other major milk-borne hazards of the time, BTB originated in particular animals' bodies. Finding these animals depended on conducting a bacteriological examination of the milk, tracing the milk back to a particular farm, and then relying on veterinary inspection to find infected individual animals on that farm. The combination of these methods allowed municipalities, during the first two decades of the twentieth century, to extend new regulatory powers— and control—over the surrounding suburban and rural zones of production.

These zones of dairy production defined the milkshed of a municipality and thus the area of epidemiological interest. In the United States in 1900, the largest cities and those in the eastern part of the nation commanded milksheds of tremendous size. For example, milk arrived in Minneapolis from up to 75 miles away every day. New York City received milk from up to 350 miles distant, and Memphis, Tennessee, from as much as 527 miles away by rail. On average, cities with populations greater than 100,000 (such as these three) received milk from up to 105 miles distant.[23] Larger milksheds, of course, provided a more daunting challenge for mapping the disease. In 1900, few American municipalities had much knowledge of the bacteriological condition of their milk supplies or of the characteristics of the animals supplying them. Exceptions included New York City, with its active medical milk commission and a 1901 Rockefeller-funded study, and Montclair, New Jersey, home of reformer Henry Coit and the certified milk movement. Similar conditions applied in the United Kingdom: milkshed size was generally in proportion to the human population of municipalities, and officials had made little systematic study of municipal milk supplies' origins. Thus at the turn of the century, American and British municipalities generally exerted little control over their suburban and rural milksheds. This situation would change drastically over the next twenty to thirty years.

A prototypical program for studying and controlling the municipal milkshed began in the English city of Manchester in the 1890s. Manchester's MOH, James Niven, and its chief bacteriologist and pathologist, Sheridan Delépine, both had a particular interest in BTB. Moreover, they worked within a supportive political and institutional environment. Manchester was well known for stringently enforcing the national Dairies, Cowsheds, and Milkshops Order along with its own municipal regulations. Furthermore, its university boasted a school of public health and a diagnostic laboratory for human and animal disease (headed by Delépine).[24] As a bacteriologist, Delépine became interested in BTB while studying its culture and inoculation properties in the 1890s. He quickly ascertained that infection within a cow's udder and body, not contamination by external sources (human or animal), was responsible for most tubercle bacilli in milk. For Delépine, this direct connection between bodily disease in the cow and the infectious agent in the milk determined the appropriate public health approach—a combination of bacteriological screening and mapping the

[23] Alvord and Pearson, "Milk Supply" (cit. n. 16), 26–9, table 1.
[24] See volumes of the *Archives of the Public Health Laboratory of the University of Manchester* (Manchester, 1906–1912), especially vol. 1 (1906), 66–8; Sheridan A. Delépine, *Report on Investigations in the Public Health Laboratory of the University of Manchester upon the Prevalence and Sources of Tubercle Bacilli in Cows' Milk: Extract from the Annual Report of the Medical Officer of the Local Government Board for 1908–09* (London, 1910).

locations of infected individual animals. Not all cases of tuberculosis could be easily identified by veterinarians' physical examinations of cows, so bacteriology was necessary to screen the milk-producing population. Milk samples taken from specific animals or whole herds were tagged to identify their exact origins before being bacteriologically analyzed. Without this documentation, samples were discarded—underscoring Delépine's determination that bacteriology, unless linked to a specific animal or location, was useless.[25]

Delépine based his public health program on his linked mapping and screening techniques. Since 1896 he had been recording the locations of BTB in the Manchester milkshed. He and Niven used this information to prompt the city to enact special model milk provisions. "All samples found to cause tuberculosis are followed to their source at the farm by the Medical Officer of Health (or his representative) and the veterinary surgeon," the provisions directed. When veterinarians identified infected cows, they encouraged the farmer to slaughter the animals on the spot or required him to isolate them. The milk of the farm's cows was then periodically retested. This type of surveillance had been in force on farms and dairies within the city limits since 1896; the 1900 milk provisions extended the control of Delépine, Niven, and J. W. Brittlebank (the chief veterinary officer) to the entire milkshed surrounding the city.[26]

Between 1900 and 1910, Delépine refined his mapping technique, making it generalizable to other locations. He divided Manchester's milkshed into sixteen-square-mile grids, marked the location of each farm on the map, and used open and closed dots to indicate whether the farm had tubercular cows on its premises. (See Figure 1.)

The maps were updated annually so Delépine could demonstrate reductions over time in the number of farms infected and isolate grids whose farms were cleared of BTB. By using grids, he could study the distribution of BTB in areas of equal size and compare otherwise unrelated areas around the country—or even in other countries. In addition to standardizing space, Delépine sought to map and standardize animal bodies. His method included standard terminology for the location of lesions in cows' bodies and a schematic drawing of a guinea pig body (into which potentially tubercular samples were injected to determine their infective power). Even the bacteriological examination results on the milk were to be recorded using a standardized notation.[27] By thus creating maps of BTB's bodily and environmental imprints, Delépine sought to understand and control the disease.

Delépine's program, by his own account at least, was successful for Manchester. In his major published summary of his results, he used his maps to demonstrate "the considerable reduction that has taken place in the number of areas in which dangerous cows have been found during the [past] 11 years."[28] For Delépine, determining whether land areas had tuberculosis-free bovine inhabitants, rather than checking bacterial counts or using other possible measures, provided the best indicator of success. BTB was viewed as a grounded, localizable, and thus controllable disease—and the data that could be collected by mapping powerfully supported arguments in favor of area-based public health policies.

[25] Delépine, *Report on Investigations* (cit. n. 24), 352–3, 355, 373–4.
[26] Ibid., 370.
[27] Ibid., 378–84.
[28] Ibid., 385; see maps 1, 2, and 3 for supporting data.

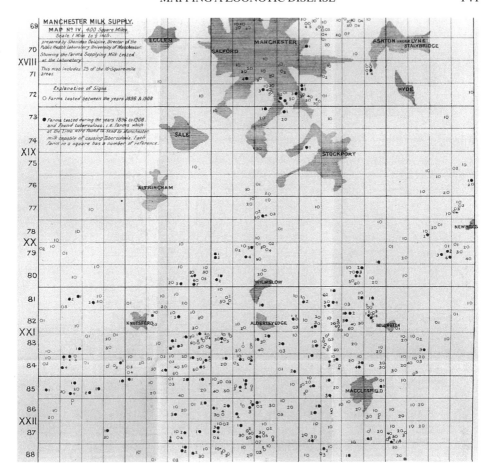

Figure 1. Map: "Manchester Milk Supply," in Sheridan A. Delépine, Report on Investigations in the Public Health Laboratory of the University of Manchester upon the Prevalence and Sources of Tubercle Bacilli in Cows' Milk: Extract from the Annual Report of the Medical Officer of the Local Government Board for 1908–1909, *383. (From* Annual Report, Medical Officer of the Local Government Board, 1908–1909 *[London, 1910]. Courtesy of Wellcome Library, London.)*

REGULATING AND CONTROLLING BOVINE TUBERCULOSIS

Policies and regulations designed to control BTB disrupted traditional linkages and forged unlikely unions between dairy producers, consumers, and public health officials. They proved difficult to negotiate prior to World War I for three major reasons. First, such a large percentage of dairy cattle in the United Kingdom and the northeastern United States were infected that neither the state nor private producers were willing to bear the economic burden of removing these animals from milk production. Second, rural producers resented urban authorities' attempts to extend regulatory control over them. Finally, concerned reformers held competing ideas about the definition of "healthy" animals, "pure" milk, and "natural" landscapes for production.

These problems intensified as the parameters of milk production changed between

1900 and 1914, influencing the shape of new regulations. First, the zones of milk production surrounding cities increased in size. As noted in the previous section, at the
turn of the century, New York City received milk from as far away as 350 miles; by
1912, this zone had increased to 400 miles, encompassing some 40,000 dairy farms.
Two-thirds of the milk supply for Washington, D.C., now came into the city from re-
mote places by rail, a pattern that characterized Manchester, Edinburgh, and London
as well.[29] Moreover, fewer producers remained within urban boundaries as municipalities increasingly sought to ban animals as nuisances and polluters. In Edinburgh,
for example, the number of city byres decreased from about 125 to 57 between 1903
and 1919. By 1914 in the United States, and the 1920s in the United Kingdom, almost
all municipalities had forbidden the keeping of livestock within their boundaries.[30]

This shift in bovine demographics meant not only that distance between producer
and consumer was increasing but also that the focus of opposition to regulation increasingly included rural as well as urban and suburban producers. Despite being
free-market competitors, urban, suburban, and rural dairy producers joined forces to
challenge the incursion of public health officials into their businesses. In 1909, Milwaukee adopted an ordinance that required dairymen selling milk in the city to file, in
the health commissioner's office, a certificate from "a duly licensed veterinary surgeon" stating that their cows were free of BTB. Every year, each cow had to be numbered, described, tested, and pronounced healthy. Under the new municipal ordinance,
milk from untested, unregistered herds would be confiscated and destroyed with no apparent recompense to the dairy producer. Outraged producers found a champion in the
pseudonymous "John Quincy Adams," a dairy farmer who carried his suit against the
Milwaukee ordinance all the way to the United States Supreme Court. "Adams" alleged, among other things, that the city health commission had no right to extend its
jurisdiction to his milk. The Court ruled against him, holding that municipalities did
have the right to insist on veterinary inspection of cows and testing of milk from rural
dairies outside their normal jurisdictions. In matters of public health, the Court reasoned, "there is no discrimination" between "milk from cows outside and milk from
cows within the city."[31] In the United Kingdom, the 1915 Milk and Dairies (Consoli-
dation) Act likewise empowered local authorities to control the sale of milk and for-
bade the sale of milk from any cow with a tubercular udder, inside or outside the city
limits.[32] Along with urban consumers and public health officials, judicial and politi-

---

[29] Rosenau, *Milk Question* (cit. n. 15), 260; Thomas Darlington, "The Methods of Dealing with the
Milk Supply of New York City," *Journal of the American Medical Association* 49 (1907): 1079; "Tu-
berculosis in Cows" (cit. n. 19), 1913; Arthur Gofton, "Veterinary Department Report, 1919," 105–6,
SL 27/1/2 (cit. n. 19); Swithinbank and Newman, *Bacteriology of Milk* (cit. n. 18), 383.

[30] Rosenau, *Milk Question* (cit. n. 15), 261; Gofton, "Veterinary Department Report" (cit. n. 29).
Cities often forced the removal of dairy animals by requiring ownership of large areas of land per stable
or animal, which was prohibitively expensive and difficult to acquire as the city grew. See C. Hamp-
son Jones, "The Present Needs of the Milk Supply of Baltimore," *Charities* 16 (1906): 499–502, 499.
This issue of *Charities,* a popular American reform journal, was largely devoted to problems of the
milk supply.

[31] See 228 U.S. 572, 580, for appellate court rulings; and 33 S. Ct. 610, 612, decided May 12, 1913,
for the Supreme Court case *John Quincy Adams v. City of Milwaukee and Gerhard A. Bading* (chief
public health official of the city); Jacqueline H. Wolf, *Don't Kill Your Baby: Public Health and the Decline of Breastfeeding in the 19th and 20th Centuries* (Columbus, Ohio, 2001), 59–61.

[32] Smith, *Retreat of Tuberculosis* (cit. n. 10), 183. Implementation was delayed in many places by
World War I (Bryder, *Below the Magic Mountain* [cit. n. 2], 133).

cal authorities recognized the permeability of urban boundaries to animals, the products of their bodies, and their diseases.

This understanding supported the increasing desire of urban health officials and concerned consumers not only to regulate producers but also to assign to them the blame for BTB deaths and disabilities in Anglo-American children (a position historians have since perpetuated).[33] The American reformist journal *World's Work* summed up the attitudes of many citizens and health officials when it placed the responsibility for BTB on owners of dairy herds and called them a "menace" to the public health.[34] This attitude was hardly universal, however; many health officials had sympathy for dairy producers.[35] Furthermore, American physicians and British MOsH often problematized their alliance with concerned consumers by asserting that dairymen would produce milk free of BTB and other dangers only if consumers would demand it and pay a fair price for it. Maryland physician John Fulton cautioned that "the weight of enlightened public opinion determines the quality of every public service." In Boston, James O. Jordan asserted that "consumers must pay for cleanliness," while veterinarian Leonard Pearson complained that hospitals and children's homes bought the cheapest (and most infected) milk they could find "as if they were buying coal."[36] In the United Kingdom, as well, MOsH and other officials decried the lack of consumer interest in supporting the cost of a clean milk supply. Wilfred Buckley wrote that farmers were only being reasonable in refusing to expend more money on producing clean milk when they received no higher price for it.[37]

The early-twentieth-century pasteurization debate perhaps best demonstrates the tangled alliances among producers, consumers, and public health officials on questions of BTB and wholesome milk. Pasteurization, when done properly, killed pathogenic bacteria (including bovine tubercle bacilli) in milk. With mandatory pasteurization in place, public health officials would no longer need to expend resources to map the locations of tubercular cows and remove their products from the milk supply. These advantages did not, however, ensure that pasteurization would quickly become a feature of municipal milk policies. Predictably, most milk producers opposed municipalities' efforts to force them to pasteurize their milk because municipalities insisted that producers bear the cost. Producers were joined in their anti-pasteurization stance by some unlikely allies. For a variety of reasons, advocates of pasteurization (such as Nathan Strauss in New York City) at first could not persuade some municipalities—in fact, the majority of them—to require the

[33] See, e.g., Smith, *Retreat of Tuberculosis* (cit. n. 10), chap. 6; and Wolf, *Don't Kill Your Baby* (cit. n. 31), 53–64, 72.

[34] "The Menace of the Dairyman," *World's Work* 13 (Nov. 1906): 8152. For British producers' opposition to regulation, see Smith, *Retreat of Tuberculosis* (cit n. 10), 183–4.

[35] See, e.g., Rosenau, *Milk Question* (cit. n. 15), 243–6; and John S. Fulton, "What the Public Ought to Know as to Milk," *Charities* 16 (1906): 479–84, 481. Some producers also supported anti-BTB measures: see Rosenau, *Milk Question* (cit. n. 15), 15, 20, 243–5; and Pearson and Ravenel, "Tuberculosis of Cattle" (cit. n. 16), 150, 181, 185.

[36] Jones, "Present Needs of the Milk Supply" (cit. n. 30), 499; Fulton, "What the Public Ought to Know" (cit. n. 35), 483; James O. Jordan, "Boston's Campaign for Clean Milk," *J. Amer. Med. Ass.* 49 (28 Sept. 1907): 1082–7, on 1087; Leonard Pearson, "Tuberculosis of Cattle: How It May Be Repressed and Its Relation to Public Health," *Charities* 16 (4 Aug. 1906): 497–9, on 498.

[37] Wilfred Buckley, discussion commentary, *Public Health* 27 (Jan. 1914): 124; F. G. Bushnell, "The Scheme for the Prevention of Tuberculosis and the Treatment of Persons Suffering from the Disease in the County Borough of Plymouth," *Public Health* 27 (Aug. 1914): 379–80.

procedure.[38] As contemporaries at the time (and historians since) have discussed, many physicians distrusted the nutritional qualities of pasteurized milk. Methods had not yet been standardized, contributing to confusion about which heating procedure would best kill bacteria while preserving nutrients. Pasteurized milk was more expensive and tasted strange to many consumers. BTB played a specific role in anti-pasteurization arguments. Following the lead of German scientist Emil von Behring, some physicians on both sides of the Atlantic speculated that ingesting small amounts of milk-borne BTB might inoculate children, safeguarding them against the pulmonary infection later in life. Pasteurization was thought to remove this "protective" effect.[39] Though widespread, these arguments represented only a part of the reasoning against pasteurization.[40]

Both the conception of BTB as an environmental disease and arguments over the definitions of "healthy" milk production and "purity" of the milk supply caused many municipal milk regulators to distrust pasteurization. As I have argued, the original model of controlling BTB stressed the monitoring and controlling of dairy cows' bodies and environments. While most members of the scientific community had accepted the idea that pasteurization killed tubercle bacilli, many authorities and consumers still felt they and their families were ingesting "diseased" milk if it had come from tubercular cows. Health officials recognized that such food, even if it could not be proved to cause disease, was nonetheless abhorrent to concerned consumers. As MOH Harold Scurfield put it, "[T]here is an aesthetic objection to consuming tubercle bacilli, either roast or boiled."[41] In the early 1900s, pasteurization, in the minds of Anglo-American physicians and veterinarians, such as Scurfield, Milton Rosenau, Samuel Prescott, Robert W. Philip, and Alonzo Melvin, served only as a last resort for salvaging milk that could not meet cleanliness requirements in its natural state. They agreed that pasteurization could not "purify" tubercular milk.[42] Influenced by popu-

---

[38] See Nathan Straus, *Disease in Milk: The Remedy, Pasteurization; the Life Work of Nathan Straus, by Lina Gutherz Straus*, 2d ed. (New York, 1917); Alvord and Pearson, "Milk Supply" (cit. n. 16), see textual information on each city for its pasteurization policies.

[39] Wolf, *Don't Kill Your Baby* (cit. n. 31), 64–5; Rosenau, *Milk Question* (cit. n. 15), 188–90; Samuel C. Prescott, "The Production of Clean Milk from a Practical Standpoint," *Charities* 16 (1906): 488–91, 489; C. W. M. Brown, "Certified Milk in Small Cities," *J. Amer. Med. Ass.* 48 (1907): 587–8, 587; Smith, *Retreat of Tuberculosis* (cit. n. 10), 183, 188, 191. On the Behring vaccination theory, see Clive Riviere, *Tuberculosis and How to Avoid It* (London, 1917), 49; Smith, *Retreat of Tuberculosis,* 187; and Georgina D. Feldberg, *Disease and Class: Tuberculosis and the Shaping of Modern North American Society* (New Brunswick, N.J., 1995), 128.

[40] See Peter J. Atkins, "The Pasteurization of England: The Science, Culture, and Health Implications of Milk Processing, 1900–1950," in *Food, Science, Policy, and Regulation in the Twentieth Century,* ed. David F. Smith and Jim Phillips (London, 2000), 37–51.

[41] Harold Scurfield, "An Anti-Tuberculosis Programme," *Journal of the Sanitary Institute* 17 (1896): 426–32, on 429; Brown, "Certified Milk in Small Cities" (cit. n. 39), 587; Manfred J. Wasserman, "Henry L. Coit and the Certified Milk Movement in the Development of Modern Pediatrics," *Bull. Hist. Med.* 46 (1972): 359–90, 386; Stuart Galishoff, *Safeguarding the Public Health: Newark, 1895–1918* (Westport, Conn., 1975), 93–4; Charles E. North, "Milk and Its Relation to Public Health," in *A Half Century of Public Health,* ed. Mazyck P. Ravenel (Lynn, Mass., 1921), 272, 277; Samuel Hopkins Adams, "Rochester's Pure Milk Campaign," *McClure's Magazine* 29 (1907): 142–9, 142.

[42] Scurfield, "Anti-Tuberculosis Programme" (cit. n. 41), 429, and see also commentary from MOsH James Niven and Francis Vacher on 434, 438; Prescott, "Production of Clean Milk" (cit. n. 39), 489; A. D. Melvin, "Commercial Classes of Milk," *J. Amer. Med. Ass.* 49 (28 Sept 1907): 1092. Robert W. Philip (Edinburgh) also preferred tuberculin-tested herds over pasteurization, and he heavily influenced the policies of the British National Association for the Prevention of Tuberculosis. Pasteurization predominated in American cities by the end of the 1920s but not in Britain until after World War II (see Smith, *Retreat of Tuberculosis* [cit. n. 10], 191–2).

lar beliefs and their own understanding of BTB as a disease locatable in particular bodies and environments, these authorities and others argued that pasteurization would only encourage milk producers to relax standards of cleanliness and animal health and guarantee an unwholesome milk supply. They believed municipal regulations should control disease notification, animal health on farms, and cleanliness of dairies prior to mandating pasteurization.[43]

The pasteurization debate thus united producers (opposed to any costly regulation), consumers (disgusted by "impure" milk), and many physicians and veterinarians on both sides of the Atlantic. It can be argued that this coalescence of opposition prevented pasteurization from being widely implemented in Anglo-American municipal regulations before the 1920s.[44] For present purposes, however, I will stress that, in the minds of consumers and their health officials, pasteurization could not replace efforts to remove the source of tubercle bacilli from the milk supply—the infected urban, suburban, or rural cow. Municipalities could accomplish this goal by regulating every cow, in every urban byre and on every farm, that produced milk consumed in cities and towns.[45]

Therefore municipalities looked to programs that would control the milksheds, such as Sheridan Delépine's plan (discussed in the previous section). Once infected animals had been removed from one area, Delépine's plan mandated the testing of all new cattle introduced into it to create a "disease-free island." He extended his model beyond Manchester to the whole United Kingdom, dividing the countryside into administrative districts. The districts with the highest prevalence of BTB would be cleansed, regulated, and isolated, followed by adjoining districts, until the whole country was under administrative control.[46]

Delépine published the details of this administrative plan in the MOH journal *Public Health* and as part of the 1908 report of the Medical Officer of the Local Government Board. The conference papers, texts, and journals on both sides of the Atlantic that reproduced or reported on Delépine's results attested to a high level of interest among public health officials. At the International Tuberculosis Congress held in Washington, D.C., in 1908, Canadian chief veterinarian J. G. Rutherford pronounced the Manchester plan "much more sensible and likely to be productive of ultimate benefit than the diffuse policy of promiscuously testing a herd here or there over an extensive territory."[47]

---

[43] M[ilton] J. Rosenau, "Pasteurization," *J. Amer. Med. Ass.* 49 (28 Sept. 1907): 1093; idem, *Milk Question* (cit. n. 15), 188–91.

[44] Ilyse C. Barkan has argued that, as with pasteurization of milk, the regulation of the meat supply in the United States could not be achieved until producers, consumers, and health officials cooperated with each other. See Barkan, "Industry Invites Regulation: The Passage of the Pure Food and Drug Act of 1906," *American Journal of Public Health* 75 (1985): 18–26.

[45] George W. Goler, "Municipal Regulation of the Milk Supply," *J. Amer. Med. Ass.* 49 (28 Sept 1907): 1077–9, 1078; Darlington, "Methods of Dealing with the Milk Supply" (cit. n. 29), 1080. MOsH and American proponents of "certified" and other regulated milk programs were well aware of each other's activities. See, e.g., the discussion following Joseph Cates, "The Sale of Milk," *Public Health* 27 (Jan. 1914): 120–8.

[46] Delépine, *Report on Investigations* (cit. n. 24); *Bibby's Book on Milk, Section IV. Bovine Tuberculosis: Cause, Cure, and Eradication* (Liverpool, [1912?]), 315, 316.

[47] J. G. Rutherford, "The Control of Bovine Tuberculosis," *Transactions of the Sixth International Congress*, vol. 4, pt. 2 (cit. n. 10), 874. See also L. M. Bowen-Jones, "The Control of the Milk Supply," *Public Health* 22 (1908–09): 170–5, especially 175; *Bibby's Book on Milk* (cit. n. 46), 308, 315, 377; Swithinbank and Newman, *Bacteriology of Milk* (cit. n. 18), 218, 457–9. Not all authorities viewed Delépine's plan favorably, however; see, e.g., Rosenau, *Milk Question* (cit. n. 15), 98.

The basic premises of Delépine's plan—combining bacteriological with epidemi-ological investigation, pinpointing BTB in individual animals and farms on maps, and certifying BTB-free areas—shaped Anglo-American schemes for BTB control for decades. At the 1908 congress, American veterinarian D. Arthur Hughes argued that efforts to eradicate BTB needed to begin with "maps . . . showing the topography of animal tuberculosis in each State."[48] The chief of the U.S. Bureau of Animal Industry, John R. Mohler, agreed and successfully advocated a national BTB eradication pro-gram, begun in 1917, that proceeded from one accredited BTB-free county to another. These local areas were joined to create accredited BTB-free districts, finally culmi-nating in a nation proclaimed almost free of the disease in 1940. Delayed by two wars and the central government's inability to bear the cost of eradication, Britain's anti-BTB efforts continued to be based in local regulation until 1950. Between then and 1960, Britain conducted a national eradication program that included systematic test-ing and removal of infected animals, creating "attested" areas. In Ireland, an anti-BTB campaign began in September 1954 with American Marshall Plan funds and local funds; the whole country was declared attested in October 1965. The British and Irish campaigns adopted many aspects of Delépine's original plan, including selecting areas in which the prevalence of BTB was low and starting there, expanding attested areas, and not allowing untested cattle into those areas afterward.[49]

CONCLUSION: RESPONSES TO ZOONOTIC DISEASE

Due to their insidious biological origins in food-producing animals, zoonotic diseases have elicited responses from public health officials that have encompassed individual animals as well as human sufferers. In the case of BTB, the response was mediated by several factors: scientists' understanding of BTB as an environmental and infectious entity, consumers' perceptions of "healthful" food, producers' anger at market inter-ference, and governments' reluctance to assume economic responsibility for eradica-tion of the disease. It remains for historians to investigate how well these factors char-acterize the overall history of Anglo-American responses to other zoonotic diseases in the past century.

For present purposes, a transnational comparison helps to demonstrate the general-izability of these factors while exposing the complexities of local contexts. For ex-ample, the eventual BTB eradication programs in the United States and the United Kingdom both included the British "disease-free island" model proposed in 1908. De-spite the availability of pasteurization after World War I, both nations carried out erad-ication campaigns in response to the demands of consumers for milk free of bovine tubercle bacilli. Local configurations of power and interest dictated the response to BTB prior to eradication. In the United States, producers fought municipal TB-testing regulations in the courts; in the United Kingdom, county medical officers of health re-

[48] D. Arthur Hughes, "Precautionary Sanitary Legislation Against Tuberculosis of the Domesticated Animals in the United States," *Transactions of the Sixth International Congress,* vol. 4, pt. 2 (cit. n. 10), 978.

[49] J. A. Kiernan and L. B. Ernest, "The Toll of Tuberculosis in Livestock," *Yearbook of the United States Department of Agriculture, 1918* (Washington, D.C., 1920), 277–88; Swithinbank and New-man, *Bacteriology of Milk* (cit. n. 18), 457–9; J. N. Ritchie, "Britain's Achievement in the Eradication of Bovine Tuberculosis," The George Scott Robertson Memorial Lecture, 1958 (Belfast, 1959); J. Arthur Myers and James H. Steele, *Bovine Tuberculosis Control in Man and Animals* (St. Louis, 1969), 76, 77, 106, 273.

fused to attempt to persuade farmers to have cows examined by a veterinary surgeon.[50] Persuasion, along with market incentives, was required to gain producers' cooperation. In the United States, officials in favor of BTB eradication cited the effects of municipal statutes condemning infected milk and the ire of neighboring livestock owners as incentives for producers to eradicate BTB from their herds.[51] Diverted by two wars and economic depression, neither BTB eradication nor widespread pasteurization was accomplished in the United Kingdom until the early 1960s.[52] While a nationwide BTB eradication program began much earlier in the United States (1917), it had only conditionally succeeded by 1940. Distrusted by consumers and opposed by producers, pasteurization remained rare in the nonurban United States, as in Britain, for at least another decade.[53]

One important change characterized both the United States and the United Kingdom before World War I: BTB was steadily diminishing as an urban problem. Compared with almost 90 percent of British cities by 1910, only two-thirds of the rural districts had any enforcement of milk laws, with the result of a high prevalence of bovine tubercle bacilli in rural milk.[54] By 1914, cities with milk regulations had become demonstrably healthier places for children to live. Physicians began to highlight a slightly increasing rate of bovine-origin tubercular infections in children outside the city limits by comparing it with a significantly decreasing rate of pulmonary infection. They asserted that the urban penalty of BTB, which had killed, crippled, and disfigured many, had been reversing since the 1890s. American physician Milton Rosenau explained in 1912 that "the tables have now been turned, for there is less chance of contracting infection in a well-ordered city than in the average country place."[55] The pastoral countryside, once viewed as the haven of escape from urban ills, had been exposed by health officials' surveillance as the unwholesome breeding ground of this abhorrent disease.[56]

No longer an "emblem of purity," milk had been shown to be a vector of disease capable of crossing the organic boundary between animal and human bodies and the geographic boundaries of urban and nonurban places. Consumer fears of exposure to BTB reflected not only fear of the disease but also an acknowledgment of people's far-reaching vulnerability. To police some of those boundaries and control the damage done to children by infected milk, public health officials constructed an identity for milk-borne BTB based on their understanding of it as an environmental disease. Their mapping techniques identified dairy cows and their milk as potentially dangerous agents in need of surveillance and control. For these officials and for concerned consumers, zones of milk production defined the local community and the area of

[50] Harold Kerr, "Some Differences in the Control of the Milk Supply," *Public Health* 22 (1908–09): 446–51, 451.
[51] See, e.g., Pearson and Ravenel, "Tuberculosis of Cattle" (cit. n. 16), 185.
[52] Bryder, *Below the Magic Mountain* (cit. n. 2), 133–4.
[53] Myers and Steele, *Bovine Tuberculosis Control* (cit. n. 49), 106.
[54] Fulton, "What the Public Ought to Know" (cit. n. 35), 480; Smith, *Retreat of Tuberculosis* (cit. n. 10), 185; Kerr, "Some Differences in the Control" (cit. n. 50), 451.
[55] Rosenau, *Milk Question* (cit. n. 15), 18; A. P. Mitchell, "A Bacteriological Study of Tuberculosis in the Lymph Glands of Children," and idem, "The Milk Question in Edinburgh," *Edinburgh Medical Journal* 13 (1914): 213–27; J. S. Fowler, "The Milk Problem and Tuberculosis in Infancy and Childhood," 27–34, and Nathan Raw, "Bovine Tuberculosis in Children," 35–42, in Kelynack, *Tuberculosis in Infancy and Childhood* (cit. n. 14).
[56] According to Worboys and Condrau, "Urban Penalty to Urban Advantage" (cit. n. 20), this was the pattern for pulmonary tuberculosis as well.

epidemiological and bacteriological interest. Those zones of milk production contained problematic people (self-serving dairy producers) and environments (filthy, unventilated, BTB-infested byres in which cows lived their whole lives), linked by milk to children's health.

While consumers and public health officials often cast dairy producers and the economic systems of milk production as the villains of the BTB narrative, in reality the situation was more complex, linked as it was to scientific debates, larger social changes, and cultural concerns. Public health officials had tools, such as pasteurization, that would have broken the causal chain of disease without any consideration of animals' bodies and environments. But because so many people understood BTB as being embedded in particular milieux, pasteurization and similar measures for a long time seemed the wrong tools for the job of controlling the disease. Indeed, the disease could not be separated from patterns of social interaction and fears of cultural decline in the early years of the twentieth century. Milton Rosenau called BTB and other milk-borne diseases artifacts of "an artificial civilization to which we have not yet adjusted ourselves," dependent for survival and expansion on the parameters of modern living.[57] The material and social ecology of BTB were embedded in particular landscapes, communities, and beliefs—factors that continue to inform our responses to zoonotic diseases a century later.[58]

[57] Rosenau, *Milk Question* (cit. n. 15), 20–2, on 21.

[58] BSE ("mad cow disease") is the most salient recent example. With mandatory pasteurization, BTB is no longer a public health problem in the United States or the United Kingdom. Current concerns focus almost solely on economic losses to livestock producers; see "Keeping the Lid on Bovine TB," *Veterinary Record* 150 (2 March 2002): 257, 260–2.

"Clever Microbes":

Bacteriology and Sanitary Technology in Manchester and Chicago During the Progressive Age

By Harold L. Platt*

ABSTRACT

A neglected aspect of the history of germ theories is its use in the purification of sewage. In the 1890s, progressive reformers rapidly developed bacteriological methods of wastewater treatment. A comparison of the United Kingdom's Manchester and the United States' Chicago shows, however, that science and technology were mediated by political culture and institutions. In Manchester, a politics of deference and strong extralocal government gave the authority of scientific expertise a decisive role in policy formation. In Chicago, devolution of power to the ward bosses meant a quarter-century of defiance against the national authority and its effort to get the city to install a modern sanitation system.

INTRODUCTION

In September 1896, the town council of Manchester confronted a decision-making crisis in uncharted waters of science, technology, and medicine. A national body, the Local Government Board (LGB), and a regional agency, the Mersey and Irwell [Rivers] Joint Committee (MIJC), were pressing the local government to choose a method of sewage treatment. In spite of town council representation on the MIJC, the majority voted to sue the city for not meeting the committee's standard of wastewater quality. At the same time, the LGB was threatening to cut off funding for all public works projects. Though under pressure to act, the council felt adrift in a sea of conflicting recommendations from engineers, chemists, biologists, and doctors.

"The question was surrounded with great difficulty, and probably there would be opposition on all sides," a member of the town's council's Rivers Committee, Nathaniel Bradley, reported to the council. "The Committee," he explained, "in great measure depended upon expert and scientific evidence, and were largely dominated by officials and professional advisers. That must necessarily be so from the very nature of the thing." Seeking shelter from this political storm, the committee proposed that the city construct a fifteen-mile conduit that would not only dump the sewage in

---

* Department of History, Loyola University of Chicago, 6526 N. Sheridan Road, Chicago, IL 60626; hplatt@luc.edu.

the tidal estuary above Liverpool but also escape the jurisdiction of the regional bureau of watershed conservation.[1]

Across the Atlantic, Chicago, by contrast, was in the process of constructing its "ultimate sink" virtually free of interference from outside governmental agencies. Seven years earlier, the city's municipal reformers had lobbied a bill through the state legislature that created the Chicago Sanitary District (CSD), which was made essentially autonomous by giving it the power to tax and to borrow. Moreover, the special purpose district was granted authority to operate a twenty-eight-mile sanitary channel between the Chicago River and the Illinois River Valley. In 1896, CSD's elected administrators were in the midst of building what was, in effect, a gigantic drainpipe that would tap Lake Michigan to wash the city's untreated, albeit diluted, sewage down to St. Louis. Although the U.S. Corps of Engineers formally exercised jurisdiction over inland waterways, it had always acted to help Chicago by sponsoring various improvement projects. The state and federal governments had an unbroken record of facilitating the growth of the city, rather than hamstringing it with regulations.[2]

In both industrial centers, policy makers attempted to combine sanitation and transportation in their plans for engineering the environment. In the case of Manchester, a ship canal to bypass Liverpool had been opened in 1894 as the remedy to perceptions of urban and economic decline. The city had also begun to respond to a long series of disastrous floods by constructing a main drain with an outfall at the top of the canal to keep it filled. While satisfying this purpose, an unintended consequence was the transformation of the turning basin into an insufferable cesspool, constantly churned by ship propellers. Even the installation of a sewage treatment facility, the Davyhulme plant, next to the canal failed to bring relief or to meet the standards set by the MIJC.[3]

In the case of Chicago, the original goal to improve public health had also been transformed but with full and open volition. Led by the engineer Lyman E. Cooley, local politicians had wrested control over the CSD from the reformers, turning their plan for a sanitary channel into a Lakes-to-the-Gulf superhighway of commerce. Cooley and fellow city boosters pointed to Manchester's ship canal as a model. Yet the amount of lake water needed to dilute the city's liquid wastes to safe levels was so large that navigation became problematic on the canal's downstream currents. To resolve this dilemma, Cooley redesigned the artificial river to be much wider and deeper, leaving little money for other infrastructure projects needed to safeguard the drinking supply from sewage contamination. An unintended consequence would be an epidemic of typhoid fever less than two years after the grandiose scheme opened to tremendous fanfare and great expectations.[4]

---

[1] Nathaniel Bradley as quoted in *Manchester Guardian* (hereafter cited as MG), 10 Sept. 1896, 7. Also see ibid., 3–29 Sept. 1896. For the best overview of the subject, see Alan Wilson, "Technology and Municipal Decision-Making: Sanitary Systems in Manchester, 1868–1910," (Ph.D. diss., Univ. of Manchester, 1990). The expression "clever microbes" can be attributed to testimony by Balfour Browne in Local Government Board (LGB) Inquiry, *Application to the Local Government Board for Sanction to Borrow Money for Purposes of Sewerage and Sewage Disposal Before Maj-Gen H. D. Crozier, R.E., and Theodore Thomspon, M.D., Inspectors, Jan. 12–13, 1899* (n.p., [1899]), 9.

[2] For more general perspectives, see Harold L. Platt, "Chicago, the Great Lakes, and the Origins of Federal Urban Environmental Policy," *Journal of the Gilded Age and Progressive Era* 1 (April 2002): 122–53; and Joel Tarr, *The Search for the Ultimate Sink* (Akron, Ohio, 1996).

[3] See I. Harford, *Manchester and Its Ship Canal Movement* (Halifax, 1994); and B. T. Leech, *History of the Manchester Ship Canal* (Manchester, 1907): 2:177–9.

[4] See Elmer Corthhell, "The Manchester Ship Canal," *Journal of the Western Society of Engineers* (hereafter cited as *JWSE*) 4 (Feb. 1899): 1–11; and Platt, "Chicago, the Great Lakes" (cit. n. 2).

In each city's case, the relatively new and fast-evolving science of bacteriology became increasingly central to the policy debate over sanitation strategies. A comparative study of the two cities helps illuminate the reception of science in the public sphere. Over the past decade, germ theories have received considerable scholarly attention in the interrelated areas of epidemiology and urban water supplies. However, historians have given much less consideration to its role in revolutionizing the other side of water management, the treatment and disposal of human and industrial wastes. By the mid-1890s, contemporaries involved in the search for the ultimate sink were closely following experiments in the United States and Great Britain to put "clever microbes" to work in finding it.[5]

As Daniel Rodgers highlights in his brilliant *Atlantic Crossings,* progressive reformers were well versed in the latest trends and inventions to ameliorate the industrial city's worst environmental conditions. Pressured from above, Manchester's officials made a thorough study of the contact beds of William Dibdin in London, the septic tanks of Exeter, the intermittent filters of Salford, and the Massachusetts Experimental Station's theories of biological sewage treatment. In 1893, Owens College (now the University of Manchester) established a department of bacteriology. During the same year, Chicago underwent a complete makeover of its public health department in the wake of its worst typhoid fever epidemic, which had threatened to cancel its World's Columbian Exposition. While loudly proclaiming their city's good health, officials quietly established a bacteriology laboratory and put a young physician and microscopist, Aldoph Gehrmann, in charge.[6]

A comparison of the reception of bacteriology in the two cities shows that institutional structures and political cultures played a pivotal, if not decisive, role in shaping the formation of public policy on sanitation technology. As Christopher Hamlin reminds us in his pioneering essay on Dibdin, the issues facing decision makers of the Progressive Era were not simple questions of scientific "progress." They "were political and pragmatic, concerned as much with appearance as with substance, and as much with persuading people as with purifying sewage." The contrast between the strong arm of regional and national authorities in the United Kingdom and the "weakened springs" of the federal government in the United States could not have been more complete. While Manchester found no safe haven from the pressure above, Chicago functioned as a virtual city-state, defying with impunity Washington's feeble efforts to enforce the law for more than a quarter-century. While the English city operated in a political atmosphere of tight-fisted paternalism and privilege, its American counterpart created a climate of open-ended wheeling and dealing that put the interests of self-serving ward bosses above those of all others. Both city councils did foster environmental inequality and social discrimination in the name of low-cost government. This rhetorical trope had little effect in either place, however, on the outcome of the application of bacteriological science to sewage disposal.[7]

---

[5] See Nancy Tomes, *The Gospel of Germs* (Cambridge, Mass., 1998); Christopher Hamlin, *Public Health and Social Justice in the Age of Chadwick* (Cambridge, 1998); and Michael Worboys, *Spreading Germs: Disease Theories and Medical Practice in Britain, 1865–1900* (Cambridge, 2000).

[6] Daniel T. Rodgers, *Atlantic Crossings: Social Politics in the Progressive Age* (Cambridge, Mass., 1998); Fred O. Tonney, "The Introduction of Bacteriology into the Service of Public Health in Chicago," *Bulletin of the Society of Medical History of Chicago* 5 (Jan. 1937): 22–3.

[7] Christopher Hamlin, "William Dibdin and the Idea of Biological Sewage Treatment," *Technology and Culture* 29 (April 1988): 218; W. D. Farnham, "The Weakened Spring of Government: A Study in 19th Century History," *American Historical Review* 68 (1965): 662–80.

Instead, political cultures and institutions gave definition to the authority of science in the decision-making process on sanitation strategies. In Manchester, England, a politics of aristocratic hierarchy and deference privileged the special knowledge of science in the formation of sewage disposal policy. Rivers committeeman Bradley's sense that "the very nature of the thing" put the policy question in the hands of scientific advisers reveals the basic assumptions of this type of political culture. In a similar manner, the various governmental bodies and parliamentary commissions gave legitimacy to the authority of science by calling prominent representatives from the academic community to give expert testimony on policy questions. In many cases, the officials and the scientists came from the same social class. In the Progressive Era, a classical education in pure science remained an honorable degree for members of the upper class trained at Cambridge and Oxford. A relatively centralized structure of government and society created a political framework that pointed Manchester toward a sanitation technology decision based on the authority of science.[8]

In Chicago, Illinois, a politics of anarchistic growth and individualism largely marginalized scientists and doctors in the formation of policy. Politicians excluded those experts from most inner circles of decision-making on water management and large-scale public works, a sharp contrast to the politicians' reliance on the municipal engineers. As Stanley Schultz has convincingly shown in *Constructing Urban Culture,* their technical methods and bureaucratic style became an integral part of the political culture of Gilded Age America. Those claiming authority based on the sciences of medicine or biology or both were forced to play the role of outside critics, warning the public of impending public health disasters that would be caused by flawed sanitary strategies. Beginning in the early 1880s, members of Chicago's flourishing scientific communities used germ theories to predict epidemics. Over and over again, they explained why the drinking water was being contaminated with sewage containing harmful microorganisms. With equal regularity, the politicians largely rejected the scientific communities' recommendations in favor of those of the engineers, who proved more cooperative in advancing the self-serving goals of their party organizations.[9]

Though their cultural frameworks and environmental settings were very different, Chicago and Manchester invite comparison because they had three important similarities. First, unlike many cities located downstream from other sources of pollution, these two were then, as now, richly endowed with pure water for their drinking supplies. Mancunians built a system of upland reservoirs and used the three rivers running through it only for wastewater. Chicagoans built their city on the banks of Lake Michigan, part of the chain of Great Lakes, which hold one-fifth of the world's fresh water. Second, the factories of both urban centers produced massive amounts of organic and chemical liquids that added a significant burden to the task of sewage disposal. Third, although Chicago in 1890 had about twice the number of inhabitants

---

[8] Bradley was no friend of science; see Nathaniel Bradley, "Manchester Sewage Problem," *Manchester Statistical Society, Transactions* (1896–1897): 135–57. On the link of class and education, see Martin J. Weiner, *English Culture and the Decline of the Industrial Spirit, 1850–1980* (Cambridge, 1981).

[9] Stanley K. Schultz, *Constructing Urban Culture* (Philadelphia, 1989). Also see Barbara Gutmann Rosenkrantz, "Cart before the Horse: Theory, Practice, and Professional Image in American Public Health, 1870–1920," *Journal of the History of Medicine and Allied Sciences* 29 (Jan. 1974): 55–73; and Rima D. Apple, "Constructing Mothers: Scientific Motherhood in the Nineteenth and Twentieth Centuries," *Society for the Social History of Medicine* (Aug. 1995): 161–78.

(nearly 1 million) as Manchester, the populations of both cities were big enough to require large-scale technologies to solve their sanitation problems. In addition, each city's search for an ultimate sink was complicated by ambitions to turn it into an engine of economic development, a booming shipping lane.

MANCHESTER AND THE POLITICS OF DEFERENCE

The Rivers Committee's report of September 1896 outlined the contours of the politics and science of sewage disposal in the heavily industrialized Mersey-Irwell watershed. An imminent judicial ruling had forced this powerful committee to ask the full town council to define the city's response. The county police court was expected to help enforce the MIJC's effluent standards by hitting the municipal government where it hurt, with a £50 fine for each day the "Corporation" failed to comply. Reminding the council of his thirty-one years of service, River Committee chairman Joseph Thompson admitted that he felt trapped and defeated, caught between the proverbial "rock and hard place."

On the one side, the MIJC was turning the screws on the Corporation to meet the minimally acceptable "limits of impurity" that had been set almost three years earlier. The exasperated alderman now admitted that the best efforts of Davyhulme's engineers and scientists had failed. Technology designed to precipitate most of the solid matter out of the liquid by adding chemicals had neither filtered out nor killed off the organic matter responsible for causing the horrid smells of putrefaction in the ship canal. Instead, the sewage treatment plant had been generating a mountain of sludge, and the city had had to buy a 1,000-ton steamer to haul the semisolid wastes out to sea. Even more problematic, Thompson lamented, the key expert in defining a series of chemical indicators of wastewater quality for the regional agency, Sir Henry E. Roscoe, was also the city's top adviser in charge of the sewage treatment plant. In effect, in his private position as expert consultant and his public positions as a member of the Parliament, the MIJC, and the Owens College faculty, the chemistry professor held veto power over any plan that might be proposed.[10] On the other side, the LGB was insisting on land filtration as a final step of wastewater disposal regardless of any other novel sanitation strategies the city might adopt.

The most promising of those novel strategies were based on germ theories. Rather than just deodorizing organic waste with chemicals, enthusiastic advocates said, bacteria could be put to work "eating" it. In 1896, however, the LGB remained unconvinced by the biologists' assertions that sewage farms represented a primitive, inefficient form of this natural process of purification. Like the MIJC, the national governing body had the power to hurt the Corporation in the pocketbook by holding its ability to borrow for public works projects hostage until LGB demands were met. Alderman Thompson recounted the committee's dismal and equally frustrating experience in trying to meet those demands.

The only way out, the chairman concluded, was to adopt the committee's new plan for a tidal conduit. It would convey the 26,000,000 gallons of effluent reaching Davyhulme each day to an outfall point of tidal flow at Randle's Sluices, about three miles

---

[10] MG, 3 Sept. 1896. For the reply of the MIJC to Alderman Thompson's position and the most complete account of Sir Roscoe's definition of the "limits of impurity." See ibid., 5 Oct. 1897. For biographical information on Roscoe, see Charles Coulston Gillispie, ed., *Dictionary of Scientific Biography* (New York, 1970), 2:536–9.

above Runcorn. The ship canal could be compensated, Thompson proposed, by using fresh water from the city's other controversial water management project, the Lake Thirlmere aqueduct. Besides a bottom-line advantage of being the lowest cost option, he argued, the conduit scheme would allow Manchester to escape the jurisdictional reach of the MIJC.[11] For Thompson, this alternative had deviously delightful prospects of turning a bitter political defeat into a final glorious vindication.

The politics of the report and the storm of public debate it engendered over the following year exposed the ways in which different groups of progressives understood the new science of bacteriology and the role it could play in reducing shocking urban mortality rates, especially among infants. At the time Thompson made his surprise announcement about the tidal conduit plan, the council was undergoing a historic shift from its first generation of ruling Liberals. Since 1884, when voting was finally extended to most working-class men, the civic sphere of political discourse had been becoming more open and contested.[12]

A broadening of the suffrage from Manchester's especially acute case of elite rule also had profound impacts on the political culture within the town council. Two of the era's most influential activists, Sidney and Beatrice Webb, recorded their impressions of Manchester during a five-week stay in the midst of these tumultuous years of civic debate on the links between public health, biological science, and the urban environment. The Webbs believed the tasks of city government had outgrown its organizational structure. Lending support to the thesis offered here, they observed that "the different parts of the machine are out of joint; it rumbles on in some sort of fashion, because it is pushed along by outside pressure."[13]

The Webbs' skepticism perfectly captures the spirit of the age in challenging those in positions of authority. During the Progressive Era, reformers pitted expert against expert in a lively exchange of ideas over the best ways to improve city life. At the September 1896 meeting of the town council, for instance, objections were immediately raised after Thompson proposed that the council adopt the committee's plan. Sir John Harwood, a venerable leader with a political and social stature to match Thompson's, demanded time for a thorough vetting of the report's recommendations. Harwood was the chair of Water Committee, hero of the successful Thirlmere aqueduct project, and member of the MIJC, along with Thompson. Mayor Lloyd, too, expressed surprise that Thompson was reverting to the days when committees routinely expected the council to rubber-stamp their proposals.[14]

The week intermission gave both sides time to prepare their speeches for a wider public than the council membership. Alderman Thompson delivered a well-rehearsed lecture on the current status of the science and technology of urban sanitation. Taking the audience step by step through the various methods of sewage disposal let him establish his technical expertise and legal command of the policy question. After laying land filtration to rest, the committee chairman accurately called the biological work

---

[11] MG, 10 Sept. 1896.

[12] See the memoirs of Medical Officer of Health James Niven, *Observations on the History of Public Health Effort in Manchester* (Manchester, 1923). For the political culture of Manchester see Derek Fraser, *Urban Politics in Victorian England* (London, 1979).

[13] Sidney Webb and Beatrice Webb, *Methods of Social Study* (1932; repr. New York, 1968), 195; Harford, *Ship Canal Movement* (cit. n. 3), 62–98, 147–66, appendix B.

[14] MG, 3 Sept. 1896. On the Thirlmere aqueduct, cf. John James Harwood, *History and Description of the Thirlmere Water Scheme* (Manchester, 1895); and Thirlmere Defense Association, *Manchester and Thirlmere Water Scheme* (Windemere, [1877?]).

of Dibdin promising but still in the experimental stage. "The whole process as to . . . call in the use and assistance of bacteria to do the work which land could do better . . . [could be proclaimed to have] succeeded very fairly, but it must be remembered that those experiments [have] only been on a small scale." He cited cost as the reason this approach was ruled out, leaving only the tidal conduit option.[15]

Although city surveyor T. De Courey Meade fell in line behind the committee chairman, his report confirmed that the bacteriological basis of sewage purification was reaching a milestone of technical understanding. In August 1895, Roscoe had converted some of his mechanical, sand filtration beds into biological ones, more or less duplicating Dibdin's work in London. Meade understood that the "slime deposit on the sand constitute[s] the real filtering material in the waterworks filter." He even conceded that Dibdin's studies showed that wastewater effluent could be purified to meet any standard of purity. Translating Dibdin's scientific theory into a practical technology, the engineer posited, would result in "the oxidation of organic matters, both those in suspension and those in solution, through the agency of living organisms. It is the preliminary establishment and subsequent cultivation of these organisms which is to be aimed at in the scientific process of purification by [artificial] filtration."[16] Yet, at the same time, Thompson did not have to distort Meade's encouraging report to portray this new approach as largely unproven and incalculable in terms of ultimate cost.

Reflecting the relative novelty of the application of germ theories to wastewater treatment, the opposition to the tidal conduit plan made no use of it or any other technology as an alternative policy approach to the problem. Instead, the opposition seemed content to lambaste the River Committee for incompetence. Rallying the small property owners, the tidal conduit plan's local political opponents joined with lobbyists from downstream interests to force a ratepayers' referendum on the national authorization act. For the first time on an important policy issue, qualified voters rejected a proposal endorsed by their representatives on the town council. After this unprecedented political defeat at the hands of their own constituents, the council members could only feel the outside pressure more intensely.[17]

Linking bacteriology and sanitation technology was not only a new idea in 1896; it was also one attracting a tremendous amount of attention among urban progressives. In many respects, Professor Roscoe saved the day for the town council by persuading the MIJC to give the city one year to come up with a complete plan of wastewater treatment and disposal that could meet the agency's minimum "limits of impurity." During this interval, Dibdin's theories appeared to sweep the field in the scientific community, leaving only practical questions of engineering and management.

Germ theories played an important role in shaping what became a surprisingly well-informed and wide-ranging debate on sewage disposal. Compared with a year earlier, the sides staked out on the council floor spilled over into the daily press, fostering a lively discourse of expert against expert and one insider version against

[15] MG, 10 Sept. 1896. For an early examination of the relationship between the science of bacteriology and the technology of sewage disposal, see F. J. Faraday, "On Some Recent Observations in Micro-Biology and Their Bearing on the Evolution of Disease and the Sewage Question," *Literary and Philosophical Society of Manchester, Proceedings* 25 (1885): 46–55.

[16] City of Manchester, *Proceedings of the City Council,* 9 Sept. 1896, 1342, 1360. For the city surveyor's full report, see ibid., 1337–89.

[17] MG, 10 Sept.–11 Dec. 1896. For the concurrent Sanitary Congress in Leeds, in which Dibdin and others were moving from scientific theory to practical engineering, see ibid., 18 Sept. 1897.

another. Most fascinating was the way in which the Davyhulme experiments became a popular metaphor for scientific and technological progress while the tidal conduit became an icon of a "policy of despair," in the words of an opposition councilman. The chief spokesman of the ratepayers, Dr. R. M. Parkhurst, also alluded to it as "a scheme at once of panic and of despair."[18]

Within this politically charged atmosphere, bacteriological science was cast as a shining beacon of knowledge that could lead the city from the dark pessimism of the council chambers to the bright light of the healthy city of tomorrow. Letters by academics on both sides of the issue underscore the point that the authority of science and technology remained privileged, albeit contested, terrain in the 1897 policy debate in Manchester. In this wide-ranging discourse, chemists rather than biologists still held center stage as the voice of science on effluent standards of quality, but their main focus of attention was increasingly dominated by the biological filter beds in London, Manchester, Salford, and similar field experiments elsewhere in England and in America. People lined up against the River Committee made good use of germ theories to offer the public an attractive and ingenious alternative to the culvert scheme.[19] The science of bacteriology promised not only to solve a major problem of the industrial city but also to advance a progressive ideal of the conservation of natural resources.

In the end, Dr. Parkhurst's mobilization of the protest vote proved decisive. Opponents of the council's plan raised legitimate questions about the effects of the diversion of so much water from the ship canal, adding to the arguments against defeatism and for faith in science and technology. In contrast, the council's position rested on too thin a base of legalistic and bureaucratic politics. Such a rationale was not strong enough to carry the weight of public opinion needed to prevail at the polls. On 31 October 1897, Mancunians voted 49,069 to 20,528 against the culvert plan. No one could question the finality of this overwhelming rejection of the council's policy choice. In a dramatic gesture of defeat, if not despair, Chairman Thompson and his colleagues resigned their positions on the Rivers Committee.[20]

Under different leadership, Manchester emerged in 1898 as a champion of the new science of bacteriology. In part, a process of elimination left it as the only viable alternative to the old science of sewage farms. However, the fast-growing consensus among sanitary experts behind biological solutions to the problem gave the reconstituted Rivers Committee the confidence it needed to make a commitment to a specific course of action. In this rapidly-moving subject of research, the various field and laboratory experiments left little doubt that the scientists were headed in the right direction. Although the precise details of an appropriate technology remained to be worked out by trial and error, the path of knowledge opened by Dibdin and company now seemed not just the most promising but also the only rational sanitation strategy.

The choice of Sir Bosdin T. Leech as the new chairman of the Rivers Committee was equally important in turning Manchester into an outspoken advocate of biological methods of sewage treatment. Along with fellow Liberal councilor Harwood, Leech had been most responsible for steering the ship canal proposal through the town council. This considerable achievement earned the yarn merchant a knighthood and a

---

[18] R. M. Parkhurst, letter to the editors, MG, 8, 14, 15 Sept. 1897; ibid, 18 Sept. 1897 (quotation).

[19] For "The Scientific Aspects of Sewage Purification," see MG, 26 Oct. 1897. For additional commentary with science-related aspects, see ibid., 14, 15, 16, 20, 27 Sept., 2, 5, 6, 25 Oct. 1897.

[20] MG, 2, 25 Sept.–31 Oct. 1897. For the vote, see ibid., 1 Nov. 1897.

directorship of the transportation company. Now he was asked to use his considerable political skills to convert the city's governmental overseers into apostles of the new science. In sharp contrast to Thompson, Leech asserted that "the Committee [is] distinctly of opinion that biological filtration [presents] at once a less costly and more effectual means of filtration than any other." In the spirit of compromise, he accepted Harwood's suggestion that the city also comply with the LGB's demand for land filtration by purchasing the required 200–300 acres while pushing forward at Davyhulme with maximum speed.[21]

The Leech-Harwood compromise plan contained the elements of a political resolution of the city's conflict with its regulatory overseers. First, Leech rallied the needed council majorities to proceed with plans for a small-scale operational test of the bacteriological method of sewage treatment. Then he turned to fostering a new partnership with the regional and national agencies. By adopting a specific technology and by embracing germ theories, Leech effectively shifted the burden of scientific proof back to those agencies. The city could now ask whether it had their official sanction to ratchet up the experimental station into a full-scale facility. On 12–13 January 1899, the showdown came at a crucial hearing of the LGB. At issue was Manchester's petition for a loan of £160,000 to expand the four acres of filter beds devoted to biological methods of sewage purification to thirty acres.

The city's solicitor, M. P. Balfour Browne, came well armed with the powerful authority of experts to bolster his case for the new science. He strove to demonstrate that Manchester's bacteriological filtration system could more than meet the MIJC's minimum standards of effluent quality. Technical data was supplied by a chemist-bacteriologist at Owens College, Gilbert John Fowler, who was emerging as the effective director of the Davyhulme experiments. The city's solicitor also brought along several heavyweight reinforcements of the scientific establishment, including Professors Percy Frankland and W. H. Perkins. Browne himself may have best captured the historic meaning of the hearing in observing that "sewage disposal at one time was simply a matter of engineering . . . and it is only recently that this matter has passed out of the hands of the chemists and passed into the hands of the biologist, who will tell us . . . that the method, and the only method, of disposing of sewage is by the bacterial method." The biology-and-land package looked increasingly attractive as a face-saving way out of the interagency conflict for all three public institutions. In less than a year, each level of government agreed to the compromise plan, ending the policy standoff over urban sanitation and watershed conservation.[22]

Under pressure from above, the town council became highly motivated to find an ultimate sink for Manchester's rapidly swelling volumes of wastewater. Between 1896 and 1900, this political struggle over science and technology policy spilled over into the larger arena of popular opinion. Urban reformers embraced germ theories as modern and "progressive"; in Manchester, they seemed eager to adopt biological methods of effluent treatment as a step toward the future. The very notion that bacteriology offered a new source of authority to challenge the old also may have appealed to some political activists of the Progressive Era.

---

[21] For Leech's assumption of leadership, see MG, 7 April 1898. For the proceedings in the courts, see ibid., 26 March 1898. For the scientific work at Davyhulme as reported by Fowler, see ibid., 21 March and 5 April 1898.

[22] LGB Inquiry, *Application* (cit. n. 1), 6–7, passim. For biographical information on Fowler, see *Chemical Society Journal* (Dec. 1953): 4191–2.

To be sure, Alan Wilson's assessment that the city's sanitation policy was driven by considerations of lowest cost has much to recommend it. He posits, for example, that the referenda of 1897 rejecting the council's culvert plan was simply a protest against higher taxes during a period of depression.[23] Yet this victory for the hard-pressed ratepayers was equally a triumph for the new science and the faith people had in its power to solve the industrial city's environmental and social problems. In the case of Manchester, the authority of experts helped shape policy formation toward a science-based approach to problem solving. The resolution of Manchester's sanitary strategy at the turn of the century offers an opportune moment to segue into the Chicago story and the opening of its dual-purpose ship canal and sewage ditch.

CHICAGO AND THE POLITICS OF DEFIANCE

On 17 January 1900, the much anticipated public works project opened with the third formal proclamation of pure water for Chicago. The two earlier promising, but ultimately disappointing, milestones of environmental engineering had been the two-mile water-intake tunnel and crib of 1866 and, five years later, the deep-cut, drainage channel and canal. Bursting with its usual civic pride, the *Chicago Tribune* declared that CSD's great achievement meant "the city at last is free from the growing menace of a contaminated water supply." Medical experts, however, responded to the city's sanitation plans, as they had over the previous twenty years, more cautiously. They continued to urge people to heed the daily newspaper bulletins from the bacteriology lab's health department that warned residents when they should boil their drinking supplies.[24]

In contrast, the sanitarians of the pre–germ theory period had no solid base of contrary knowledge upon which to raise reservations about the city's environmental planning. Beginning in the early 1880s, however, the science of bacteriology gave chemists and microscopists from the medical colleges new perspectives that cast doubt on the basic design of city hall's water management system. Their investigations of water quality near the two-mile crib raised serious questions about a strategy based on the belief that Lake Michigan furnished a "fountain inexhaustible" of pure water. Formulated into policy by Chicago's first prominent municipal engineer, Ellis S. Chesbrough, in the 1850s, this notion of nature's boundlessness led him to propose that the lake could be tapped as a virtually free source of pure water both for the city's drinking supplies and for diluting its liquid wastes to safe levels. Over the succeeding years, however, scientific evidence that linked contamination at the intake cribs to discharges from the sewerage system mounted.[25]

[23] Manchester Corporation, Rivers Dept., *Experts' Report on Treatment of Manchester Sewage, Oct. 30 1899* (Manchester, 1899); MG, 15 Oct. 1901; Wilson, "Technology and Municipal Decision-Making" (cit. n. 1), 259–329.

[24] *Chicago Tribune* (hereafter cited as CT), 18 Jan. 1900. For a range of contemporary opinions, see ibid. For an engineer's perspective, see George M. Wisner, "A Description of the Opening of the Chicago Drainage Canal," *JWSE* 5 (Feb. 1900): 8–11.

[25] City of Chicago, Dept. of Public Works [E. S. Chesbrough], *Annual Report* (1877), 6. For greater detail, see Harold L. Platt, " 'A Fountain Inexhaustible': Environmental Perspectives on Water Management in Chicago, 1840–1980" (paper presented at the annual meeting of the Society of the History of Technology, London, June, 1996). On the reception of germ theory, see notes 5 and 9. For an introduction to Chicago medical history, see Thomas Neville Bonner, *Medicine in Chicago, 1850–1950,* 2d ed. (Urbana, Ill., 1991).

In America's political culture, unlike in Britain's, the authority of science held a tenuous place, especially compared with the nearly sacred space reserved for engineers in planning large-scale infrastructure projects to improve the quality of urban life. Though this conflict forms part of a broader one between Chicago's municipal reformers and its ward bosses, attention here will stay focused as much as possible on the influence of germ theories in the struggle for control of water management policy. In spite of growing confidence in the science of bacteriology during the Progressive Era, Chicago's doctors and academics would remain frustrated, ignored, and isolated on the fringes of decision-making. As they came to see the urban environment as a world teaming with microbes, they kept pointing to the fatal flaw in the city's sanitary strategy. Yet the engineers were able to retain their privileged positions by proposing ever-bigger technological fixes that generated more and more jobs and contracts for the politicians to dole out. Without any effective outside pressure from the state or federal governments until the mid-1920s, Chicago's ward bosses promoted their own self-serving interests at the expense of the people and the environment.

Ironically, left on the sidelines of policy formation, the city's scientific community found itself free to create one of the world's great centers of experimental work on biological methods of sewage disposal. Between 1908 and 1925, they and their allies among the municipal engineers became key pioneers in the development of methods for treating and disposing of wastewater, methods still in use today. During this same period, Chicago's policy makers stubbornly refused to install even a single water filtration station or full-scale sewage treatment plant. Instead, they continued to build one gigantic project of hydraulic engineering after another, justifying their actions with the outmoded notion of nature's boundlessness. Despite overwhelming scientific evidence to the contrary, the politicians in charge of Chicago's water management system continued to act on the assumption that they could always count on the Great Lakes for an unlimited amount of pure water.

The 1890s, as discussed in the previous section, was a period of ferment in bacteriology and epidemiology. Chicago's scientific community became an active participant in the transatlantic crossings of urban progressives. While Cooley and the engineers were directing the construction of the ship canal, the scientists and doctors were gaining fresh insight into the relationship between the world of the microorganism and the environment of the industrial city. Most important here was the growing realization of the pervasiveness of the former in the latter. In other words, germs were everywhere, and efforts to contain them would prove difficult, if not futile.

In 1889, for instance, the state's leading public health official, Dr. John H. Rauch, explained, "[I]n Chicago the sewage undergoes decomposition in the mains. . . . In rain or floods, sweeping everything out rapidly [into the lake], there is danger of [it] being carried a long distance away and infecting the water supply." Five years later, a popular account of the city's sanitary history exclaimed that "in the operation of these minute beings [microorganisms] a new world is brought to light. Their number, even in a defined space, is inconceivable. In a single gramme of butter . . . there are said to be 2,465,555 micro-organisms." More to the point, in 1895 the city's bacteriologist, Adolph Gerhmann, warned that "there is an area of *continually* contaminated water along the lake front. . . . To attempt to obtain pure water by locating cribs beyond this line of permanent contamination leads to a false security." His tests found sewage pollution twelve miles out and beyond.[26]

Yet until the typhoid epidemic of 1902, Chicago's elected officials could brazenly ignore—and did—the critics of the city's water management policies. When the number of deaths from the disease suddenly jumped to 471 during August and September 1902, Dr. Gerhmann's prescient advice came back to haunt the local defenders of the Chicago Sanitary District. The *Tribune,* for example, now confessed, "[T]he entire water supply is of inferior quality. The only consolation is that if it were not for the drainage canal the water would be inconceivably worse. . . . It would be rank poison."[27]

Among those investigating the causes of the epidemic was Edwin Oakes Jordan, an assistant professor of bacteriology at the University of Chicago. A recent graduate of the M.I.T/Massachusetts Experimental Station program, he put the epidemic in broad, national perspective by highlighting the city's role as a rail hub, the place where train passengers supplied with Chicago water were dispersed to points across the country. His analysis was remarkable in several respects. Not the least was its timely publication in the December 1902 issue of the prestigious *Journal of the American Medical Association.* That the new public health should reach mature form in only ten years is testimony to the revolutionary pace of the paradigm shift in the etiology of disease.[28]

The report represents a comprehensive understanding of the environmental implications of the new science for the industrial city. Jordan immediately acknowledged the benefits of the drainage channel. Since the channel's opening, he said, deaths from typhoid had been reduced to the lowest rate in the city's history. However, he was equally quick to take the channel's planners to task for failure to divert all the city's sewage away from the lake and into the waterway. "Through a lack of foresight and coordinated endeavor on the part of the responsible authorities," he complained, "a large part of the sewage system of Chicago remains at this date unconnected with the Drainage Canal. . . . It is certainly singular that the present situation should not have been foreseen and guarded against. The excuse for . . . pour[ing] fresh sewage into the lake for upward of three years after the completion of a great and enormously expensive sanitary undertaking can hardly be adequate." Jordan produced a map, which showed that the wastewater of more than a quarter-million people was still flushing into the lake. This was undoubtedly the source of the problem "since there seems to be no instance on record where a large city possessing a pure or purified water supply has experienced an epidemic of typhoid fever of the proportions of the one that has just visited Chicago."[29]

To the scientist, the logic of the equation between the protection of the environment and the health of the city now appeared to be self-evident. After making a comparative, statistical analysis of the links between water quality and public health in Amer-

[26] John H. Rauch, *Preliminary Report to the Illinois State Board of Health* (Springfield, 1889), xxii; G. P. Brown, *Drainage Channel and Waterway* (Chicago, 1894), 37; and [Adolph Gerhmann], "Report of the Municipal Laboratory," in City of Chicago, Dept. of Health, *Annual Report* (1895): 178 (my italics).

[27] CT, 8 Aug. 1902.

[28] Edwin Oakes Jordan, "Typhoid Fever and Water Supply in Chicago," *Journal of the American Medical Association* 39 (20 Dec. 1902): 1561–6. Also see Carolyn G. Shapiro-Shapin, "'A Really Excellent Scientific Contribution,'" *Bulletin of the History of Medicine* 71 (1997): 385–411.

[29] Jordan, "Typhoid Fever" (cit. n. 28), 1563–5.

ican and European cities, Jordan turned to a discussion of all the alternative theories that placed blame on more localized causes, such as rotten food, flying insects, and infected dust. While conceding the scientific possibility, he reasoned that the probability in the present situation of transmission by these agents was extremely low as opposed to that of transmission due to sewage contamination of the water supply during the two months of heavy rains that preceded the August outbreak. "Since there is an explanation so simple, so in accord with the general experience regarding extensive epidemics of typhoid, and so consonant with the past experience of Chicago itself, it would seem logically unnecessary to seek for another cause."[30]

According to Jordan, completing the lakefront interceptor sewer project was imperative, but it would be a serious mistake to believe that it represented a final solution to the problem. On the contrary, he argued, all plans must be based on the assumption that Lake Michigan would remain a source of pollution, not purity, into the foreseeable future. "No one familiar with the general sanitary history of water supplies can expect that all chances for water pollution will cease with the completion of the sewage system," Jordan concluded. He painted a picture of the world as a place filled with microbes. Reflecting the fascination of science with statistics, the bacteriologist explained that as many as 172,000 typhoid germs could be found in just one cubic centimeter of urine. Although a single cruise ship or bather could cause the next crisis, the real threat was rapid industrial and suburban growth, a fact that made any water management plan based on the use of the lake without some method of purification problematic.[31]

The scientific revolution in the theory of disease causation led Jordan inexorably toward a dual approach of water filtration and sewage treatment for inland lakes and streams. He concurred with engineers already calling for a major expansion of the CSD to integrate the affluent North Shore suburbs and the Calumet District into the metropolitan system. But he sharply disagreed with their plans for two more heroic canal-building projects, especially the one for the industrial district because it would still leave those dependent on water from the 68th Street intake crib in the Hyde Park neighborhood vulnerable to contaminated supplies. Jordan thought that instead of spending an estimated $12 millions on the so-called Cal-Sag project, the city should build a water filtration plant, which at a cost of $2 million would be not only far less expensive but far more effective in protecting the public health. In fact, Jordan calculated, it would cost only $8.5 million to install filtration works to safeguard the city's entire water supply. "It would prove most discouraging," he predicted, "to discover after the expenditure of seven or eight million dollars for the construction of a drainage canal for the Calumet region that the pollution of the Hyde Park water supply from towns in Indiana south of the Calumet and from other sources was still so great that the amount of typhoid fever in that portion of the community served by this supply remained excessive."[32]

But Cooley and the engineers had already run up a bill of more than $48 million, twice the projected cost. So neither Chicago's ward bosses, who controlled city hall, nor the CSD had any intention of heeding the authority of science—in spite of the

[30] Ibid., 1064–5.
[31] Ibid. For a full analysis of this epidemic, see Harold L. Platt, "Jane Addams and the Ward Boss Revisited," *Environmental History* 5 (April 2000): 194–222.
[32] Jordan, "Typhoid Fever" (cit. n. 28), 1566.

canal's glaring failure in terms of public health and commercial development. The city did not have pure water, and it did not have a superhighway of commerce. In fact, the elementary incompatibility of the two goals had become immediately evident. The rushing flow of lake water into the artificial river had been so great that it had created a hazard for the navigation of its cumbersome barges, repeating the history of the 1871 deep-cut.

This time, however, the Army Corps of Engineers intervened. In May 1901, it ordered the local agency to reduce the flow by almost 60 percent, from 6.5 to 2.7 billion gallons a day. By now, the federal government had also taken an unyielding stand against paying an astronomical amount to deepen the 278 miles of the Illinois River from Joliet to the Mississippi River, to accommodate ocean-going ships—to say nothing of the cost of dredging the additional 800 miles to reach the Gulf of Mexico.[33]

Moreover, the CSD's massive withdrawals lowered the Great Lakes by as much as six inches, resulting in Canada's equally unrelenting opposition to Cooley's megalomaniac dreams. As corps engineers had predicted as early as 1887 in debating Cooley's proposals, such a drop would cause serious problems for navigation through the system of locks and canals connecting the Great Lakes. Furthermore, docking ships in the lakes' shallow harbors would result in economic losses, estimated at $50 million a year. Here, then, lay the origins of the legal dispute between Washington and Chicago that would languish in the courts until the mid-1920s. Since an enduring tradition of American federalism has been the appointment of federal prosecutors and jurists with strong local attachments, without constant diplomatic pressure from the Canadians even this case of justice long-delayed might never have been brought to a resolution. During the interval between 1910 and 1912, when chlorination was introduced, the health of the people of Chicago remained at risk from drinking water contaminated with sewage.[34]

CONCLUSIONS:
POLITICAL CULTURES OF SCIENCE AND TECHNOLOGY

By the turn of the twentieth century, the germ theories of disease causation had triumphed over previous approaches in Manchester and Chicago.[35] In both places as well, the biological basis of sewage purification had become the common understanding. Although the new science was widely accepted, the technology of wastewater treatment and disposal to make best use of this knowledge remained to be worked out. Closely related were political questions because the costs to advance experimental studies, install large-scale facilities, and operate them year-round would

[33] CT, 16–28 July 1871.

[34] International Waterways Commission, "Reports of the International Waterways Commission 1906," in Report of the Minister of Public Works, Canada, *Sessional Paper no. 19a. A. 1907* (Ottawa, 1907). For an account of Canadian diplomacy and federal law, see Platt, "Chicago, The Great Lakes" (cit. n. 2), which includes a bibliography of historical and legal citations to the controversy. For the earliest expression of concern about the proposed sanitary canal's impact on the water levels of the Great Lakes, see the remarks of the Army Corps of Engineers' Major Handbury at the 1887 Rivers Convention in Peoria, Illinois, as reported in *Chicago Morning News,* 12 Oct. 1887. For a similar debate, see Western Society of Engineers, *The Levels of the Lakes as Affected by the Proposed Lakes and Gulf Waterway* (Chicago, 1889). The two key cases were *Sanitary District of Chicago v United States* 266 US 405 (1925), and *Wisconsin v Illinois,* 278 US 367 (1929).

[35] See, e.g., Arthur N. Talbot, "Recent Progress in Sewage Purification," *JWSE* 5 (Dec. 1900): 543–60.

be significant. For Manchester, a politics of deference meant complying with the rulings of the regional and central agencies while searching for ways to reduce operational expenses. For Chicago, a politics of defiance meant disobeying the orders of the national government while enhancing the self-serving goals of the politicians. Nonetheless, both cities would play key roles in the development of an advanced technology, the activated sludge method, which remains in general use today.

In the case of Chicago, adoption of the new science by sanitary engineers brought them into ever-increasing conflict with policy makers at city hall and CSD headquarters. After the great annexation of 1889, the burden of paying for sewer extensions had shifted from common taxes to special assessments on affected property owners. The profits, or so-called surpluses, from the waterworks were no longer diverted into subsidizing sewer construction, giving the ward bosses in control of the finance committee a huge slush fund for patronage jobs and pet projects. As Table 1 demonstrates, these were very substantial sums.

The aldermen knew that these obscene "surpluses" were generated by a flat-rate system of charges for water service as opposed to meter-based billing. They also understood from engineering reports dating back to the 1870s that universal metering was the only practical way to curb the profligate waste of more than half of the water pumped through an underground network of mains that leaked like a sieve. Much of this water found its way by osmosis into the brick sewers. Chronic low pressure and periodic shortages justified an endless round of public works projects.[36]

Any effort to filter the water supply or to purify the resulting sewage discharges would be extremely expensive unless this gross abuse of the fountain inexhaustible were curbed. City engineer John Ericson, appointed in 1901, would spend his entire career (the next quarter-century) trying to do so. One panel of outside experts after another would reinforce his recommendation of universal metering. In 1915, for example, the Chicago Real Estate Board sponsored a study of Chicago's water management strategies. Among its findings of fault, the blue ribbon panel, which included several experts from England, reported that "the rate for 1913 of 218 gallons per head per day is excessive even for American cities, whose generous use of water appears to European engineers to be lavish and inexplicable." At that time, the city's waterworks were pumping, on average, more than 600,000,000 gallons a day. Like Ericson's personal efforts, however, those of this panel and others ultimately proved of no avail. Chicago's politicians were not willing to address the problem of curbing water use.[37]

Politicians in charge of the CSD also had no intention of treating sewage other than by dilution. Its professional staff was given relatively free rein, however, to conduct small-scale experiments and came up with an alternative solution for delivering drinking water free from dangerous germs. In Chicago, the 8-10 million gallons a day of liquid wastes laden with organic matter pouring out of Packingtown posed the single

---

[36] For the first recommendation for meters as the only way to curb waste, see City of Chicago, Board of Public Works, *Annual Report* (1874): 13.

[37] George A. Soper, John D. Watson, and Arthur J. Martin, *A Report to the Chicago Real Estate Board on the Disposal of the Sewage and Protection of the Water Supply of Chicago, Illinois* (Chicago, 1915), 65. For Ericson's travails, see John Ericson, "The Water Works System of Chicago" (paper presented at a meeting of the Western Society of Engineers, Chicago, 15 May 1901), as reported in *JWSE* 4 (1901): 231–304; idem, "Chicago Water Works," ibid., 18 (Oct. 1913): 763–96; and idem, "An Improved Water Supply for Chicago and the Relation of Metering to Service," ibid., 14 (Oct. 1924): 1–8.

Table 1: Chicago Waterworks Revenues

Year	Total Revenue	Costs				"Surplus"	
		Salaries	Fuel	Other*	Total Costs	Amount	Profit Ratio
1893	$2,950,000	$252,800	$250,000	*$63,200	*$566,000	$2,384,000	421.2%
1903	$3,690,000	$277,900	$392,100	$275,000	*$945,300	$2,744,700	290.4%
1913	$6,500,000	$403,200	$437,000	$780,000	$1,621,000	$4,879,000	301.0%

*Includes repair, maintenance, and new construction other than the distribution system.
SOURCES: Chicago, Department of Public Works, Annual Report (1893, 1903, 1913, 1914), passim.

greatest challenge. In 1908, chief engineer George M. Wiser supervised the use of chlorine as a disinfectant at a research station located near the junction of the stockyard's infamous "Bubbly Creek" and the Chicago River. Wiser and other engineers, including George A. Johnson, concluded that chlorine could also be used to kill germs in water supplies. Johnson soon applied this lesson in Jersey City, New Jersey, which resulted in a well-publicized court case decided in favor of the novel method.

Soon cities across the country without filtration systems were adding the chemical to their drinking supplies. In 1911, an outbreak of typhoid fever struck consumers of water from the Hyde Park intake crib, just as Professor Jordan had feared. As a result, Chicago, too, adopted chlorination. Although phenols dumped by steel makers into Lake Michigan meant extra heavy doses of the additive were required to protect the public health, the politicians believed they had found the perfect answer to the demand for "pure" water. Chlorination would prevent future epidemic crises from waterborne disease without endangering the politicians' "surplus" fund.[38]

In the case of Manchester, the main challenge facing scientists and engineers was reducing the costs of sewage treatment and disposal. After 1901, various technologies were strung together to produce a more or less acceptable effluent, but the bill for dumping the sludge at sea kept mounting. As more and more working-class homes were finally allowed to install indoor plumbing and hooked up to the main drain, the Davyhulme facility fell behind in adding enough new capacity to handle the everlarger quantities of wastewater to be processed. Open septic tanks were installed because they proved better than contact beds in reducing the amount of sludge, pointing the way toward using bacteria more efficiently to consume the waste. However, the increase in the volume of wastewater to be purified more than offset these gains.[39]

In 1912, Professor Fowler visited the Lawrence Experimental Station of Massachusetts and observed studies of aeration of sewage in bottles. Inspired, he returned home to continue this line of inquiry, enlisting the help of two of Manchester's engi-

[38] For the most complete account, see George A. Johnson, "The Purification of Public Water Supplies," in Water Supply Papers, U.S. Dept. of the Interior, U.S. Geological Survey, No. 315 (Washington, D.C., 1913). Also see George M. Wisner, Report on Industrial Wastes from the Stockyards and Packinghouses of Chicago (Chicago, Sept. 1914). On the chlorination of Chicago's water and its problems with phenols, see City of Chicago, Dept. of Public Works, Annual Report (1913–1916); and idem, Annual Report (1926–1930), 470–6.

[39] Wilson, "Technology and Municipal Decision-Making" (cit. n. 1), 259–329.

neers, Edward Ardern and W. T. Lockett. They found that allowing the microorganisms to accumulate in the sewage would purify it much faster and more completely than separating the solids from the liquids as the first stage of the treatment process. Some of this "activated sludge" could then be added to the next patch of raw sewage, producing even better results. Moreover, the much-reduced waste by-product had pecuniary value as fertilizer. First published in 1914, news of the Manchester breakthrough spread throughout the professional community with remarkable speed.[40]

In less than a year, Professor Fowler was providing advice on how to duplicate the new method to sanitarians in Chicago and Milwaukee as well as at the University of Illinois at Urbana. Information on the results of this work were reported in the trade journals and discussed at the regional professional organization, the Western Society of Engineers. The CSD's Langdon W. Pearse quickly began tests at the Packingtown station, but large-scale trials that advanced the science and technology of activated sludge were only conducted in the other two cities. Over a relatively short period of three years, Manchester, Urbana, and Milwaukee became international leaders in this field. They established the new method as far superior to all previous methods of sewage treatment and disposal.[41]

In contrast, Chicago would lag further and further behind, stubbornly holding on to notions of the fountain inexhaustible to underpin its sanitary strategy. In 1922, the CSD would open its first sewage treatment plant in the heavily industrialized area of Calumet. It would use older technologies rather than taking advantage of the new method. Only the forceful intervention of a U.S. Supreme Court three years later finally began to bring Chicago up to modern standards of water management. Except for the installation of a water filtration plant, almost twenty years more would pass before the city was brought into basic compliance with the court's orders on sewage disposal.[42]

Comparative studies of Manchester and Chicago lend strong support for Daniel Rodgers's assertion that progressivism took place within a transatlantic context. The ascendancy of germ theories provides an especially useful test case because the timing closely parallels the rise of the impulse for urban reform. European medical practitioners may have been ahead of their American counterparts in adopting microbiological theories of disease causation. Yet their American colleagues appear to have kept abreast of the rapid development of bacteriology, at least in Chicago. Beginning in the early 1880s, scientists from Chicago's medical schools applied the lessons of the microscope to answer questions about the safety and quality of the water supply. With the establishment of the Massachusetts Experimental Station in 1893, important research findings began flowing back to Manchester, informing its debate over sanitation policy. The movement of information and people across the Atlantic in both

[40]See Edward Arden and W. T. Lockett, "Activated Sludge Experiments," *Journal of the Society of the Chemical Industry* 33 (1914): 523–39; and A. Redford and I. S. Russell, *The History of Local Government in Manchester* (London, 1940), 3: 112–20.

[41]For the earliest American reports, see Leslie C. Frank, "English Experiments on Sewage Aeration Reviewed as Preliminary to Baltimore Tests," *Engineering Record* 71 (6 March 1915): 288–9; and Edward Bartow and F. W. Mohlman, "Sewage-Treatment Experiments with Aeration and Activated Sludge," *Engineering News* 73 (1 April 1915): 647–8. For the triumph of the activated sludge method, see George W. Fuller, "Current Tendencies in Sewage Disposal Practice," *JWSE* 26 (Aug. 1921): 273–88; and T. Chalkley Hatton, "Activated-Sludge Process Has Come to Stay," *Engineering New-Record* 93 (Oct. 2, 1924): 538–9.

[42]See note 34 above.

directions made the search for improved methods of biological treatment and disposal of sewage truly international. There was no gap between the two cities in knowledge about this area of science, technology, and medicine. Emblematic of this process were the close interchanges among experts in the Manchester and the Chicago areas that led to the development of the activated sludge method.

A comparative approach also reinforces Christopher Hamlin's claim that political culture played a crucial role in shaping public health and sanitation policy during the Progressive Era. This transnational case study illuminates the ways in which sharp contrasts in the authority of science and experts, the structure of government and society, and the style of partisan organization and mobilization affected decision-making. In the United Kingdom, the central state had already accumulated a long record of social investigation and direct intervention in the affairs of its cities. Equally important in the formation of a politics of deference were the intimate bonds of social class among government officials and academic scientists. Working together, they took incremental steps toward finding solutions to the interrelated problems of urban sanitation and river conservancy.

In the United States, the federalist ethos of local self-government worked against the creation of national urban and environmental policies until the crisis of the Great Depression. In Chicago, an extreme version of this devolution of power gave the ward bosses extraordinary leverage in defining public policy. While the engineers in their city building roles also played a part in the configuration of municipal administration, their authority was always subservient to the self-serving goals of the professional politicians. After the turn of the century, they defied not only the national government but also their own sanitation experts. Only highly exceptional circumstances—persistent Canadian diplomacy—eventually prodded Chicago to begin to conform to modern standards of public health and sanitation. Until then, those forced by low incomes into the slums bordering the industrial corridor of the river paid the price of environmental degradation and diminished lives.

Harold Knapp and the Geography of Normal Controversy:

Radioiodine in the Historical Environment

By Scott Kirsch[*]

ABSTRACT

In 1962, after high levels of the isotope Iodine-131 were detected in Utah milk supplies, Dr. Harold Knapp, a mathematician working for the AEC's Division of Biology and Medicine, developed a new model for estimating, first, the relation between a single deposition of radioactive fallout on pasturage and the levels of Iodine-131 in fresh milk and, second, the total dose to human thyroids resulting from daily intake of the contaminated milk. The implications of Knapp's findings were enormous. They suggested that short-living radioiodine, rather than long-living nuclides such as radiostrontium, posed the greatest hazard from nuclear test fallout and that children raised in Nevada and Utah during the 1950s had been exposed to internal radiation doses far in excess of recommended guidelines. This paper explores the explicit historical revisionism of Knapp's study, his refusal, contra normal AEC practices of knowledge production and spatial representation, to distance himself from the people and places downwind from the Nevada Test Site, and the reactions his work provoked among his AEC colleagues.

INTRODUCTION

In summer 1962, Dr. Harold Knapp, a mathematician working for the Fallout Branch of the Atomic Energy Commission's Division of Biology and Medicine in Washington, was, in his own words, "trying to put the finishing touches on a report on *Radiation Exposure in the United States from Nuclear Test Fallout* and discovered at the last minute that the most important problems concerning the most important nuclide hadn't been thought about."[1] During the 1940s and 1950s, external emitters of radiation, especially long-living radionuclides such as Strontium-90 and Cesium-137, had been the health hazards of primary concern in connection with atmospheric nuclear testing. It was not until the early 1960s that many public health scientists, reacting in part to renewed testing and new data, began to look more seriously at ingested internal emitters from short-living radioiodine as potentially a greater threat. Health physi-

[*] Department of Geography, Campus Box 3220, University of North Carolina at Chapel Hill, Chapel Hill, NC 27599; kirsch@email.unc.edu.

I am grateful to Michael Brown, Wil Gesler, Gregg Mitman, and the participants of the 2002 Madison conference "Environment, Health, and Place in Global Perspective" for their constructive comments on earlier versions of this paper.

[1] Harold Knapp to Charles Dunham, 27 June 1963, DOE doc. #0153306, U.S. Dept. of Energy Coordination and Information Center (hereafter cited as CIC), Las Vegas, Nevada.

cists, ecologists, and various radiation safety scientists and professionals were beginning to understand that though the half-life of Iodine-131 (I-131) was less than eight days, the isotope moved quite rapidly through the food chain as it was deposited on pasturage (as fallout), taken up by cattle, and in turn, ingested by people drinking fresh milk. In the latter some of the radioiodine would accumulate in the thyroid gland, raising the risk of cancer. It quickly became clear that their small thyroid glands and high levels of milk consumption made infants and children especially susceptible to thyroid cancer. By 1963, the release of iodine had become, as one internal Atomic Energy Commission (AEC) report put it, "one of the principal, if not the principal, controlling factors in terms of environmental contamination."[2] And yet, as Harold Knapp would comment, at the Nevada Test Site alone, "over 1000 kilotons equivalent of I-131 were released *before* we obtained any reliable data on the I-131 levels in milk in Off-Site communities."[3]

This was not, however, an insurmountable problem for Knapp. In response to high levels of I-131 observed in Utah in July, 1962, he had begun, on his own initiative, to raise questions about how the radioiodine hazard could be measured, and further, he sought to "retrofit" new concerns about short-living radionuclides to the incomplete data of past environments.[4] Basing his model on previous AEC studies predicting the biological effects of a nuclear war, Knapp developed techniques for estimating first, the relation between a single deposition of fallout on pasturage and the levels of I-131 in fresh milk, and second, the total dose to the thyroid resulting from daily intake of the contaminated milk.[5] According to Knapp's study, the thyroid dose to a one-year-old child from ingested I-131 could be 50 to 250 times the whole body external (gamma) dose to the child, if the child consumed at least one liter per day of fresh milk.[6] From a single, relatively small nuclear test in June 1962, Knapp estimated that the thyroid dose to a one-year-old child in Fruitland, Utah, could have been more than 130 times existing annual radiation protection guidelines.[7] Perhaps even more disturbing were the questions he raised about radioiodine in what might be called the "historical environment" of fallout in the 1950s, before significant offsite measurements of it had been taken. By extrapolating I-131 measurements from existing external radiation measurements from these tests and working these converted data through his model for estimating the uptake of I-131 in milk, Knapp showed that, from just one 1953 test, infants who had been living in a radiation hotspot around

---

[2] Gordon Dunning to E. J. Bloch, "Recommendations of the Ad Hoc Working Group on Radioiodine in the Environment," 25 April 1963, U.S. Dept. of Energy doc. #43348, CIC.

[3] Knapp to Dunham, 27 June 1963 (cit. n. 1) (my italics).

[4] In this sense, Knapp's project was broadly akin to more contemporary "dose reconstruction" estimates. See, e.g., F. Warner and R. J. C. Kirchmann, eds., *Nuclear Test Explosions: Environmental and Human Impacts* (New York, 2000); National Cancer Institute, "I-131 Fallout from NTS: Informing the Public" (workshop proceedings, 19–21 Jan. 2000), http://rex.nci.nih.gov/ INTRFCE_GIFS/radiation _fallout/day_one.html (accessed 22 Feb. 2002).

[5] Harold Knapp, "Iodine-131 in Fresh Milk and Human Thyroids Following a Single Deposition of Nuclear Test Fallout," 1 June 1963, Division of Biology and Medicine, AEC, Washington, D.C., DOE doc. #0153314, CIC. See also Knapp to Charles Dunham, "Recommendations for Additional Measures Related to the Evaluation and Control of Radioiodine from Nevada Nuclear Tests," DOE doc. #27355, CIC. As will be discussed below, the AEC models for estimating I-131 uptake after a nuclear war were based partly on data from the 1957 reactor fire at the United Kingdom's Sellafield nuclear reprocessing plant.

[6] Knapp, "Iodine-131 in Fresh Milk and Human Thyroids" (cit. n. 5).

[7] Ibid.

St. George, Utah, might well have received I-131 doses anywhere from 150 to 750 times existing annual permissible doses.[8]

This paper explores the unambiguous historical revisionism of Knapp's work, and that of several other public health scientists during the Utah radioiodine controversies of 1962–1963, and the reactions to that work within the AEC. In part, it is a study of what Bruno Latour calls the characteristic "historicity" of scientific knowledge, referring not only "to the passage of time—1999 after 1998—but to the fact that something happens in time, that history not only passes but transforms."[9] Just as Latour answers the question of "where were the microbes before Pasteur?" by professing "without contradiction" that, *after* 1864, airborne germs had been there *all along,* we might say that after 1962, ingested I-131 had posed the most serious radioactivity hazard to infants and children living downwind from nuclear testing *throughout* the previous decade. Of course, there are several possible responses to such a claim. One might be to insist that the radionuclides had been there "all along," in the bodies of those living downwind, anyway, and that to suggest otherwise is therefore to buy into a kind of constructivist fallacy that systematically devalues the materiality of nature, artifice, and disease itself. But an alternate response, and a more productive one I argue, insists that we need not deny such materialities—especially not the slow production of undetected thyroid cancers—to take advantage of Latour's heuristic for identifying both the historically embedded and the historically constitutive aspects of scientific knowledge, qualities that have been especially prominent in the emergence and assessment of new environmental health risks. In this sense, it is precisely the struggles of Knapp and others to name, quantify, and make visible the radioiodine hazard, along with the strategies of Knapp's erstwhile AEC colleagues for maintaining its invisibility, that we need to better understand. For Knapp, we will see that this quality, the historicity of his radioiodine study, was in a sense empowering, politically cum morally, in that it compelled the mathematician to allege, quite publicly, that the AEC had likely been wrong in its previous assurances of safety. What is more, he persisted in arguing that his report should be published, one way or another, and that, uncertainties notwithstanding, his findings should be used to inform existing "operational safety" practices.

Yet if knowledge is both "timeless" and historical in this sense, as the geographer Nigel Thrift has argued, it must have important spatial (or spatiotemporal) characteristics as well, since "knowledge of germs does not stretch everywhere."[10] Because knowledge, as well as the lack of it, is an obvious resource and constraint for all manner of social relations, Thrift sketches a "geography of social knowing and unknowing" to examine how knowledge is socially distributed in space, including questions of access to knowledge among social groups, and of the spatial variation of different kinds of knowledge and "unawareness."[11] Although in trying to broadly locate knowledge (rather than knowledge *production* or science as such) in a theory of situated

---

[8] Harold Knapp, "Conclusions of a Report on Observed Relations between the Deposition Level of Fresh Fission Products from Nevada Tests and the Resulting Levels of I-131 in Fresh Milk," included in Charles Dunham to W. Langham, C. Comar, J. Gofman et al., 7 June 1963, DOE doc. #0153320, CIC.

[9] Bruno Latour, *Pandora's Hope* (Cambridge, Mass, 1999), 306.

[10] Nigel Thrift, "Flies and Germs: A Geography of Knowledge," in *Spatial Formations* (Beverly Hills, Calif., 1996), 96. Essay first published in 1985.

[11] Ibid., 108. On "unawareness," see Ulrich Beck, "Knowledge or Unawareness? Two Perspectives on 'Reflexive Modernization,'" in *World Risk Society* (Cambridge, 1999), 109–32.

social action Thrift provides a rather static sense of knowledge itself (i.e., a group either has access to a particular "stock of empirical knowledge" or not), the questions he raises—about how the spatiality of knowledge has been articulated with the institutionalization and stabilization of different forms of knowledge in society—might easily be asked of knowledge producers such as Harold Knapp and the AEC. If the historicity of knowledge is most likely to surface, and to challenge prior norms and ideas, with the construction of new knowledge and the settlement of controversies, then how do these controversies come to depend on particular geographies of social knowing and unknowing? (Such geographies include not only the geographic and institutional settings for knowledge production but also the strategies of spatial representation through which scientific and technical information is circulated.) Furthermore how, as Knapp's work on the problem of radioiodine in the historical environment will help us to understand, have these geographies at times been disrupted?

I begin to address these questions in the following section by providing an overview of radiation and fallout controversies in the United States during the 1950s. I examine how, on one hand, particular institutional sites of knowledge production persisted despite "outside" scientific opposition and how, on the other, the scientific-technical work of nuclear testing and radiation safety depended on particular spatial (and scalar) perspectives, and cartographies of socially empty space, that helped preserve the invisibility of the fallout hazard.[12] Next, starting with Knapp's attempt to publish his report and the AEC's decision, as a stopgap, to censor it, I explore in greater detail the nature of Knapp's "internal" opposition and how Knapp's emphasis on people and locality, in contrast to the AEC's more distanced view, disrupted the boundaries of public-private and classified-unclassified that structured much AEC science in a moment of regulatory transition.

RADIATION CONTROVERSIES AT A DISTANCE

The AEC never carried out nuclear tests without public safety plans; indeed, the very notion that safety from fallout hazards *could* be planned and accounted for was the central assumption of the AEC's self-regulation. For example, among the AEC's site selection criteria in the search for a continental nuclear proving ground in 1950 had been the presence of a "90 degree possible fallout sector, to a radius of 125 miles downwind from site."[13] The history of "radiation safety," however, like that of many other areas of public and worker health regulation, had been largely a reactionary one, characterized by changing standards developed over time in response to new scientific knowledge of environmental health risks.[14] While AEC scientists and engineers felt that a properly run experimental program could keep human radiation exposures within "permissible dose" limits, these limits were revised downward repeatedly during the period of atmospheric and underwater testing by the Federal Radiation Coun-

---

[12] For a related point linking this geography of unknowing to the production of invisibility, see Michelle Murphy, "Uncertain Exposures and the Privilege of Imperception: Activist Scientists and Race at the U.S. Environmental Protection Agency" (this volume).

[13] U.S. AEC, "Selection of a Continental Atomic Test Site; Report by the Director of Military Application," as attached to "Atomic Energy Commission: Location of Proving Ground for Atomic Weapons," 13 Dec. 1950, DOE doc. #30419, CIC.

[14] See Peter Bacon Hales, *Atomic Spaces* (Urbana, Ill., 1997), 273–300; Christopher Sellers, *Hazards of the Job* (Chapel Hill, N. C., 1997).

cil, and in turn, the AEC's advisory committee to the Division of Biology and Medicine (in 1951, 1955, 1957, 1960, and 1961).[15]

As numerous legal, activist, and academic critics have argued, the AEC sought to protect itself from public scrutiny during this period by using security classification procedures developed for protecting the new secrets of state.[16] However, for this technocratic form of expertise—defined partly by access to special stocks of secret (but putatively public) information—to maintain legitimacy in the face of resistances from the scientific community was clearly another matter. By the early 1950s, a number of scientists, drawn mainly from universities, entered public debate on the problem of fallout hazards and public health. While AEC scientists—from physicists at the weapons laboratories to physicians at the Division of Biology and Medicine (DBM)—continued to argue that low-level radiation below certain "thresholds" was insignificant compared with the amount of radioactivity received through existing natural and artificial sources, most geneticists flatly disagreed.[17] By 1954, according to A. H. Sturtevant of the California Institute of Technology (CIT), it was "widely confirmed" that (1) high-energy radiation produced mutations in reproductive cells; (2) the frequency of mutations was *proportional* to the dosage of radiation; (3) there was "almost certainly" no "threshold value" below which radiation was insignificant; and (4) the genetic effects of exposure to radiation were cumulative and permanent.[18] Special committee studies commissioned by the National Academy of Sciences, however, produced mixed results that largely conformed to disciplinary and institutional boundaries.[19]

Controversies over the genetic effects of radiation, on germ or reproductive cells,

[15] U.S. AEC, "Some Thoughts on Radiation Protection Criteria at the Nevada Test Site," n.d., DOE doc. # 68015, CIC. In this sense, the ways scientists, or institutions, behave with respect to the dynamism and necessary uncertainties of the health sciences take on a central significance. Yet this is precisely what Barton Hacker, in his comprehensive (and generally quite useful) history of the AEC's radiation safety program, *Elements of Controversy,* leaves out of the story. The stated purpose of that volume is worth repeating here, for it is one, in its implicit acceptance of "permissible dose" levels as "timeless" (at a given point in time), that is only possible by ignoring the "historicity" of testing practices as discussed above. For Hacker, "Those responsible for radiation safety in nuclear weapons testing under the auspices of the Atomic Energy Commission were competent, diligent, and cautious. They understood the hazards and took every precaution within their power to avoid injuring either test participants or bystanders. Testing, of course, meant taking risks, and safety could never be the highest priority. Those in charge sometimes made mistakes, but for the most part they managed to ensure that neither test participants nor bystanders suffered any apparent damage from fallout. Describing how they did so, in terms as neutral and as fully as the sources will allow, defines this book's ultimate purpose." Hacker, *Elements of Controversy: The Atomic Energy Commission and Radiation Safety in Nuclear Weapons Testing, 1947–1974* (Berkeley, Calif., 1994), 6; on the radioiodine controversy, see 219–30.

[16] See, e.g., Howard Ball, *Justice Downwind: America's Atomic Testing Program in the 1950s* (New York, 1986); Phillip Fradkin, *Fallout: An American Nuclear Tragedy* (Tucson, Ariz., 1989); J. Fuller, *The Day We Bombed Utah: America's Most Lethal Secret* (New York, 1984); Peter Goin, *Nuclear Landscapes* (Baltimore, 1991); S. Hilgartner, R. Bell, and R. O'Connor, *Nukespeak: Nuclear Language, Visions, and Mindset* (San Francisco, 1982); and J. Turley, *Facing Reality: A Guide to Citizen Law Enforcement; Fighting Environmental Crime at Facilities of the U.S. Departments of Energy and Defense* (San Francisco, 1996).

[17] For a useful summary of the debate, see Carolyn Kopp, "The Origins of the American Scientific Debate over Fallout Hazards," *Social Studies of Science* 9 (1979): 403–22.

[18] As Sturtevant put it in his 1954 presidential address to the Pacific Division of the American Association for the Advancement of Science, these observations were "so widely confirmed that we may confidently assert that they apply to all higher organisms including man." In ibid., 405–6.

[19] "Nuclear Weapons Tests," *Science,* 9 Nov. 1956, 925–6. While the National Academy of Sciences' Pathology Committee, composed mainly of medical scientists and chaired by a former director of the DBM, concluded that nuclear test explosions could be increased *tenfold* "without causing any serious genetic danger." Sturtevant, a member of the same body's Genetic Effects Committee, contested this characterization in a forceful letter to the *Washington Post,* 26 Oct. 1956 (cited in the *Science* article).

were soon matched by disputed claims over the pathological effects of radiation, on somatic cells involved in the formation of bodily organs and tissues, focusing on the relationship between leukemia and the isotope Strontium-90 (Sr-90). Commonly taken up through the ingestion of milk and vegetables, Sr-90, due to its chemical similarity to calcium, tended to accumulate in bone marrow, where it continued to emit radiation. In May 1957, the geneticist Linus Pauling, also at CIT, estimated that 10,000 people had *already* died or were dying of leukemia because of nuclear testing. Biologist E. B. Lewis, another CIT scientist, supported Pauling's estimate in a quantitative analysis published in *Science*.[20]

Yet in the same year the AEC would proclaim, in a pamphlet produced for southern Nevada and Utah residents, that its research had "confirmed that Nevada test fallout has not caused illness or injured the health of anyone living near the test site."[21] In this sense, what we might call (with apologies to Kuhn) a state of "normal controversy" over the health effects of fallout persisted, allowing the AEC to continue its nuclear testing program for a time, along with related research at its laboratories and test sites. During this same period, though, testing practices and normal controversy engendered still more oppositional scientific work, with more scientists seeking to end or limit the tests joining public politics.[22] The debate among scientists hinged on issues of public information and decision-making. While the AEC hoped to remove the threat of "unreasoning" public participation from its operations, scientifically oriented activist organizations such as the Greater St. Louis Committee for Nuclear Information (CNI) and the Federation of American Scientists (formerly the atomic scientists' movement) were after just the opposite. They sought to expand public knowledge of fallout and radiation science; in doing so, they appealed to the public for the leverage necessary to enact political change. CNI, in one striking example of science in public education, even enlisted the public as data collectors in a scientific project, a campaign to collect 50,000 baby teeth to investigate how children's bodies, in particular, were susceptible to bone-seeking radionuclides such as Sr-90.[23]

In one sense, this condition of normal controversy could be plotted as an uneven *geography* of knowledge production, in which normative disciplinary and institutional conventions structured scientific and technical work at particular sites, whether Livermore or Cal Tech, therein enabling certain kinds of knowledge production and limiting others.[24] By focusing on the spatiality of contested knowledge production—the

---

[20] E. B. Lewis, "Leukemia and Ionizing Radiation," *Science,* 17 May 1957, 965–72.

[21] U.S. AEC, *Atomic Tests in Nevada* (Washington, D.C., 1957), 15. On AEC public information practices, see Hilgartner, Bell, and O'Connor, *Nukespeak* (cit. n. 16); and Scott Kirsch, "Watching the Bombs Go Off: Photography, Nuclear Landscapes, and Spectator Democracy," *Antipode* 29 (1997): 227–55.

[22] This, of course, was after the much more ambitious politics of the postwar atomic scientists' movement had largely been reduced, under anticommunist legal and political attacks, to a defense of scientists' civil liberties. See Jessica Wang, *American Science in an Age of Anxiety: Scientists, Anti-Communism, and the Cold War* (Chapel Hill, N. C., 1999). See also Mary Jo Nye, "What Price Politics? Scientists and Political Controversy," *Endeavour* 23 (1999): 148–54.

[23] W. Wyant, "50,000 Baby Teeth," *Nation,* 13 June 1959, 535–7.

[24] This is not meant, however, to draw an absolute distinction between the national laboratories and the universities nor to claim, reductively, that where someone worked ultimately *determined* his or her stance on the fallout question (as the case of Harold Knapp makes plain). On the relationships between universities and the military establishment during the cold war, see Stuart W. Leslie, *The Cold War and American Science: The Military-Industrial-Academic Complex at MIT and Stanford* (New York, 1993); see also Noam Chomsky, Laura Nader, Immanuel Wallerstein et al., *The Cold War and the University* (New York, 1997).

geography of normal controversy—we call attention to the simultaneity of different truths generated in differently-fashioned scientific spaces. Furthermore, we can deepen this focus on the physical and institutional settings wherein scientists worked by exploring how scientists used spatial representations, and other "distancing" strategies, in assessing the potential environmental and health hazards from nuclear testing and the damage already done.

In his critique of diffusionist medical geographies of AIDS, geographer Michael Brown writes that, in the mappings of spatial science, "gay men with AIDS inter alia seem to be important . . . only as data points or nodes by which the virus spreads across space."[25] This reduction of social space to measurable, physical distance, Brown argues, facilitates both a validation of the scientists' objectivity and a *social* distancing from the people with HIV/AIDS, their bodies, and their localities. In a similar manner, many of the spatial representations used in AEC science (and in its communication to public audiences) worked by distancing the scientists from more direct questions about people and their local environments and from the actual spatial variation of fallout hazards that could be masked by generalizing the data at different spatial scales. For example, physicists and testing advocates Edward Teller and Albert Latter insisted that for the "average American," the actual dangers from cancer- and leukemia-causing radionuclides such as Sr-90 and Cesium-137 were minimal, amounting to a dosage of only a small fraction of one roentgen per person.[26] "If the tests continue at the present rate, radiation levels might increase as much as fivefold," the physicists argued in *Life* magazine, yet "even in this situation it is extremely unlikely that anyone would receive a lifetime dosage of as much as five roentguns [*sic*]. If radiation in this small amount actually does increase a person's chance of getting bone cancer or leukemia, the increase is so slight it cannot be measured."[27] Responding in a letter to the editor, Linus Pauling challenged many of Teller and Latter's points as either false or misleading. The emphasis on the "average American," Pauling insisted, obscured the geographic variation and specificity of fallout patterns, including the hotspots that, though by no means limited to areas immediately downwind, were nevertheless disproportionate to Nevada and Utah. Pauling argued that it was not at all unlikely that lifetime doses would exceed five roentgens, and he cited as evidence an AEC report listing towns in Nevada and Utah where thousands of inhabitants had already been exposed to more than five roentgens of fallout from a single test series in Nevada.[28]

The mapping of predicted and observed fallout patterns also facilitated social distancing between testing scientists, on the one hand, and "we 'relatively unpopulateds,'" on the other (as one frustrated Utah public health scientist would describe his home state, refracting the AEC's standard characterization of the "fallout sector").[29] If

[25] Michael Brown, "Ironies of Distance: An Ongoing Critique of the Geographies of AIDS," *Environment and Planning D: Society and Space* 13 (1995): 159–83, on 167.

[26] Edward Teller and Albert Latter, *Our Nuclear Future: . . . Facts, Dangers, and Opportunities* (New York, 1958); and idem, "The Compelling Need for Nuclear Tests," *Life*, 19 Feb. 1958, 64–72.

[27] Teller and Latter, "Compelling Need for Nuclear Tests" (cit. n. 26), 65.

[28] Linus Pauling, letter to the editor, *Life*, 17 March 1958, 21. Teller and Latter countered that they had meant that such lifetime doses were unlikely outside the "immediate downwind area"—they wrote off the immediate downwind populations, in other words, as statistically insignificant—as they reiterated their mission to "allay fears aroused by exaggerated and frightening statements." Teller and Latter, "Doctors Teller and Latter Reply," ibid.

[29] R. C. Pendleton to Harold Knapp, 6 May 1963, DOE doc. #0027353, CIC. For Pendleton, a University of Utah radiation ecologist who by this time had become a forceful critic of the AEC, the "relatively unpopulateds" had become a "new category of citizenship."

mapping, as historians of cartography have argued, works as a technology of both inclusion and exclusion—that is, by portraying some spatial distributions or landscape features and not others—then the ideal fallout map, for maximizing social distance, was one excluding both people and place and in the case of the map of predicted fallout reductions reprinted in Figure 1, by reducing not only the biological environment but also the cardinal directions to physical categories of distance and distribution.[30]

Such distancing from the human consequences of nuclear testing was also reflected in many of the questions asked (and those not asked) by radiation safety scientists and other professionals. In 1957, when the Windscale reactor fire at the United Kingdom's Sellafield nuclear reprocessing plant had generated substantial new data on the problem of I-131 contamination in milk, career DBM scientist Dr. Gordon Dunning produced a new study of the hypothetical biological consequences of a nuclear war but did not apply the same techniques used in that work to existing test fallout data. In fact, despite the ample time for reflection allowed by the 1958–1961 test moratorium, it would be five years before the implications of Dunning's I-131 study would be explored in connection with fallout in Utah. After a number of tests in July 1962 had produced alarming levels of the isotope in some of the state's milk and vegetable monitoring stations, Harold Knapp felt compelled to take up the problem himself, extending Dunning's I-131 analysis from the possible to the real.

"THE OBVIOUS IN A HAYSTACK":
I-131 IN THE HISTORICAL ENVIRONMENT

The AEC was eager to publicize neither Knapp's findings on I-131 in the environment following the July 1962 *Sedan* and *Small Boy* nuclear tests nor his retrofitted estimates for the 1953 test noted above.[31] When Knapp sought to publish his findings in fall 1962, AEC officials initially determined that his report contained "Restricted Data"—at least while they scrambled to assemble a special technical review committee and "Ad Hoc Working Group on Radioiodine and the Environment" to assess the paper and its implications for radiation safety practices.[32] Gordon Dunning, who had become deputy director of the AEC's Division of Operational Safety, was part of the ad hoc working

---

[30] For geographic histories of cartography, see J. B. Harley, "Maps, Knowledge, and Power," in *The Iconography of Landscape*, ed. Denis Cosgrove and S. Daniels (Cambridge, 1988); and idem, "Deconstructing the Map," *Cartographica* 26 (1989): 1–20. See also Denis Cosgrove, ed., *Mappings* (London, 1999); and Matthew Edney, *Mapping an Empire: The Geographical Construction of British India, 1765–1843* (Chicago, 1997).

[31] The *Sedan* and *Small Boy* nuclear tests, conducted 7 and 14 July 1962, respectively, at the Nevada Test Site, figured prominently in the production of the high levels of radioiodine detected in Utah in July and August 1962. *Sedan* was a shallowly buried, 100-kiloton "nuclear-cratering" experiment and the largest nuclear test to date in North America. On *Sedan*, see Fradkin, *Fallout* (cit. n. 16); and on its role in Project Plowshare, the AEC's "peaceful nuclear explosives" program, see Scott Kirsch, "Experiments in Progress: Edward Teller's Controversial Geographies," *Ecumene* 5 (1998): 267–85. *Small Boy* was a low-yield, tower-type atmospheric test.

[32] The review committee was composed of scientists from Livermore and Los Alamos laboratories, Hanford, the U.S. Public Health Service, and Cornell University. As historian Peter Westwick argues, review committees such as this one were one common means for the AEC's "classified community" to maintain some elements of the critical review and debate that had always been central aspects of science as a rational public space. Westwick, "Secret Science: A Classified Community in the National Laboratories," *Minerva* 38 (2000): 363–91.

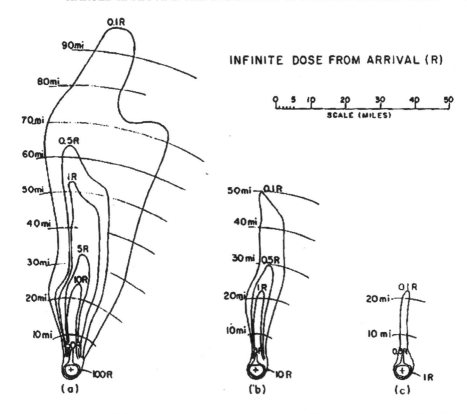

Figure 1. *External dose estimates for "nuclear cratering" explosions presented to Congress in 1965. The exclusions of people, place, and even geographic direction and location from the maps were characteristic distancing strategies in AEC radiation safety practices. (From John Kelly,* Statement before the JCAE, *January 5, 1965, Office of History and Historical Records, Lawrence Livermore National Laboratory, Livermore, Calif.)*

group, and it was he who emerged as Knapp's harshest critic during what amounted (as Knapp saw it) to some nine months of censorship. Of Knapp's report, Dunning argued:

> There are two basic problems involved, one, the questionable technical validity . . . and two, a policy question of the Commission publishing the paper. In response to a question asked as to the motive for publishing the paper, a member (not the author) of the Commission's staff said, "The Commission has been telling the world for years that it has been conducting its operations safely, now it appears that this may not be so." If a member of the staff says this about the paper, what reaction may we expect from the press and the public?[33]

Dunning's main technical criticism of the paper was that Knapp's equations for relating I-131 deposition to its uptake in fresh milk had too many uncertainties: he had extrapolated from too few data points, and there were potential errors in data measurements and reliability.[34] Dunning thus called into question some of the

---

[33] Gordon Dunning to N.H. Woodruff, 14 June 1963, "Comments on 'Iodine-131 in Fresh Milk and Human Thyroids following a Single Deposition of Nuclear Test Fallout' by Dr. Harold Knapp," included in Woodruff to Charles Dunham, 17 June 1963, DOE doc. #0153315.

[34] Cf. ibid.; Gary Higgins to Charles Dunham, 30 March 1963, "Preliminary Comments on the Knapp I-131 Study," DOE doc. #0160850, CIC.

mathematical techniques *he* had developed (in the context of assessing the environmental health risks of nuclear war) and the adequacy of *AEC's* data measurement,
hence challenging the reliability of the same data used by the commission before the
I-131 controversy to defend the safety of nuclear testing in Nevada.[35]

"All of science," Knapp responded, in comments to the DBM's Charles Dunham,
Dunning's superior, "involves relating one set of measurements to an entirely different set of measurements." In addition to insisting that he was using the best *available*
data, Knapp questioned Dunning's emergent skepticism and, perhaps more to the
point, the reason the AEC had such poor data in the first place:

> Why, when the Division of Operational Safety had a hot spot it didn't expect, didn't it
> find out what the gamma levels were? All one can say is that some measurements were
> made and the results were reported and used in a straightforward fashion, and all the
> rough pieces of data fit together very well and consistently to give exactly the levels of
> I-131 in milk that Dr. Dunning would have estimated on the basis of his own calcula-
> tions. It looks mighty silly to suddenly become so doubtful just because when the the-
> ory is applied, we suddenly realize we have all been missing the obvious in a haystack
> for over 10 years, and that the public relations impact of this oversight are painful to
> ponder.[36]

In fact, until the July 1962 tests, as Knapp would later argue, there had been no systematic effort by the AEC to obtain local fallout levels at the same place and the same
time as I-131 measurements in milk were obtained.[37] What questions *were* being asked
about radioiodine at the time? For the AEC's Nevada Test Organization, the problem
of how to obtain reliable data on I-131 contamination was understood largely as a matter for experimentation. Shortly before Knapp began his 1962 investigations of I-131,
the 100-kiloton *Sedan* test in Nevada offered the opportunity for an experiment, contracted through a U.S. Public Health Service laboratory, involving beagle dogs placed
in cages at 31 and 42 miles from ground zero.[38] The experiment was meant to compare
physical sampling devices with biological ones—beagle thyroids—with the *Sedan* explosion providing "a fission source enabling some of these factors to be studied."[39] As
it turned out, the experiment suggested that the iodine inhalation hazard was greater
than expected and, indeed, greater than existing equipment was capable of measuring.
If anything, the study revealed the general inadequacy of technical instrumentation in
fallout monitoring.[40] Though the analysis of the data yielded few firm conclusions, it

---

[35] See Gordon Dunning, "Shorter-Lived Fission Products in Fallout," *Health Physics* 4 (1960):
35–41.

[36] Knapp to Dunham, 27 June 1963 (cit. n. 1).

[37] "The most direct information available," as Knapp later described it, "is limited to some coinci-
dental measurements of iodine-131 levels in milk from two dairy herds and to the open-field external
[gamma]-dose rate reported by Public Health Service monitors in the vicinity of the area where the
cows were grazing." These measurements were made at Alamo and Caliente, Nevada, following the
14 July 1962 *Small Boy* test. Harold Knapp, "Iodine-131 in Fresh Milk and Human Thyroids following a Single Deposition of Nuclear Test Fall-Out, " *Nature,* 9 May 1964, 534–7, on 535.

[38] "Radiation Safety Plan for Field Operations Portion of the Iodine Inhalation Study to be Conducted by the U.S. Public Health Service in Connection with the Sedan Event of the Plowshare
Program," n.d. (pre-shot planning document), DOE doc. #3763, CIC; "Iodine Inhalation Study
for Project Sedan," Southwestern Radiological Health Laboratory, Public Health Service, Received at Lawrence Radiation Laboratory, Livermore, 3 Aug. 1964, LLNL, Plowshare Collection,
Office of History and Historical Records, Lawrence Livermore National Laboratory, Livermore,
California.

[39] "Iodine Inhalation Study for Project Sedan" (cit. n. 38).

was obvious to the experimentalists that there was "an urgent need for additional studies directed toward the determination of radioiodine concentration in air which will lead to the development of truly quantitative sampling methodology."[41]

Of course, Knapp, too, was interested in developing better quantitative sampling methodologies. But the decision to enact the beagle thyroid experiment is still telling in terms of research priorities, unasked questions, and the *distancing* from local places and communities I have characterized as part of the AEC's geography of normal controversy. Rather than monitoring the human environment downwind more extensively, to better establish the sorts of relationships Knapp outlined—that is, between external (gamma) radiation levels and I-131 measured in milk—the AEC had contracted out an experiment that worked by irradiating and harvesting beagle thyroids for scientific and technical information, even as the same *Sedan* test was significantly irradiating the dairying, ranching, and gardening ecosystems downwind in central and northeastern Utah. Small wonder that the cover for the laboratory's project report on the beagle thyroid experiment resorted to gallows humor to provide a *comic* distancing from the event, doubly distancing the report, for its scientist-readers, from the people and places who were actually caught, whether directly or indirectly, in *Sedan*'s dust cloud (see Figure 2).

By spring 1963, it was clear to Gordon Dunning that the boundary of classification placed over Knapp's study was only a partial and temporary solution to the problems posed by the mathematician's new techniques for estimating I-131 uptake. This was in part because Knapp, in his efforts to gather as much information as possible for his analysis, had *already* disrupted the boundary, metaphorically and geographically, by consulting with scientists from the Public Health Service and the University of Utah, as well as with people in downwind communities. As Dunning complained:

> Through meetings and exchange of letters between the author [Knapp] and others outside the Commission, the paper probably is well known. The author spent several days traveling and talking with many people around the Nevada Test Site, followed by exchange of letters with them. This has just goaded them on. We have spent years of hard, patient effort to establish good and calm relations with the public around NTS. Such action as the author's has been harmful.[42]

Whether Knapp's actions or the AEC's public information practices were more harmful was clearly a matter of perspective. What troubled Dunning most, though, was Knapp's breaching of the boundary between the classified and the public sphere and the fundamental challenge posed by this disruption in the normal procedures for the movement of information.

"Consider the author's problem," Knapp replied to Charles Dunham in a biting set of comments on Dunning's latest review of Knapp's work. Because Dunning, Knapp insisted, had denied him access to relevant fallout data, he had had to turn to Utah's public health scientists, such as University of Utah radiation ecologist Robert Pendleton and Grant Winn of the state's Department of Health, who had shared their data with him.[43]

---

[40] The report concluded: "The physical sampling gear currently considered optimum for sampling radioactive iodine effluents is actually very inefficient, being on the order of 10% or less." The "biological sampler" of the beagle thyroid, however, "takes a more accurate representation of the iodine activities than does physical sampling equipment because the animal 'does its own chemistry' and deposits these activities in strictly correct ratios in the thyroid gland." Ibid.

[41] Ibid.

[42] Dunning to Woodruff, 14 June 1963, "Comments" (cit. n. 33).

Figure 2. Cover sheet for Southwestern Radiological Health Laboratory, "Iodine Inhalation Study for Project Sedan," 3 Aug. 1965. (From Office of History and Historical Records, Lawrence Livermore National Laboratory, Livermore, Calif.)

In fact, the public health scientists had informed Knapp that the Utah Department of Health had "found it necessary to build duplicate monitoring facilities beside those of

---

[43] R. C. Pendleton to Harold Knapp, 26 March 1963, in *Fallout, Radiation Standards, and Counter-measures: Hearings before the Subcommittee on Research, Development, and Radiation of the Joint Committee on Atomic Energy, Congress of the United States, Eighty-Eighth Congress, August 20, 21, 22, and 27, 1963* (Washington, D.C., 1963), pt. 2, 1034–7; Pendleton to Knapp, 6 May 1963 (cit. n. 29); Grant S. Winn to Knapp, 28 March 1963 and 1 April 1963, in *Fallout, Radiation Standards, and Countermeasures,* pt. 2, 1039–46.

the Atomic Energy Commission's Off-Site Radiological Safety Organization, because data pertaining to the safety of the citizens of Utah was not forthcoming from the AEC."[44] Knapp had also spent a week visiting sites where data had been collected to talk with the monitors about collection procedures, and he sought out additional firsthand observations from nearby Nevada ranchers. Quite unlike AEC fallout maps that excluded people from spatial analysis, Knapp not only endeavored to depict human *content* in the "fallout sector" but also relied on both experts and laypeople living in the region to provide additional data, to actively contribute to his evaluation of the hazard.

Characterized by a different perspective on place and community as well was Knapp's eight-page set of "Recommendations for Additional Measures Related to the Evaluation and Control of Radioiodine from Nevada Nuclear Tests" (submitted in March 1963 as his revised paper was still being held "under review" by the AEC). The recommendations also reflected a fundamentally different understanding of the scientist's role in the evaluation of uncertainty and risk than had heretofore typified AEC operational safety and research.[45] The obvious concern, for Knapp, was for the infants and children living in rural areas downwind from the test site. He argued for the necessity of conducting new inventories—of dairy herds within 500 miles of the test site and of the people in this area who took their milk from family cows. His insistence on bringing people and places into the problems of data collection, and thus into the scientist's sense of social responsibilities and values, was also evident in his discussion of the kind of care required in monitoring children's health in downwind areas.

> It may also be noted that the population group of greatest concern may be small children in isolated rural communities who consume fresh milk. Such children and their families often see outsiders only once a week, or less. If a truck with fancy counting equipment turns up and starts a series of mysterious measurements on small children, it's not at all clear what perturbations might be introduced into the diets or habits of the children being counted, or whether the parents permission to count will be readily obtained, or whether the child will hold still long enough to be properly counted. . . . It is important that the scientist who is actually going to analyze the data and write the report have a large say in where and when and how the data is to be collected, and that great care go into the design of the program for collecting the data.[46]

In additional departures from the norms of AEC radiation safety practices, Knapp called for thorough descriptions of the pasturage and an inventory of the names and ages of people consuming fresh milk from cows who grazed on the pastures. He

[44] Harold Knapp, "Comments on Gordon Dunning to N.H. Woodruff," 14 June 1963, DOE doc. #0161119, CIC; idem, "Iodine-131 in Fresh Milk and Human Thyroids" (cit. n. 5); see discussion of the Knapp-Pendleton correspondence in Fradkin, *Fallout* (cit. n. 16). Both scientists believed the AEC was holding up publication of their papers in *Science* (Pendleton) and *Nature* (Knapp). Knapp was especially irritated by one particular stalling tactic on the part of the AEC: it repeatedly treated his report, and a revised version, as "drafts." "The report transmitted is not a draft," Knapp insisted. "It is a formal, complete document, ready for Commission action." Knapp to Dunham, 27 June 1963 (cit. n. 1). See Gordon Dunning to E. J. Bloch, "Report of the Ad Hoc Working Group on Radioiodine in the Environment," 26 March 1963, DOE doc. #43350, CIC.

[45] Knapp, "Recommendations for Additional Measures" (cit. n. 5). Knapp had originally planned to include the recommendations with his report on the relations between fission products and the levels of I-131 in fresh milk, but decided that "since the recommendations were addressed to a more restricted audience and contain suggestions and observations of a more controversial nature than the factual data of the report, their inclusion would only delay distribution of the report." Knapp to Charles Dunham, 12 March 1963.

[46] Ibid., 5.

stressed that the cow owners must be advised of the I-131 content of their milk when radioactivity was found to be above certain levels.

While some members of the reviewing committee saw no reason to hold up publication of Knapp's report, Dunning worried that if the AEC published it—even with a "disclaimer" by the ad hoc working group that would expose the report's many uncertainties and qualifications—then the commission would be placing itself "in the untenable position of condemning itself" while adding "dignity and prestige to [Knapp's] report."[47] Dunning proposed instead "one possible 'out' to the situation":

> Let the Commission tell Dr. Knapp in a matter-of-fact and bland manner that the Commission interposes no objection if he, as an individual scientist, wishes to publish his paper.
> If Dr. Knapp can find a reputable scientific journal to accept the manuscript, we may expect letters and inquiries. We can treat these in the same professional and unemotional way that we have many others in the past. By doing so, we can make it clear that the Commission has been conducting its operations in a responsible manner and that the Commission has a highly competent staff that has considered Dr. Knapp's ideas.[48]

The AEC did ultimately agree to publish Knapp's report (along with the comments of its "ad hoc committee review").[49] It had, in effect, little choice but to do so because by this time Knapp had already disrupted the commission's geography of social knowing and unknowing in his cooperation with "outside" scientists and in his refusal to distance himself from the living places that were his objects of study.

SCIENCE'S REVISIONIST HISTORIES

When, in August 1963, the Joint Committee on Atomic Energy (JCAE) reconvened hearings on fallout, radiation standards, and countermeasures begun earlier that summer, Harold Knapp had already left his position in the DBM's Fallout Studies Branch to become a systems analyst at the Pentagon. From there he completed his I-131 study. Born classified into an internal AEC controversy, Knapp's report at last entered the public record as an appendix in the JCAE published *Hearings on Fallout, Radiation Standards, and Countermeasures,* along with both the review team's comments and Knapp's comments on his reviewers.[50] Thus as the public debate over the effects of hotspots and low-level radiation from fallout articulated with debates over the ratifi-

---

[47] Ibid.

[48] Ibid.

[49] Harold Knapp, "Iodine-131 in Fresh Milk and Human Thyroids Following a Single Deposition of Nuclear Test Fallout," *TLD-19266, Health and Safety,* TID-4500, 24th ed. (Washington, D.C., 1 June 1963). Knapp did manage to find a reputable outlet for a summary of his I-131 study; Knapp, "Iodine-131 in Fresh Milk and Human Thyroids," *Nature* (cit. n. 37).

[50] Although the report is dated June 1, 1963 (probably the date that Knapp transmitted his final version from the Institute for Defense Analysis at the Pentagon), it had in fact not yet been published when it was submitted to the JCAE on 16 August. "Report by Dr. H. A. Knapp, AEC, on 'Iodine 131 in Fresh Milk and Human Thyroids Following a Single Deposition of Nuclear Test Fallout,'" in *Fallout, Radiation Standards, and Countermeasures* (cit. n. 43), 915–1031; "Ad Hoc Committee Review of Report on 'Iodine 131 in Fresh Milk and Human Thyroids Following a Single Deposition of Nuclear Test Fallout,'" in ibid., 1076–8; "Letter Concerning Comments Relative to the Committee Review of the Knapp Report from Dr. Harold A. Knapp, to John T. Conway, Executive Director, JCAE, Dated September 9, 1963," in ibid., 1078–82. These latter comments overlap with, but are not the same as, Knapp's comments on Dunning quoted in this paper.

cation of the Partial Test Ban Treaty, the historicity of Knapp's study—the grim revisionism of estimating risks for dangers to which we have already been exposed—became part of a political discourse of regulation and reform that pushed nuclear testing underground soon after.[51]

The larger story of nuclear testing "radiation safety" in the United States is an archetype for the technocratic administration of hazards in which, as the sociologist Ulrich Beck would argue, we have passed "the monopoly of interpretation [of danger] to those who caused it, of all people."[52] The case of Harold Knapp and the 1962–1963 Utah radioiodine controversy complicates Beck's notion of the "unity of culprits and judges" in certain ways, however. It shows that even under normative institutional arrangements, and across normalized geographies of social knowing and unknowing, such unities are never total and at times have been resisted "from within" on moral and epistemological grounds. Yet it is one thing to suggest that the historicity of scientific knowledge, and the recognition of past errors underlying it, can empower scientists and others to become political subjects, bringing new perspectives and new social bases into scientific practice and public debate. It is another to ask what *kinds* of politics or political changes science's revisionist histories can help to inform. This is the structural issue to which Beck speaks in his calls for a fundamental division of powers, and a shift in the burden of proof, governing those nuclear, chemical, and biogenetic experiments that are, like the effects of I-131 in the historical environment, "testable only after they are built."[53]

"Adverse criticism in retrospect," Gordon Dunning would argue in 1964, in another iteration of the radioiodine controversy, "is a comparatively easy task. I would respectfully submit that we utilize our time and energies now to getting on with solving technical problems since decisions must continue to be made."[54] By this time, however, adverse criticism had begun to change the very frameworks through which technical problems could be solved and decisions made in Dunning's field of operations. In the emergence and assessment of environmental health risks, it is often precisely in the resistances generated in reaction—"in retrospect"—that new political forms of the scientific take shape.

[51] This is not meant to imply, however, that the Partial Test Ban Treaty (PTBT) resolved all the environmental and health problems associated with nuclear testing. While the PTBT was certainly good news for most people living downwind from nuclear tests, it also facilitated developments in underground testing technologies that allowed for the *acceleration* of nuclear weapons development and with it both an expanded labor force and the displacement of social and environmental costs across the AEC (and later the Dept. of Energy) network of production, from Rocky Flats to Paducah, Kentucky, to Hanford, Oak Ridge, and other sites, while debates over the effects of low-level radiation continued. Kirsch, "Watching the Bombs Go Off" (cit. n. 21).

[52] Beck, *World Risk Society* (cit. n. 11), 58; see also idem, *Risk Society: Towards a New Modernity,* trans. M. Ritter (Beverly Hills, Calif., 1992).

[53] Beck, *World Risk Society* (cit. n. 11), 60. Beck refers specifically to nuclear reactors in this passage.

[54] Gordon Dunning, "AEC Official Protests," *Bulletin of the Atomic Scientists,* Sept. 1964: 29–30. Dunning's letter was a response to Lindsay Mattison and Richard Daly, "Nevada Fallout: Past and Present Hazards," *Bulletin of the Atomic Scientists,* April 1964, 41–5; see also the response to Dunning by Mattison and Daly, *Bulletin of the Atomic Scientists,* Sept. 1964, 30.

The Artificial Nature of Fluoridated Water:

Between Nations, Knowledge, and Material Flows

*By Christopher Sellers**

ABSTRACT

An exercise in "historical ontology," this paper charts the contrasting ways fluoridated water and its effects crystallized as objects of knowledge and concern in three quite different realms over the mid twentieth century. Among U.S. health officials and experts, fluoridated water emerged and stabilized as a public health goal, preventing tooth decay. Indian doctors and scientists defined it as a public health problem, causing "skeletal fluorosis." Fluoridated water also acquired an intense presence among laypeople in the United States, especially those voting in local referenda on fluoridation. More often than not rejecting it, suspecting bias and myopia in profluoridation expertise, they cobbled together a lay ontology that proved predictive of the varied and changing flows of fluoridated water itself. The paper concludes by suggesting a principle of environmental symmetry as an aid to this kind of comparative ontology.

INTRODUCTION

On January 25, 1945, the first fluoride-loaded drops of drinking water splashed through the faucets of Grand Rapids, Michigan. They bore the rising hopes of a handful of public health officials that fluoridation would prove a "magic bullet" remedy against tooth decay. Since the germ theory of the late nineteenth century inaugurated a new public health in the early twentieth century, few measures have so readily found their way into the American health officials' pantheon of global remedies. Fluoride's inhibiting effects on cavities had only become recognized by public health scientists in the 1930s, yet by the early fifties, years before the Grand Rapids experiment was over, water fluoridation had received endorsements from the American Medical Association, the American Dental Association, the American Public Health Association, and the U.S. Public Health Service, and a campaign was in full swing to bring it to the rest of the nation and world.[1]

* History Department, Stony Brook University, Stony Brook, NY 11794; csellers@notes.cc. sunysb.edu.

For their comments on earlier versions of this essay, I would like to thank my coeditors, the participants in the volume-based workshop at Madison, Wisconsin, and those at the Yale Colloquium in the History of Medicine.

[1] Donald R. McNeil, *The Fight for Fluoridation* (New York, 1957); M. W. Easley, "Celebrating 50 Years of Fluoridation: A Public Health Success Story," *British Dental Journal* 178 (1995): 72–5; Greg

The achievements of the fluoridation campaign have fallen significantly short of those heady aspirations, however. The science justifying fluoridation, like the practice itself, has remained predominantly American. By the early sixties, the United States had fluoridated more water than the rest of the world combined (serving 46 million versus 25 million people). By the 1990s, that was no longer the case, but Americans still made up nearly half of those receiving fluoridated water worldwide (132.2 million of 298.2 million people).[2] Why have so many people been so hesitant about fluoridation? Social scientific answers to this question abound, but in this paper I want to suggest another answer involving the perceived ontology of fluoride itself. The "fluoride" of fluoridation scientists and advocates differed from the "fluoride" of many others around the world who ignored or blocked the practice.

Scientists outside the United States had concerns about fluoridation's universal value. For instance, prominent experts saw fluoride as a veritable disease menace in a place such as India. At the same time American investigators were making the case for fluoridation in drinking water as a health benefit, Indian scientists and doctors began linking fluoridated water to a mysterious, crippling bone malady, skeletal fluorosis. In the 1950s, as American public health officials strove for public support to build fluoridation plants, their Indian counterparts were desperately seeking ways of defluoridating local water supplies.

Skepticism about the universal value of fluoridation was not limited to non-American scientists. Despite the quick embrace by America's national organizations of health professionals, fluoridation stirred up an unusually vocal and effective lay opposition in the U.S. scientists' backyard. By the late 1950s, lay voters had quashed the idea in some 70–80 percent of fluoride referenda across the United States.[3] The fate of fluoridation offers a dramatic counterpoint to many historical characterizations of this midcentury era of American health history—such as Paul Starr's claim that doctors had achieved "cultural authority," or John Burnham's that this period capped a "Golden Age of Medicine."[4]

Setting the story of fluoridation in these wider contexts, global as well as inexpert, illuminates the parochialism of those midcentury American experts who made the case for fluoridation. They did so, in part, by bracketing out the questions about fluoride others continued to ask. Extending a project in science and medical history that Ian Hacking dubs "historical ontology," I chart the bumpy and divergent ways fluoridated water and its effects emerged and stabilized as contrasting objects of scientific concern in two very different national traditions of public health, only partly reconciled by the later mediation of the World Health Organization (WHO). Projected transnationally, the historical ontology approach to fluoride suggests a model for

Field, "Flushing Poisons from the Body Politic: The Fluoride Controversy and American Political Culture," typed manuscript. See also Brian Martin, *Scientific Knowledge in Controversy: The Social Dynamics of the Fluoridation Debate* (Albany, N.Y., 1991).

[2] G. H. Leatherman and J. Ellis, "Fluoridation round the World" (1963), Folder "Fluoridation around the World," Box 1, Fluoride Papers, Epidemiology Dept. of National Institute of Dental Research, History of Medicine Division, National Library of Medicine, Bethesda, Md., Institute of Social Sciences, *Fluorosis in India* (New Delhi, 1992), 27; Public Health Service, Ad Hoc Subcommittee on Fluorine of the Committee to Coordinate Environmental Health and Related Programs, *Review of Fluorine Benefits and Risks* (Washington, D.C., 1991), 22.

[3] Robert Crain, Elihu Katz, and Donald Rosenthal, *The Politics of Community Conflict: The Fluoridation Decision* (Indianapolis, 1967), 22.

[4] Paul Starr, *The Social Transformation of American Medicine* (New York, 1982), especially 134–40; John C. Burnham, "American Medicine's Golden Age: What Happened to It?" *Science,* 19 March 1982, 1474–9.

health history akin to, yet distinct from, the "disease-frame" approach so popular among social and cultural historians. Instead of revolving around a particular disease, this story pivots around pathological interpretations of an environmental "risk factor," which disease-centered histories often marginalize or ignore.[5] In adapting this project of historical ontology to the history of environment and health, this story of fluoride pushes it in two other new directions.

While historical ontologists such as Hacking, Lorraine Daston, and Bruno Latour confine themselves to the historical "cocreation" of natural objects such as "bacteria" by experts and laboratories, the fate of fluoridation suggests the importance of extending this project beyond the purview of past scientists. What Daston terms the "quotidian," the realm of inexpert material experience turned out to be considerably less "obdurate," "enduring," or "obvious" than she asserts, for Indian villagers as well as lay American anti-fluoridationists. For these laypeople, fluoridated water remained, as Daston puts it for her scientists, "simultaneously real *and* historical," natural and constructed.[6] Indeed, what distinguished so many laypeople from profluoridation scientists and health professionals was that, for them, the artifice of putting fluoride in their faucets threatened realities that were more diverse and dire than these experts thought "realistic."

In reconstructing these historical ontologies of fluoride side by side, I also want to suggest a more eclectic approach to fluoride's own historical agency. Much as we learn from Latour and Daston about the "historicity of things," their strict focus on established expertise of a given era, in combination with a studied refusal to consider "travel *through* time of an already-existing *substance*," steers them away from questions that help set fluoridation within a broader and longer history. How, for instance, might germs or fluoridated water, as historical "actors" or "influences," actually circumvent the frames of past experts, yet be seen, asserted, or imagined by nonexperts? How might the history of this environmental substance's actual distribution, flow, and changing human contacts have intersected with the ways fluoridated water has been conceived and debated? Historians of science and medicine may shrug off Latour's blanket attack on these questions, which says that by taking fluoride as an "already existing substance," they "accept too much of what the giants demand."[7] Still, the scholarly scope and emphasis of this edited volume reflect how difficult it has become to address such questions within the fields of science and medical history alone.

Positing some historical continuity to fluoride not only helps recover its diverse and changing flows but also raises more insistent questions about what laypeople knew. I draw here from environmental history, in which much recent scholarship has highlighted a materialist ecology of "flows," from streaming wheat and lumber into Chicago in William Cronon's work to Martin Melosi's history of urban channels of water and sewage to Joel Tarr's depiction of industrialization as "the search for the ultimate sink."[8] Material reconstruction of fluoride's shifting distribution—in part by

---

[5] Charles Rosenberg and Janet Golden, eds., *Framing Disease: Studies in Cultural History* (New Brunswick, N.J., 1993).

[6] Ian Hacking, *Historical Ontology* (Cambridge, Mass., 2002); Lorraine Daston, ed., *Biographies of Scientific Objects* (Chicago, 2000), 2–3; Bruno Latour, *Pandora's Hope: Essays on the Reality of Science Studies* (Cambridge, Mass., 1999), especially 145–73.

[7] Latour, *Pandora's Hope* (cit. n. 6), 162.

[8] William Cronon, *Nature's Metropolis: Chicago and the Great West* (New York, 1991); Martin Melosi, *The Sanitary City: Urban Infrastructure in America from Colonial Times to the Present* (Baltimore, 2000); Joel Tarr, *The Search for the Ultimate Sink: Urban Pollution in Historical Perspective* (Akron, Ohio, 1996).

drawing on more recent, "anachronistic" science—spotlights historical contingencies missed by a focus on past expert perceptions alone: stark differences in the "natural" fluoride concentrations between nations, as well as contrasts in the engineering of human water supplies. An ecological perspective on fluoride's history brings out longer-term, larger-scale consequences of the fluoridation campaign, including the worldwide rise in fluoride exposure over the late twentieth century. It also helps move nonscientists toward the center of the historical stage by reorienting our perspective toward encounters with "fluoridated" water that were less mediated by past fluoride experts.

THE SCIENCE OF FLUORIDE FOR U.S. WATERS: FROM NATURAL TO NORMAL

Fluoride-bearing minerals originate in the tumultuous heat of volcanoes and lava flows. They make up 1 percent of the earth's crust, mostly solidified into fluorspar or cryolite. Coming into contact with water, fluoride starts to flow again, to a degree that depends on the type and porosity of the rocks and the pH, temperature, and other chemical characteristics of the water, as well as flush rates. Not surprisingly, the natural concentration of fluoride in water fluctuates widely by climate and regional as well as local geography.[9] Within the United States, groundwater in the eastern and central sections of the country, where most people lived into the early twentieth century, often averaged less than 0.1 parts per million (ppm) of fluoride. In the rapidly growing western states more recently settled by whites, measured amounts of fluoride in the drinking water often ranged higher than 1ppm, in some places to as much as 16 ppm.

A combination of natural and built hydrology helped steer early U.S. fluoride researchers toward viewing fluoridated water as manipulable, whether as problem or remedy. By the time American health officials started investigating fluoride levels of drinking water, they disregarded "unreliable" water sources such as "isolate potholes" or streams, where sediments could elevate fluoride concentrations. Most of the communities under study had "reliable water suppl[ies]": deep wells and "sanitary" systems that filtered their drinking water and piped it into homes.[10]

The first American investigations into the fluoride content of water came as part of a concerted effort during the 1910s and '20s to discern the cause of a strange, unremovable brown stain on teeth, known as mottling. As eastern-trained Colorado Springs dentist Frederick McKay discovered, Colorado Brown Stain seemed endemic to dental patients in this and other recently-established western towns. He posed few or no questions about the existence of the stain among longest-standing locals such as Native Americans, instead seeking a "retrofit" for this condition mainly in the European medical literature. A 1901 report out of Naples, Italy, on volcano-contaminated groundwater led him to discount contributions from food or air early on. Following models in industrial hygiene and vitamin deficiency studies, correlating clinical with environmental information, McKay and a few like-minded investigators—chemists and physicians as well as dentists—concluded that waterborne fluoride was responsible. Only through extensive public debate did they convince lay citizens in towns

[9] Subcommittee on Fluorine, *Fluorine Benefits and Risks* (cit. n. 2), 2; on fluoride's history, see also Kaj Roholm, *Fluorine Intoxication: A Clinical-Hygienic Study* (Copenhagen, 1937), 7–60.
[10] World Health Organization (WHO), *Fluorides and Human Health* (Geneva, 1970), 274–87. Waterworks served 94 million Americans in 1945 and 160 million in 1965. Melosi, *The Sanitary City* (cit. n. 8), 298.

such as Colorado Springs that their sparkling tap water was to blame.[11] From the start, American efforts to link fluoridated water with disease confronted fluoride's imperceptibility to lay water drinkers, along with the acquired reputation of piped water as "sanitary."

The broader backdrop for this move was what I have described elsewhere as an environmental turn in the American health science community. Especially over the 1920s and '30s, health scientists applied the methods of analytical chemistry ever more ambitiously and rigorously to the "natural" and environmental exposure levels of substances known or suspected to cause human ills. Investigators of mottled enamel followed lead poisoning researchers in undertaking some of the most extensive of these early initiatives. Yet despite detailed scrutiny of its geographic variations that increasingly focused on a drinking water link, fluoride only emerged as a causal candidate in 1929. Unlike lead scientists, who focused mainly on that toxin's occupational origins, investigators of the link between fluoride and mottling built on the assumptions of this earlier epidemiology that "natural" groundwater, rather than industry, was to blame. Indeed, the corporate ties of this new U.S. environmental health science, in an era before extensive federal funding of health research, helped steer laboratory confirmation of the link between fluoridated water and mottling. Henry Churchill, who produced the first analytical evidence for the connection between Brown Stain and the fluoride content of faucets, was the chief chemist of ALCOA, the Aluminum Corporation of America. Driving chemist Churchill's investigation were his concerns that the stained teeth of inhabitants of Bauxite, Arkansas, an ALCOA town, would feed consumer fears about the health dangers of aluminum cooking ware. As with Robert Kehoe's demonstrations that measurable lead levels in many Americans were natural and normal, Churchill's proof that fluoride was the culprit for the browning of Bauxite children's teeth quieted people's fears about industry.[12]

Beyond quantifying fluoride's ties to mottling, what secured the uniqueness of the American science of fluoridated water during this era was its establishment of benefits. During the 1930s, investigators at the Public Health Service (PHS), led by H. Trendley Dean, established elevated "natural" levels of fluoride as a cause not just of mottling but also of reduced rates of cavities. Trendley Dean and other PHS investigators first set out to study mottling across larger populations and a wider geography, then focused on determining what fluoride levels could suppress cavities without mottling children's teeth. Dean acknowledged strong evidence that other environmental factors outside of water were likely contributors to cavities—diet ("of major importance") and possibly latitude and sunlight intensity. To bring out just how influential water was, however, he and his colleagues designed studies in which these other factors remained relatively constant. The "magnitude of the sample" presumably washed out any dietary contributions among communities whose proximity equalized other

[11] J. M. Eager, "Denti di Chiaie (Chiae Teeth)," *Public Health Reports* 16 (1901): 2576; among McKay's studies, see Grover Kempf and Frederick McKay, "Mottled Enamel in a Segregated Population," *Public Health Reports* 45 (1930): 2923–40; McNeil, *Fight for Fluoridation* (cit. n. 1), especially 19–21.

[12] H. V. Churchill to F. S. McKay, 15 Nov. 1931, McKay papers, as cited in McNeil, *Fight for Fluoridation* (cit. n. 1), 27 n. 40; Christopher Sellers, *Hazards of the Job: From Industrial Disease to Environmental Health Science* (Chapel Hill, N. C., 1997); Kaj Roholm, "Fog Disaster in Meuse Valley, 1930: Fluorine Intoxication," *Journal of Industrial Hygiene and Toxicology* 19 (March 1937): 126–37.

climatological and geographic factors.[13] No clear line emerged to denote where mottling ceased and cavities began. One part per million of fluoride showed a definite decrease in cavities but linked with the mildest form of mottling in 10 percent of the population—a percentage that rose gradually with fluoride levels of more than 1 ppm. Given the possible benefits, public health officials in the 1940s decided to intervene on an experimental basis using a prospective epidemiological design, a design that smoking studies would, in the following decade, make the gold standard of epidemiological practice. They chose two communities with low natural levels of fluoride in the water—Grand Rapids, Michigan, and Newburgh, New York—and in early 1945 began adding one part per million of this chemical to the towns' water supplies.[14] The dose-response curves constructed by PHS epidemiologists under Trendley Dean made them confident that with between 0.7 and 1.5 ppm of fluoride in the water, cavity reduction could be bought with minimal mottling.[15]

"Inoculating" an entire community's water supply with fluoride meant transporting benefits enjoyed by some western children to the cavity-prone majority in the East. The decision came at a meeting in Washington, D.C., within the Epidemiology Branch of the National Institute of Dental Research, on November 11, 1942. The PHS's Stream Pollution Division had just declared the idea technically and economically feasible. Nevertheless, uncertainties and worries plagued early PHS advocates of this initiative even as they pressed forward.

They anticipated that fluoridated water was going to be a hard sell with nonscientists. One doctor commented that "the psychology of introducing fluorine into the water supply unless handled properly, might produce condemnation of the procedure on the part of the public." Beyond psychology, they fretted about the physiology: What might happen "when minute amounts of fluorine are ingested over a long period of time"? What were the "possible undesirable effects on the human system not yet detected"? By no means "essential," fluoride, then known as a pesticide component and an air pollution culprit, was "universally recognized as a toxic element." Noting that tolerance levels had been set for fluoride insecticide residues, also that "human error" could somehow result in " high proportions of fluorine (e.g., 1,000 parts per million)," conferees pointed out that "several foreign studies (inadequately controlled) [reported] that osteosclerosis or other bone changes are found in people that had used fluorine water." PHS doctors had taken such reports seriously enough to launch a "small study (ten cases)" of their own, which failed to support these "foreign" findings.[16] American health experts were not convinced that fluoride intoxication from water was possible.

At the 1942 PHS meeting in Washington, there was little discussion of fluoride exposures outside drinking water. The lofty epistemological standards attained by Dean wrought a convincing environmental reductionism: the dose-response relationship he

[13] See, e.g., H. Trendley Dean, Philip Jay, Francis Arnold et al., "Domestic Water and Dental Caries, Including Certain Epidemiological Aspects of Oral L. Acidopholus," *Public Health Reports* 54 (1939): 862–8, as reprinted in F. J. McClure, ed., *Fluoride Drinking Waters* (Bethesda, Md., 1962), 90–101, on 90.

[14] McNeil, *Fight for Fluoridation* (cit. n. 1), 3–43.

[15] H. Trendley Dean, "Domestic Water and Dental Caries," *Journal of the American Water Works Association* 35 (Sept. 1943): 1168–9.

[16] Dr. W. J. Pelton to Dr. J. W. Mountin, "Report of Conference held on November 6, 1942, in the Bureau, regarding public health aspects of fluorine inoculation in public water supplies," Folder "Fluoridation Memo 1942," Box 1, Fluoride Papers (cit. n. 2).

had abstracted between water levels and cavities became the sole consideration. The stability of this reduction was still vulnerable; PHS officials did resolve to undertake a further, more comprehensive investigation of what happened to adults who drank "fluorine water" over a period of "many years." However, their collective eagerness to test the fluoride remedy itself remained virtually undeterred. The sole nonmedical or -dental participant at these hearings, a sanitary engineer by the name of Hoskins, voiced the greatest skepticism.[17] The group consensus was that this study of long-term exposure could accompany, rather than precede, the fluoridation of an entire community's water supply. After all, whatever the "foreign" claims, in America "populations undoubtedly move from high fluorine areas to low fluorine areas and vice versa, without any noticeable or reported undesirable effects."[18]

The PHS dismissed reports of skeletal fluorosis elsewhere, and in the United States the naturally fluoridated water found in a few parts was soon transmogrified into a new, population-wide standard for water—and dental—normality. Increasingly over the previous couple of decades, chlorine as well as water softeners had been added to urban Americans' drinking water, so public health officials themselves saw nothing unprecedented in this proposal. Now also penetrating into midsize and smaller cities, chlorination had given rise to scattered complaints about the aftertaste but no vocal or organized resistance.[19] Fluoride might even stir up even fewer problems, being both tasteless and odorless. Fluoride's American investigators soon shifted their sights to the unfluoridated parts of their nation. By the late forties and early fifties, investigators found the diminishing tooth decay in their two experimental communities (32.5 percent less in Newburgh by 1951 and as much as 43 percent less in Grand Rapids) rapidly approached that in naturally fluoridated communities (nearly 60 percent lower), unaccompanied by more dire disease. PHS investigators had initially envisioned the Newburgh and Grand Rapids studies as modest "experiments" that would last ten to fifteen years, but eager dentists began to make the case that nearly every community should consider fluoridating their water, across the United States and the world.

Over this same period, PHS efforts to *de*fluoridate drinking water also got underway in the relatively few western American towns with naturally high fluoridation levels, but these efforts remained vastly overshadowed in the United States by the scientific scrutiny and promotion of fluoridation.[20] Meanwhile, the authors of those foreign studies that PHS scientists dismissed as "uncontrolled" and unconfirmed, Indian public health scientists, had developed different priorities.

THE INDIAN SCIENCE OF FLUORIDATED WATERS:
ARTIFICE OF THE NATURAL

As in the United States, much groundwater used for drinking in India had negligible fluoride content. But larger slices of the Indian subcontinent had elevated (greater than 1ppm) fluoride waters, including long and densely settled rural areas. In places

[17] E. Hunt, "Water and Turf: Fluoridation and the 20th-Century Fate of Waterworks Engineers," *American Journal of Public Health* 87 (1987): 1235–6.

[18] Pelton, "Report of Conference held on November 6, 1942" (cit. n. 16).

[19] Only about one-third of American waterworks had chlorinated their water by 1939. Melosi, *The Sanitary City* (cit. n. 8), 223–4.

[20] F. J. Maier, "Defluoridation of Municipal Water Supplies," *Journal of the American Waterworks Association* 45 (1953): 879–88.

the content could reach much higher than that in any American drinking water supply, upward of 70 ppm. Importantly, these differences in concentration, reported in the first WHO bulletin on fluoride and health in 1970, reflected more than regional contrasts in groundwater sources. They also involved differences in the technology of water supply. Whereas the Americans measured fluoride levels along water pipes even in the smaller towns, their Indian counterparts, when outside the largest cities, analyzed water from springs, "spring pits dug out of the bed of streams," and hand-dug village wells. Such water was often laced not just with dissolved fluoride but also with fluoride-containing sediments—the kind of particles screened out of most "sanitary" waterworks in the United States by this time.[21] Indian health scientists and officials, however, not only faced intensive levels of fluoride exposure but also had to deal with problems in measuring the many sources of village drinking water.

Underdeveloped as India's water supply was, Indian health scientists themselves kept in touch with the new methods and trends in Western medical science. As in the West, their health science undertook an environmental turn: analyses of fluoride in drinking water began only slightly later on the Indian subcontinent than in America. During the mid-1930s, Indian scientists' studies of fluoride began like those of their American counterparts with a mysterious ailment, reported by health inspectors as well as a doctor from the Baptist mission in the Nellore District of southern India. The ailment was as localized as mottled enamel but far more severe. Its adult victims experienced a crippling of the bones and joints that, at its worst, rendered them immobile. The local villagers attributed the problem to drinking water. One of the first steps taken by early investigators—one not taken by their American counterparts looking into mottling—was to bring some of the victims into a hospital, the King Institute of Preventive Medicine in Guindy, near Madras. In 1936, institute researcher C. G. Pandit, "in the ordinary course of reading" European medical journals, ran across a report of similar skeletal disease found among workers in a Danish cryolite factory, identified as "chronic fluorine poisoning." Remembering Nellore's fluoride-rich mica mines, Pandit and his colleagues proceeded to investigate fluoride levels in local well water and the parallel occurrence of mottled enamel asserted in American publications. The syndrome of "skeletal fluorosis" that the Indian investigators went on to define had been unknown among Indian diseases as well as those found in the international medical literature. With the new fluoride pathology came an additional mystery. The water fluoride levels first measured in India were surprisingly close to those measured in America where mottling, but no more severe disease, had been identified.[22]

Stabilizing as a disease, skeletal fluorosis quite suddenly acquired an intense yet puzzling reality for these scientists and physicians in Guindy. They did use experi-

[21] Institute of Social Sciences, *Fluorosis in India* (cit. n. 2); WHO, *Fluorides and Human Health* (cit. n. 10), 17ff.; C. G. Pandit, T. N. S. Raghavachari, D. Subba Rao et al., "Endemic Fluorosis in South India: A Study of the Factors Involved . . . ," *Indian Journal of Medical Research* 28 (1940): 540. On Indian water systems, see Anil Agarwal and Sunita Narain, "Dying Wisdom: The Decline and Revival of Traditional Water Harvesting Systems in India," *The Ecologist* 27 (1997): 112ff.

[22] H. E. Shortt and C. G. Pandit, "Endemic Fluorosis in the Nellore District of South India," *Indian Medical Gazette* 72 (1937): 396–8, on 396; Rao Sahib, T. N. S. Raghavachari, and K. Venkataramanan, "The Occurrence of Fluorides in Drinking Water-Supplies with a Note on Their Removal," *Indian Journal of Medical Research* 28 (Oct. 1940): 517–32; Pandit et al., "Endemic Fluorosis in South India" (cit. n. 21). On the larger Indian medical context, see Mark Harrison, *Public Health in British India: Anglo-Indian Preventive Medicine, 1859–1914* (Cambridge, 1994); Roger Jeffrey, *The Politics of Health in India* (Berkeley, Calif., 1988).

mental controls, but only to test the contribution of vitamin C to the presumed intoxication in monkeys. They were convinced that fluoride was the culprit because of their patients' resemblance to the Dutch cryolite workers as well as the ubiquity and severity of the ailment around highly fluoridated wells. Some 74 percent of those examined in high-exposure areas showed some "bone affections"; 14 percent, mostly long-term residents, showed the severest form, "a complete rigidity of the spine . . . immobility of the joints of both extremities, [and] the fixation of the thoracic wall so that breathing became entirely abdominal." While U.S. PHS researchers dismissed early reports from India, the Indian researchers struggled to reconcile their findings with those of the Americans. Initially, they surmised that ailing, elderly Indian villagers in their hospital had simply been exposed much longer than any American counterparts, periods of forty years or more. When field surveys of implicated villages showed how widespread the syndrome was among the middle-aged as well as the elderly, investigators were forced into a multifactorial approach, with a much less stable focus on water than that of their American counterparts. They looked into the fluoride and other content of local diets and soil, into vitamin intake, nutrition, and even economics.[23]

Those who first defined skeletal fluorosis in India also turned to reconstructing its local past. Unlike either McKay or Latour's Pasteur, they began with a sweeping and unvarying "retrofit" of this disease on to the local past and only subsequently turned more historicist.[24] The clinicians who saw the first victims in a Guindy hospital in 1936 reasoned broadly that the disease, "since it comes in the first instance from the soil, might truly be described as being in this area as 'old as the hills.'"[25] But as other researchers visited the stricken villages to gather information, and the discrepancies with American and other studies became clear, these extrapolations quickly gave way to more limited and nuanced surmises about the disease's past. By 1940, investigators had realized how recent changes in the water supply might have contributed to the ailment's scope, even as they had gained a new appreciation of the prescientific, "quotidian" knowledge about it.

Prior to the official medical discovery of this disease, the connection between bone ailments and drinking water had become so persuasive and intense to many rural Indians that when stricken, entire villages moved, in search of less contaminated water sources. Though they also attributed other ailments to the water and faulted the micaceous rocks of the region rather than an invisible fluoride, the villagers' empirical choice of one source of water over another often correlated with a lower fluoride content, investigators were surprised to discover. The scientists' own chemical knowledge was challenged since fluoride was supposed to be imperceptible to the human senses. The investigators soon established why the villagers' quotidian surmises often proved reliable: high fluorine content correlated with "high total solids," what the villagers could see.[26] Moreover, Indian villages by the 1930s were seeking water in new ways, departing from traditional water-harvest methods partly in response to warnings

[23] Sahib, Raghavachari, and Venkataramanan, "Occurrence of Fluorides in Drinking Water-Supplies" (cit. n. 22); Pandit et al., "Endemic Fluorosis in South India" (cit. n. 21), 538; see also H. E. Shortt, G. R. McRobert, T. W. Barnard et al., "Endemic Fluorosis in Madras Presidency," *Indian Journal of Medical Research* 25 (Oct. 1937): 553–68.

[24] Latour, *Pandora's Hope* (cit. n. 6), 168–71.

[25] Shortt et al., "Endemic Fluorosis in the Madras Presidency" (cit. n. 23), 554.

[26] Shortt and Pandit, "Endemic Fluorosis in the Nellore District," 397; Sahib, Raghavachari, and Venkataramanan, "Occurrence of Fluorides in Drinking Water-Supplies," 521. (Both cit. n. 22.)

about the infectious dangers of surface waters. In areas of Nellore hardest hit by skeletal fluorosis, high fluoride was found both in springs and in wells of 15–30 feet, "in
actual daily use for several years." But in other districts, the fluoride content of wells
50–200 feet deep, dug since 1938, turned out to be considerably higher than in surface sources. With the new recognition of fluoride's risks, Indian villagers and public
health officials found themselves on the horns of a dilemma: deeper wells considered
"safe from the sanitary point of view" poured forth new, risky levels of fluoride.[27]

Indian investigators establishing this disease as a medical entity continued to think
of it as long-standing and endemic to Nellore but came to realize additional complexities to its history—the way in which it had been influenced by changing water
supply as well as the villagers themselves. From the outset, Indian public health scientists and officials also considered defluoridation but operated in a far different
technological and sociopolitical context than their American counterparts. While Indian public health was not shaped as much by corporate interests, it was also not as
well endowed; in addition, it remained overwhelmingly oriented toward infectious
disease. Even "sanitary" measures well established in western American towns—
chlorination as well as filtered and piped water—remained forestalled in the Indian
countryside where fluorosis had been found. Nevertheless, as more Indian investigators looked for this chemically defined disease, its estimated distribution grew, along
with the population at risk. By the early 1950s, middle-age adults and children had
been found to have skeletal fluorosis, in northern as well as southern India, sometimes from levels of water exposure as low as 1–2 ppm. Though India's first defluoridation plant went online in 1961, for most afflicted regions the only feasible solution was what McKay prescribed in the 1920s for overfluoridated American towns:
switching water supplies.[28] As Indian public health fitfully grappled with the subcontinent's excess of fluoride, in other parts of the world, the fluoridation campaign
itself gained momentum.

THE INTERNATIONAL SPREAD OF FLUORIDATION
AND SKELETAL FLUOROSIS

Beyond India, especially in other Western countries, the introduction of fluoride into
drinking water confirmed the uniformity of fluoride's effects on tooth decay in children, at least across the major English-speaking nations.[29] In Canada and in Britain,
campaigns for fluoridation got underway during the 1950s. International agencies
also launched single-city fluoridation projects in the third world. By 1963, fluoridation advocates at the Fédération Dentaire Internationale (FDI) could boast that water
fluoridation plants were in operation or "will start in the near future in 41 countries."
Yet nowhere had fluoridation caught on to the extent it had in the United States. While
46 million Americans drank fluoridated water, fewer than 4 million Canadians did;

[27] Sahib, Raghavachari, and Venkataramanan, "Occurrence of Fluorides in Drinking Water-
Supplies" (cit. n. 22), 518. More generally, see Agarwal and Narain, "Dying Wisdom" (cit. n. 21).
[28] A. H. Siddiqui, "Fluorosis in Nalgonda District, Hyderabad-Deccan," *British Medical Journal* 2
(10 Dec. 1955): 1408–13; P. Venkateswarlu, D. Narayana Rao, and K. Ranganatha Rao, "Studies in
Endemic Fluorosis: Visakhapatnam and Suburban Areas," *Indian Journal of Medical Research* 40
(1952): 535–47; Amarjit Singh, S. S. Jolly, B. C. Bansal et al., "Endemic Fluorosis: Epidemiological,
Clinical, and Biochemical Study of Chronic Fluorine Intoxication in Panjab (India),"*Medicine* 42
(1963): 229–46.
[29] Subcommittee on Fluorine, *Fluorine Benefits and Risks* (cit. n. 2), 26.

next in line were Colombia, Chile, and Hong Kong. Striking exceptions appeared in Europe, where only 100,000 people in Britain and none in Germany had water artificially fluoridated.[30]

Around this same time, international controversy over just what the fluoride content of water should be led to a FDI appeal to the World Health Organization to compile "an authoritative and up-to-date report on the metabolism of fluorine." Though spurred by fluoridation advocates, the resulting 1970 publication, *Fluorides and Human Health,* nevertheless aspired to an "impartial review of the scientific literature" that had accumulated across the globe. Heavily reliant on, even dominated by, U.S.-based scientists, the publication became an international showcase for Indian researchers as well. Two Punjabi scientists, S. S. Jolly and A. Singh, wrote most of the section on "Toxic Effects of Larger Doses of Fluoride."[31] An official expression of the global ambitions of fluoridation investigators, the report also confirmed the arrival of waterborne skeletal fluorosis as a stable, recognized reality for scientific and medical elites in other parts of the world, including America itself.

Reconciliation between American and Indian depictions of fluoridated water came in part through working hypotheses by Indian researchers about what gave fluoridated water in the United States a less pathological punch: lower temperatures, ingestion of smaller quantities of water and better nutritional conditions, along with the lack of undissolved sediments in American water, and the lesser physical strain involved in most Americans' work. U.S.-based researchers, for their part, conceded that even in America, "optimal" levels of water fluoride might vary by temperature. Important as well were the ways fluoride researchers in both India and America responded to a traveling network of antifluoridationists, including doctors such as George Walbott. Walbott, who began spreading word about the Indian findings in the United States in the 1950s, claimed to have found skeletal fluorosis in his American patients.[32] To refute such claims, American researchers investigated U.S. communities exposed to high-fluoride drinking water and found little hint of skeletal fluorosis, while Indian medical scientists, sealing their credibility with the Americans, questioned whether the fluoride pathology claimed by Walbott was the same as that seen in highly fluoridated Indian patients. "The evidence from the clinical studies of fluorosis with regard to systemic intoxication," wrote Singh and Jolly, "is mostly of a negative nature, with the exception of dental, skeletal and neurological effects."[33]

Questions still hovered about just why Indian villagers seemed so much more vulnerable to fluoridated water than their American counterparts. But by weaving both national versions of fluoride health science into the same official WHO publication, Indian and American health elites crafted a new level of international stability for fluoridated water and its effects. Simultaneously, they joined forces against those people whose apparent impact most irked fluoridation advocates. While poverty, technical obstacles, and the continuing dangers of infectious disease understandably

---

[30] Leatherman and Ellis, "Fluoridation round the World," 27; Subcommittee on Fluorine, *Fluorine Benefits and Risks,* 22. (Both cit. n. 2.)

[31] WHO, *Fluorides and Human Health* (cit. n. 10), 11. A majority of the authors held American posts.

[32] George Walbott, with Albert W. Burgstahler and H. Lewis McKinney, *Fluoridation: The Great Dilemma* (Lawrence, Kans., 1978), 106.

[33] WHO, *Fluorides and Human Health* (cit. n. 10), 274–84, 287, on 263.

diverted the rest of the world from fluoride's virtues, what truly dumbfounded America's fluoridation experts was the public opposition that sprouted in their own backyards.

AGAINST FLUORIDE IN AMERICA:
WATER NATURALISM ON A LANDSCAPE IN FLUX

In the Stanley Kubrick film *Dr. Strangelove,* General Jack T. Ripper obsessed about the "communist" threat of fluoride to his "precious bodily fluids." If Kubrick crystallized the link in American popular culture between antifluoridation and cold war paranoia, many social scientists of the time looked to American society itself for explanations of the movement's "antiscientific" attitudes. Drawing on national statistical aggregates that showed antifluoridationism most rife among the uneducated and lower classes, researchers attributed those attitudes to a feeling of "loss of personal control" in a world of large and complex organizations caused by "relative deprivation" or viewed the attitudes as a "political protest by the powerless." The social scientists often emphasized the effectiveness of traveling antifluoridationists in getting their message across to lay audiences. This group of antifluoridationists comprised some individuals with medical or other expert credentials such as George Walbott and Howard Spira, though their numbers were inveterately cast as "small" by social scientists.[34] In investigating antifluoridationism, however, those scientists did not ask many questions about the ontology of water, disease, and health that led lay majorities to turn down fluoridation referenda in so many cities and towns, and the local and personal histories in which that ontology was grounded.

While in India some of the laypeople's ontology of fluoridated water comes through the investigators' reports of fluoride, in America, it is the public debates over fluoride referenda that offer a revealing window into the perceptions of the antifluoridationist laity. On May 9, 1958, the Manhasset-Lakeville water district in Nassau County, N.Y., rejected fluoridation by a whopping four-to-one margin. Manhasset-Lakeville defied social scientific generalizations about antifluoridation as lower class; an older, wealthier slice of Long Island, it had the highest assessed property valuation of any Nassau County water district. Unlike with health scientists in India, no particular ailment dominated the perception of fluoridated water among its many Manhasset opponents. Local antifluoridationist sentiment in this, as other areas, hinged on what the contemporary political scientist Morris Davis somewhat derisively termed "a *drive toward naturalism* or a *naturalist syndrome.*"[35] The "natural" served a purpose comparable to scientists' "retrofits"; it traced backward what, in this case, local memories confirmed as solid and real, what laypeople felt they knew about "quotidian" substances such as food and water and their health impacts. Constituting a modern version of Hippocratic ontology (oriented toward *Airs, Waters, and Places*), Manhassetites deemed familiar substances and practices natural, even if only recently (and by

---

[34] A. Green, "Ideology of Anti-Fluoridation Leaders," *Journal of Social Issues* 17 (1961), 13–25, on 16; W. Gamsom, "The Fluoridation Dialogue: Is It an Ideological Conflict?" *Public Opinion Quarterly* 25 (1961): 526–37; Arnold Simmel, "A Signpost for Research on Fluoridation Conflicts: The Concept of Relative Deprivation," *Journal of Social Issues* 17 (1961): 26–44; Benjamin Paul, "Fluoridation and the Social Scientist," ibid., 9.

[35] Morris Davis, "Community Attitudes toward Fluoridation," *Public Opinion Quarterly* 23 (1959): 478.

their own admissions) contrived. Lay antifluoridationism in the Manhasset-Lakeville district asserted what was "natural" not just in reaction to the fluoridation campaign, but also to longer-standing local developments that traveling advocates, pro and con, rarely addressed, from groundwater pollution to fundraising for disease research.

As with Indian villagers, Manhasset-Lakeville residents' contentions about fluoride were tightly bound to the history of their water supply. Since before the district's first waterworks, in 1910, drinking water had come from underground aquifers. These absorbed a rising influx of human waste as Nassau County suburbanized. Beginning in the 1930s, public health officials began warning residents of Manhasset-Lakeville about threats of groundwater contamination and depletion. In the following decade, nitrates began to appear in the upper layers of the aquifer. During and, especially, after World War II, new influxes of people into Nassau County—a 65.4 percent population gain between 1940 and 1950 and a 93.3 percent gain in the succeeding decade— placed additional strains on the county's underground water supply, strains only partly alleviated by the opening of a county sewage plant in 1949. The biggest problem continued to be the septic tanks and cesspools on which more than half of Nassau County homes relied, which leached wastewater into the local earth. Manhassetites' attunement to "chemical tastes" and pollutants in their water pushed officials there and in neighboring districts to dig deeper, so that chlorination was only intermittently necessary. Wells delved downward as far as 780 feet by the early 1960s, ensuring an official "natural purity" to the drinking water.[36]

Throughout the fifties, discussions of waterborne as well as other environmental threats from rapid postwar growth often sidestepped any mention of pathology or toxin through a less medicalized language of "contaminants," "stink," and "pollution." When disease terminology did surge into local headlines it was because of lay groups targeting single, specific diseases. In Manhasset as elsewhere in postwar America, local chapters of the American Cancer Society, the Cerebral Palsy Auxiliary, and the March of Dimes (against polio) emerged as prominent forms of civic engagement. Mostly centered around pathologies for which there were as yet no antibiotic or other decisive cures, these organizations reflected a long-term shift in the United States from infectious to chronic diseases (with the exception of polio) as major causes of mortality and predominant targets of medical concern. Unlike that of the Western age of infection, as well as that of contemporary India, the "preventive" component of midcentury American disease agitation was much more confined: report any suspicious symptoms to one's doctor. In stark contrast to today's breast cancer movement, the groups were silent about whether laypeople themselves could do anything to avoid these diseases; the unifying purpose of these organizations was to raise funds for research in distant medical centers and laboratories. Though the *raison d'être* for these disease groups appears to have been the cultivation of medical trust, they inadvertently brought home just how little doctors knew about these ailments.

The push for fluoride in Manhasset-Lakeville, preventive in orientation, also fix-

[36] Dick Wetterau, "Editorial Musings," *Manhasset Press,* 3 April 3 1958; "County Board of Supervisors Gets Sewer Report . . . ," *Manhasset Press,* 3 Jan. 1936; Department of Rural Sociology, Cornell University Agricultural Experiment Station, "The People of Nassau County, New York, 1900–1960," *Bulletin* no. 62–28 (Aug. 1963): 2; Nassau-Suffolk Regional Planning Board, *Utilities Inventory and Analysis* (1969), 23; Manhasset-Lakeville Water and Fire Districts, (1963?) typed manuscript Folder "L. I.–Water Supply," Manhasset Public Library, Manhasset, N.Y.

ated on a single disease and on expert control. District officials seized upon the idea sometime in 1957. Keenly aware of the controversy fluoridation had aroused elsewhere, they agreed to a local referendum, the first on Long Island. The ensuing public debate penetrated into the local social fabric through town newspapers as well as meetings of several civic associations. In contrast to a successful petition drive by a newspaper editor and other advocates in Levittown, another Nassau County community, these public presentations were structured as debates rather than expert "education." Spokespeople for the opposition, given time and opportunity comparable to those of fluoridation advocates, shattered the insistent focus of the latter on alleviating cavities.[37]

Local advocates in Manhasset included a dentist and presidents of the county medical and dental societies. Unlike their opponents, they did not write letters to the local papers; in debates, they probably presented arguments similar to those of Levittown's fluoride promoters. In Levittown, dental effects were depicted as contributing to overall healthiness—and savings. "If you have a child under 12 years of age, as most Levittowners do, there is something that can be done which will make him healthier, improve his appearance, add to his happiness—and save you hundreds of dollars in dental bills." Endorsement by "many dental societies" and other professional groups was critical. Quotes from local dentists were paraded alongside assertions of proven benefits, shown by "tests that were conducted over periods of years under the supervision of government, state and local health authorities, and official dental organizations." Advocates admitted that fluoride was a "poison" but insisted there were no "adverse effects" at the levels being proposed. Crucial to their case was their account of "natural" fluoride. Discovery of fluoride's beneficial effects came through study of natural levels that were too high in some places, causing mottling, and hence, also in need of adjustment. As many a science studies scholar has recognized, this kind of argument naturalized experts' politics alongside many of their reductions and results; claims about the "nature" of water served as arguments for an expert-controlled "normalizing" of water supplies.[38]

Fluoride's lay opponents rarely questioned whether fluoride was effective against tooth decay. They emphasized what else it did and, especially, *might do*—the shadowy possibilities of harm that PHS scientists by this time considered their own studies to have mostly dismissed. "In addition there is a host of other possible disorders such as gastrointestinal disturbances, spotty baldnes [sic], eczema, cardiac and respiratory disorders, etc. that may be traced to fluoride ingestion." For the Manhasset Pure Water Committee, fluoride was a "cumulative poison . . . involved in a great number of diseases such as arthritis, heart [and] kidney disease, diabetes and allergic disorders." The multiple pathologies invoked here were far more dire than tooth decay, yet they were also more tenuous and fleeting than that found in the traveling antifluoridationists' literature. Only disease names appeared; allusion to skeletal fluorosis was confined to fluoride-related "bone

[37] See, e.g., "Lake Success Debates Fluoridation," *Manhasset Press,* 27 March 1958; see also Sheila Jasanoff, "Science, Politics, and the Renegotiation of Expertise at EPA," *Osiris* 7 (1992): 195–217.
[38] "Fluorine in Our Water? Local Dentists Favor Plan," *Levittown Tribune,* 26 July 1951; "Fluorine in Our Water? It's Safe and Tasteless," *Levittown Tribune,* 9 Aug. 1951; Sidney Sussman, "Think Then Vote," *Manhasset Press,* 31 April 1958; Sheldon Levinson, "Approves Fluoridation," *Manhasset Press,* 8 May 1958.

changes" mentioned in a single letter. This disease had only the wispiest reality for Manhasset's fluoride opponents.[39]

Fluoride itself sounded disturbingly evasive to its opponents, especially in comparison with the more familiar chlorine. Most troubling, fluoride offered ordinary water drinkers fewer clues about when it reached excessive amounts. While high chlorine levels gave rise to an "irritant reaction on the eyes and nose," fluoride's odorlessness and tastelessness made it "not as easily discoverable except by qualitative and quantitative tests," that is, tests by experts.[40] In contrast to adding chlorine to water, and the natural presence of fluoride in the water of far-off Nellore, putting fluoride into faucets would erode laypeoples' ability to taste the risks in their drinking water. The Manhasset Pure Water Committee, broaching local anxieties about sewer seepage into groundwater, placed chlorine itself squarely on the side of water purity: "Sure, chlorine in the water to kill germs is necessary . . . pasteurization of food to kill germs is necessary . . . BUT FLUORIDE FOR MEDICAL PURPOSES IS ANOTHER STORY. . . . *It is mass medication for a non-communicable disease.*"[41] Forthright about their water's historical and human shaping, they readily admitted established and continuing public health measures against infectious diseases within the scope of their "natural," while attacking the unnaturalness of fluoridation.

Repeated accusations of "mass medication" exemplify how lay antifluoridationists slid from water ontology to explicit sociopolitical critique far more readily than fluoridation's advocates. The phrase echoed a widespread targeting of "mass society" in the national media, which took fluoridated Levittown as its suburban exemplar. Antifluoridationists indicted what they saw as an overstepping of proper jurisdictional boundaries between medical and lay decision-making. "Fluoridation is *not a medical problem*," stated the Manhasset Pure Water Committee. "It is simply a question of whether or not we want our water supply used for medication." Conceding that there were "scientific" and "medical" realms in which experts rather than laypeople were most competent to know and decide, many antifluoridationists were perfectly willing to accept fluoride through traditional clinical channels (e.g., medical prescription of fluoride tablets). As Mrs. George Conway argued, however, there were "many other aspects to this question [of fluoridation] besides the scientific aspects which were within the competence of lay persons to evaluate."[42] For her, it was the fluoridation scientists who were transgressing established boundaries by asserting that all or most aspects of water supply should be left to experts. Still worse was the precedent it would set for the local civics of specific diseases. The public water system would become a "happy hunting grounds for every group with a pet illness to cure."[43]

Antifluoridation talk was full of critical commentary about science and scientists: how in this case "professional opinion" seemed "divided," how arrogant and "brazen"

[39] Letters to the editor: Dinah Dever, "Twenty Years Ago . . . ," *Manhasset Press,* 27 March 1958; Mrs. George Conway, "Hits Fluoridation," *Manhasset Press,* 24 April 1958; Cecilia Schwartz, "Fluoridation Question," *Manhasset Press,* 17 April 1958; Dinah Dever, "Fluoridation Facts," *Manhasset Press* 8 May 1958 ("bone changes").

[40] Schwartz, "Fluoridation Question" (cit. n. 39).

[41] Manhasset-Lakeville Pure Water Committee, "Let's Really Protect Our Children's Teeth," *Manhasset Mail,* 1 May 1958.

[42] Ibid.; Conway, "Hits Fluoridation" (cit. n. 39).

[43] Pure Water Committee, "Protect Our Children's Teeth" (cit. n. 41); Schwartz, "Fluoridation Question" (cit. n. 39).

the scientific advocates of fluoride seemed.[44] But it was through appeal to their own memories—as a Mrs. Tangredi put it, "the practical, down to earth experience of a person who is unbiased"—that lay antifluoridationists reached to the heart of skepticism about fluoride. Al Fisher noted how "my wife and four children have gotten along very well without the magic chemical."[45] In claiming their own ability to recognize what was best when it came to fluoridated water, many were inclined to emphasize homebound ways of contending with cavities that they had long known and practiced. According to the Manhasset Pure Water Committee, prevention was "a matter of proper diet, dental care, cleanliness and heredity," mostly the mother's realm. The committee, "composed largely of mothers," saw fluoridation as a medical and governmental imposition on the private home, domesticity, and for Conway, women's prerogatives. "Should we not conceivably expect that the time will come when our government will supervise our children's meals and the hours at which we put them to bed?"[46]

A crucial step in many antifluoridationists' arguments was casting the assault on this naturally domestic and private realm as a threat to democracy: "[a] violation of the democratic right of every individual to drink what he wants and to take the medicine he needs." The rights fluoridation failed to respect were, quite literally, consumers' rights and, by extension, rights to bodily control. As Conway eloquently summed it up, "Let us at least retain the mastery of our own bodies if by so doing we improve or injure no one but ourselves!"[47] The political ideology of Conway and others, which inclined them to accuse fluoridationists of "socialism," seemed to confirm an imagined alliance of antifluoridationists to a global struggle against communism. Yet the wider embrace of antistatist arguments about body mastery, beyond a McCarthyesque right, suggests that in 1950s Nassau County, shared local history and experiences were at least as important. A few miles from Manhasset, Marjorie Spock schemed unsuccessfully to convince her brother Benjamin, the renowned pediatrician, to speak out against fluoridation and organized a lawsuit against a state-run DDT spray campaign. This first public trial against DDT provided evidentiary bases for Rachel Carson, then commencing research for what would become the book *Silent Spring*.[48] Not just the Goldwater right but postwar environmentalists owed important debts to the antifluoridation campaigns.

In view of this legacy, it is striking how, without any Carsonian appeal to ecology, Manhasset's antifluoridationists worried about environmental ramifications of fluoridation that ranged beyond the faucets on which its advocates fixated. Leery of the stable, tidy formulae of fluoridation science, opponents imagined fluoride flows and accumulations, especially in and around their home turfs, about which fluoride's advocates had little to say. The Manhasset Pure Water Committee noted how "fluoride becomes more concentrated when boiled or used for cooking . . . to add to the hazard." For Clara Harban, the prospect of widespread fluoridation conjured up a vision of innumerable other pathways by which this "cumulative poison" might undermine even the best intentions for a "wholesome" diet. "Can you imagine if the entire coun-

[44] Conway, "Hits Fluoridation" (cit. n. 39); Mrs. S. Tangredi, letter to editor, *Manhasset Mail,* 1 May 1958.

[45] Tangredi (cit. n. 44); Al Fisher, "Dear Editor," *Levittown Eagle,* 14 Feb. 1952.

[46] Conway, "Hits Fluoridation" (cit. n. 39).

[47] Ibid.

[48] Christopher Sellers, "Body, Place, and the State: The Makings of an 'Environmentalist' Imaginary in the Post–WWII U.S.," *Radical History Review* 74 (winter 1999): 31–64.

try becomes fluoridated there won't be a canned or frozen fruit, vegetable or soft drink that won't have fluorides in them. So again, how do you control the dose?"[49] Just as the Indian investigators of fluorosis looked to diet to help explain effects for which fluoridated water alone seemed insufficient, so antifluoridationists fretted that fluoridated water would spill over into dietary flows.

The practices that antifluoridationists defended as natural were time-tested, but far from eternal: how medical and dental clinics administered their medicine, how chlorine's bitter taste clued people in to elevated water levels, how dietetic and hygienic supervision could allay cavities in children's teeth. Touting the "naturalness" of these practices not only pointed up the patent artifice of fluoridated water but also steered the arguments away from terms and terrain on which fluoridation science had evolved: dose-response curves between fluoridated water and pathology, forged on populations elsewhere. All of the practices antifluoridationists deemed natural had histories, likely including recommendations by earlier experts; the timeless, static cast of their appeals to "nature" was thus deceptive, to contemporary social scientists and many antifluoridationists themselves. Naturalism absolved them from delving into the personal histories that solidified their confidence in certain ways of handling their families' water, food, and medicine; at the same time, it affirmed the importance of *these* histories to the fluoridation issue. Even if extending back to childhoods elsewhere, in contrast to the messages of either fluoride scientists or traveling antifluoridationists, these histories were more localized, more these suburbanites' own.

Antifluoridationists' more ecologically-based questions about fluoridation science proved prescient. Braking the spread of fluoride through public water pipes, antifluoridation channeled a flood of fluoride into Americans' bodies through private consumption—drugstores and supermarkets rather than water faucets. In toothpastes, mouthwashes, and vitamin supplements, fluoride rapidly evolved from a marketing tool to a near prerequisite for American manufacturers. By the mid-1980s, 90 percent of toothpaste sold in America contained fluoride. The fluoride choices available to American consumers when it came to toothpaste verged on the compulsory—just what antifluoridationists had feared from state-fluoridated water.[50] Though controlled community studies have continued to confirm the preventive effects of mildly fluoridated water on tooth decay, the impacts measured more recently have been two-thirds to one-half of what PHS scientists measured in the 1940s. The reasons cited are largely what early antifluoridationists anticipated. With so many toothpastes and mouthwashes now prepared with fluoride supplements, and with so many foods prepared with fluoridated water, even those Americans without fluoridated drinking water take in much more fluoride than did their 1950s counterparts.[51]

CONCLUSION

Examining the historical ontology of fluoridated water among these three separate groups over the mid twentieth century highlights not so much their borrowing from as

[49] Pure Water Committee, "Protect Our Children's Teeth" (cit. n. 41); Clara Harban, letter to editor, "Why Fluoridation?" *Manhasset Press,* 10 April 1958.
[50] E. D. Beltran and S. M. Szpunar, "Fluoride in Toothpastes for Children: Suggestion for Change," *Pediatric Dentistry* 10 (1988): 185–8.
[51] Alan Hinman, Gene Sterritt, and Thomas Reeves, "The US Experience with Fluoridation," *Community Dental Health* 13 (1996): S5–9.

their indifference to one another. If American fluoridation science drew heavily upon the European literature early on, it was largely by ignoring, rather than addressing, the findings of Indian scientists that U.S. scientists concocted such a strong case for fluoridation. Lay Manhassetites, bolstered by confidence in their own personal experiences—often local, longstanding, and homebound—turned a deaf ear not just to American fluoridation experts but to an Indian science of skeletal fluorosis, despite the number of traveling antifluoridationists who took it as ammunition for their side. Most responsive among these three groups were the Indian scientists, whose findings ran against the grain of the quickly proliferating studies of American profluoride scientists. The recalcitrant, widening presence of skeletal fluorosis nevertheless helped sustain the Indian ontology of fluoridated water in a very different orientation from its American counterpart, as more problem than solution.

If a geographically circumscribed historicism illuminates the peculiar ontology of each group, careful readers will note how I have brought their indifferences alongside one another precisely by rooting each within a larger environmental history of fluoride's varied distribution and flow. What we might term an environmental symmetry linked their separate, distant locations, between an America "naturally" fluoridated only in pockets, where fluoridation took shape as a public health project, and the highly fluoridated countryside where Indian scientists uncovered skeletal fluorosis. Even in Manhasset-Lakeville itself, among lay antifluoridationists who played such key roles in steering fluoride's spread, minimal fluoride levels shaped the strengths as well as the weaknesses of their arguments. Their ontology from unfluoridated personal experience invited self-consciously "natural" labels for diet, drink, and spillage, along with an integrative analysis that we can, in retrospect, identify as ecological. But fluoride's diseases proved more fleeting and unstable in this lay ontology than for either group of scientists; without local or personal memories of fluoride's actual effects, Manhasset-Lakeville and other communities were forced to choose between more or less distant and competing echoes of the effects of fluoride ingestion. That so many did embrace fluoride's merits, unlike Manhassetites, had profound longer-term consequences on fluoride's global circulatory system. Over the late twentieth century, a gathering stream of fluoride passed through supermarkets and water pipes within America and beyond, raising fluoride exposures worldwide.[52]

The impacts of this human-mediated tide have been felt among investigators in both nations. American scientists' version of fluoride has destabilized in important ways, forcing questions that their predecessors had thrust aside. As rates of dental fluorosis have grown in many fluoridated communities, as environmental laws, agencies, and health disciplines have also expanded, health scientists and advisory committees have turned to emphasizing "the influence of sources of fluoride other than water," especially diet. A few confirmed cases of skeletal fluorosis in the United States, along with studies of fluoride's effects on cancer and reproduction, helped spur the U.S. Environmental Protection Agency to set new maximum drinking-water standards for fluoridated water. In specialized publications, some U.S. as well as Indian scientists still puzzle over why there were so many fewer cases of skeletal fluorosis in America than in India when exposures were comparable. Yet only starting in the late 1990s did

---

[52] C. E. Renson, "Changing Patterns of Dental Caries: A Survey of 20 Countries," *Annals of the Academy of Medicine of Singapore* 15 (1986): 284–98.

editions of prominent textbooks in environmental medicine and preventive dentistry make any mention of crippling pathologies of fluoride beyond American shores.[53]

In India, despite redoubled efforts to defluoridate water supplies, the risk of skeletal fluorosis has grown and spread as more people partake of dentifrices, beverages, and other goods from the fluoridated West. Outside the medical centers specializing in skeletal fluorosis, Indian public health officials are frustrated in their efforts by the limited awareness of this disease among local doctors and health officers. Not just material flows of fluoride, but the cultural flows of textbooks and other medical and dental literature from the West, which ignore this endemic disease and tout water fluoridation, stand in the way of taming its ravages.[54] On the far side of the globe, the most fearful imaginings of antifluoridationists have become a dreadful reality.

[53] Ibid., 46–47; National Research Council, Subcommittee on Health Effects of Ingested Fluoride, *Health Effects of Ingested Fluoride* (Washington, D.C., 1993); William Rom, *Environmental and Occupational Medicine* (Philadelphia, 1998), 1078; Norman Harris and Franklin Garcia-Godoy, *Primary Preventive Dentistry* (Stamford, Conn., 1999), 169; Institute of Social Sciences, *Fluorosis in India* (cit. n. 2).

[54] Institute of Social Sciences, *Fluorosis in India* (cit. n. 2), 38–9, 91.

EXPOSURE AND INVISIBILITY

The Fruits of Ill-Health:
Pesticides and Workers' Bodies in Post–World War II California

By Linda Nash*

ABSTRACT

In the postwar period, modernist frameworks of the human body, which described the body as both cosmopolitan and separated from its environment, competed with ecological frameworks that constructed the body as inherently porous and tightly linked to the surrounding world. The history of pesticide-related illness among farmworkers, and the gradual recognition that pesticides posed a new kind of public health problem, illustrates how these competing understandings were adopted, mobilized, and applied by different groups, as well as how politics shaped the emergence of new medical facts. New forms of illness generated new knowledge about the modern landscape and made visible material links between bodies and their environments.

INTRODUCTION

The orchard has long been associated with health in California's promotional literature. Nineteenth-century agriculturists and boosters alike extolled horticulture as a means for improving the landscape and the bodies of those who labored there. For consumers, the link between fruit and health would be sustained down to the present day through the advertising campaigns of the Sunkist cooperative and other producers. But as farming practices themselves changed, the reputed healthfulness of the farm and orchard environment no longer matched the reality if, in fact, it ever had. By the late twentieth century, dangerous machinery, harsh working and living conditions, and the introduction of greater and greater quantities of chemicals had rendered California orchards and fields among the most hazardous places to work. Certainly for those who labored in agriculture in the decades after World War II, the modern orchard became a location that constantly threatened their physical well-being. While postwar consumers were likely to associate oranges with vitamin C and a healthy diet, those who harvested the fruit had a radically different perspective. As one farmworker told an interviewer in 1969, "Whenever I pick oranges, I feel so bad; my mouth feels sour and dry from the nose all the way down to the stomach. It is so bad that I can't even eat."[1]

* Department of History, Box 353560, University of Washington, Seattle, WA 98195; lnash@
u.washington.edu.
   I would like to acknowledge the comments of Christopher Sellers and Simon Werrett, as well as those of the participants in the seminar "Environment, Health, and Place in Global Perspective," convened at the University of Wisconsin–Madison in April 2002.
   [1] California Department of Public Health (CDPH), *Community Studies on Pesticides,* 15 Dec. 1970, 14.

Tracing the history of the pesticide-related illness in the postwar decades makes visible the parallel effects of modernization on the landscape and the bodies of working people. In doing so, this essay considers not only the agency of humans but also the agency of nature, not only the vulnerability of the environment but also the vulnerability of human bodies. As laboring bodies transformed the orchard through pruning, picking, planting, and spraying, the orchard transformed those same bodies, often in less visible, but no less material, ways.

The principal objective of this essay, however, is to consider the ways in which politics and cultures of knowledge make visible or invisible a link between the health of human bodies and the condition of the surrounding environment. The "fact" of pesticide toxicity to agricultural workers was one that emerged into popular and political consciousness quite slowly. Though the first fatal poisonings were reported in the late 1940s, the issue received little attention until the late 1960s and no serious regulation until the 1970s. Even then, the risks faced by farmworkers received little attention in comparison with other environmental health issues.

Why did the effects of pesticides on working people remain invisible to so many for so long? Although the invisibility of pesticide poisoning had multiple sources, in this essay I focus on the discursive strategies and conceptual models of modern medicine, particularly that of occupational health, the discipline in which the problem of pesticide poisoning was first articulated. Over the course of the 1950s and 1960s, occupational health specialists would assume primary responsibility for defining and regulating the relationship between bodies and environments on the modern farm. Paradoxically, however, the methods and assumptions of their discipline both revealed and obscured the connection between the modern agricultural environment and the illnesses of workers. Visibility and invisibility were produced together.

In speaking about the post–World War II boom in pesticide production, the historical as well as the popular emphasis has been on DDT and the organic chlorinated hydrocarbons made household words in Rachel Carson's *Silent Spring*. Popular thinking still holds that the rapid introduction of pesticides after the war hinged on the fact that little or nothing was known about their harmful effects. In this view, it is the biochemical characteristics of DDT that account for its wholesale introduction and adoption. Despite its ability to accumulate and concentrate in certain plants and animals, and even in human tissue, it has a relatively low *acute* toxicity. DDT had been used on U.S. troops during the war, after all, and was credited with saving as many as 5 million lives. Toxicological ignorance, while a questionable explanation for the government's failure to restrict the use of DDT, cannot account for the simultaneous introduction of organophosphate (OP) pesticides. German researchers discovered and produced these compounds during World War II as part of that nation's chemical weapons program; scientists immediately recognized that OP chemicals were powerful human neurotoxins.[2]

---

[2] Edmund Russell, *War and Nature: Fighting Humans and Insects with Chemicals from World War I to Silent Spring* (New York, 2001); Thomas R. Dunlap, *DDT: Scientists, Citizens, and Public Policy* (Princeton, N.J., 1981); Margaret Humphreys, "Kicking a Dying Dog: DDT and the Demise of Malaria in the American South, 1942–1950," *Isis* 87 (1996): 1–17; John H. Perkins, *Insects, Experts, and the Insecticide Crisis: The Quest for New Pest Management Strategies* (New York, 1982); Christopher J. Bosso, *Pesticides and Politics: The Life Cycle of a Public Issue* (Pittsburgh, Pa., 1987); and Edmund Russell, "The Strange Career of DDT: Experts, Federal Capacity, and Environmentalism After World War II," *Technology and Culture* 40 (1999): 770–96.

As with DDT, chemical manufacturers introduced organophosphate compounds to the domestic market immediately after the fighting stopped, and American farmers were eager customers. By 1949, growers in California were applying parathion and tetraethyl phosphate (TEPP) to their fields. In 1951, they added demeton (systox) and EPN; in 1953, malathion and chlorthion; and in 1956, dipterex and the extremely toxic metacide. In 1958, several serious poisonings accompanied the introduction of thimet into cotton production. That year also saw the introduction of tetram, disulfuton (di-syston), mevinphos (phosdrin), azinphosmethyl (guthion), and carbophenothion (trithion). By 1963, more than 16,000 pesticides had been registered in California, and farmers had become increasingly reliant upon multiple applications of multiple chemicals. By the late 1960s, a typical California walnut farmer was spraying her trees with multiple chemicals several times a year: once with a copper sulfate compound to control blight; once with guthion to control codling moths; and once with trithion or parathion to control mites.[3] Synthetic chemicals quickly became a critical component of the environment on a modern California farm.

The toxicity of OP compounds resides in their ability to inhibit the action of cholinesterase, an enzyme critical to the normal functioning of the nervous system. The result is hyperexcitability, which may be observed as muscle twitches, tremors, convulsions, bronchial spasms, constriction of the pupils, abdominal cramps and vomiting, irregularities in heart beat, and in extreme cases, respiratory paralysis and cardiac arrest. Little is known about the chronic effects of these compounds, though mounting evidence suggests they may produce delayed neurological problems. In addition, both farmers and farmworkers are at increased risk for several types of cancer, which some researchers suggest is linked to their high rates of pesticide exposure.[4]

What is known is that systemic poisonings and death went hand in hand with the introduction of these pesticides into peacetime agriculture. Although poisonings would quickly become, and remain, a transnational problem, most information on the health dangers of pesticides in the 1950s and 1960s would emerge from California, more particularly from the Central Valley. In 1949, in a single incident near Marysville, California, twenty-five pear pickers became seriously ill after entering an orchard that had been sprayed with parathion twelve days earlier. By September of that year, the California Department of Public Health (CDPH) was aware of 300 cases of poisoning by agricultural chemicals within the state, including two deaths.[5]

From the beginning, poisonings had been linked primarily with fruit crops, which partly accounts for their prevalence in the Central Valley. Nature and history have combined to make California one of the premier regions in the world for growing fruits and vegetables. By 1899, California was already growing more fruit than any

[3] CDPH, *Community Studies on Pesticides,* 2 Feb. 1972, table 4.

[4] Howard W. Chambers, "Organophosphorus Compounds: An Overview," *Organophosphates: Chemistry, Fate, Effects,* ed. J. E. Chambers and P. E. Levi (San Diego, Calif., 1992), 3–17. The first evidence that OP pesticides could create long-term neuropsychological problems was reported in 1961 by Australian researchers. Lakshman Karalliedde, Stanley Feldman, John Henry et al., eds., *Organophosphates and Health* (London, 2001), xxiii. See also Marion Moses, "Pesticide-Related Health Problems and Farmworkers," *AAOHN Journal* 37 (1989): 115–30; Devra Lee Davis, Aaron Blair, and David G. Hoel, "Agricultural Exposures and Cancer Trends in Developed Countries," *Environmental Health Perspectives* 100 (April 1992): 39–44; and Aaron Blair and Shelia Hoar Zahm, "Cancer among Farmers," *Occupational Medicine: State of the Art Reviews* 6 (1991): 335–54.

[5] Herbert K. Abrams, "Occupational Illness Due to Agricultural Chemicals, 1949," *California's Health,* 15 Sept. 1950, 35.

other state in the nation. By the end of World War II, the state was producing 33 percent of the nation's pears, 42 percent of the peaches, 50 percent of the oranges, 90 percent of the grapes, and 100 percent of the lemons, olives, avocados, apricots, almonds, and artichokes. The specialization of California farms in horticulture would grow only more pronounced over the following decades.[6]

As farmers introduced a multitude of new plant varieties into the state, they also introduced new insects, which they struggled to control. In California, many introduced insects were not subject to the same ecological limits that constrained their populations in their native locales. The absence of typical predators, the mild climate, and abundant food supplies allowed many insect species to proliferate. By the early twentieth century, California fruit growers were contending with scales, beetles, thrips, moths, phylloxera, aphids, spider mites, and peach-tree borers. The prevalence of insect problems prompted many to turn to chemical control as early as the 1880s. Growers' adoption of strict standards for cosmetic quality in the early twentieth century exacerbated the rapid turn toward chemical compounds. In the 1950s and 1960s, California farmers rapidly converted to the new organic chemicals that appeared on the market. Throughout this period, they applied pesticides on more acres and in larger quantities than did their counterparts elsewhere in the country. Estimates from the 1960s typically put the state's share of pesticide usage at 20 percent of the national total, but no one knew the actual amounts applied.[7]

From 1901 until the 1970s, the department solely in charge of researching and regulating the use of pesticides in California was the state Department of Agriculture. The concerns of this agency, however, lay primarily with the efficacy of pesticides rather than their effects on human health. Staffed with agronomists, entomologists, and, increasingly, chemists, the agriculture department maintained no expertise in public health or medicine. Meanwhile, twentieth-century medicine had largely rejected the environmentalist focus of earlier eras, focusing instead on the isolated human body and particular infectious agents. The only disciplinary space in which a strong link between environment and human health remained was in the relatively marginalized field of occupational health. It was within this disciplinary space that the health effects of OP pesticides were articulated with growing concern after World War II.[8] It was also within this space that the fact of pesticide poisoning would remain confined until the late 1960s.

Though not typically considered as such, occupational health (or industrial hygiene as it was referred to in the early part of the century) was one of the earliest of the "environmental sciences." Since its inception, the discipline had highlighted the link be-

---

[6] Steven Stoll, *The Fruits of Natural Advantage: Making the Industrial Countryside in California* (Berkeley, Calif., 1998); Miriam J. Wells, *Strawberry Fields: Politics, Class, and Work in California Agriculture* (Ithaca, N.Y., 1996), 19–37; and Paul Rhode, "Learning, Capital Accumulation, and the Transformation of California Agriculture," *Journal of Economic History* 55 (1995): 773–800.

[7] Stoll, *Fruits of Natural Advantage* (cit. n. 6), 98; Martin Brown, "An Orange Is an Orange," *Environment* 17 (1975): 6–11. California fruit growers, under the auspices of their marketing cooperative (Sunkist), first developed standards for cosmetic quality in the 1910s. See Richard C. Sawyer, *To Make a Spotless Orange: Biological Control in California* (Ames, Iowa, 1996), 34–6. For 20 percent number, see Robert Z. Rollins, "Federal and State Regulation of Pesticides," *American Journal of Public Health* 53 (1963): 1427–31.

[8] For a history of pesticide regulation in California, see http://www.calepa.ca.gov/About/History01/dpr.htm (accessed 27 May 2003); Rollins, "Federal and State Regulation of Pesticides" (cit. n. 7). Comments on occupational health are drawn in part from Thomas H. Milby, M.D., former CDPH director, telephone interview by author, 1 Oct. 2002.

tween the health of bodies and the condition of their environment. Emerging as a progressive response to nineteenth-century industrialization, industrial hygiene focused on the diseases peculiar to certain trades, concerning itself not merely with the body of the worker but also with the work environment: the modern factory. It owed much to nineteenth-century theories of environmental disease, which had assumed that environment shaped the physical and moral qualities of individuals. As the discipline evolved, practitioners and researchers followed the lead of medical bacteriologists and focused increasingly on identifying and monitoring specific contaminants in the workplace. In the 1920s and 1930s, industrial hygienists turned toward laboratory research, using animal studies to establish maximum safe-concentration levels for particular chemicals. Yet even as the scope of the discipline narrowed and it dissociated itself from its earlier relationship to environmental medicine, industrial hygienists continued to focus on how the (work) environment affected health. In this endeavor, occupational health stood apart from most other realms of medical understanding in the mid-twentieth century.[9]

In the late 1940s, the problem of pesticide poisoning gradually drew a few of California's occupational health specialists onto the unfamiliar terrain of modern agriculture. Not part of the traditional domain of industrial medicine, agriculture was perceived, however inaccurately, as work that took place in a natural, rather than a man-made, environment, that was inherently healthful, not hazardous. Yet as agriculture became industrialized, it took on many of the characteristics of the factory: mass production, standardized products, wage labor, workforce segmentation, and corporate ownership. Among these industrial characteristics was the increasing rate of work-related injuries and illnesses. The mass-poisoning incident in Marysville in 1949 prompted California's Bureau of Adult Health to begin publishing annual statistics on the incidence of occupational disease attributable to agricultural chemicals. Throughout the 1950s, reports of pesticide-induced occupational disease increased within the state. By 1963, CDPH was reporting that agriculture had the highest rate of occupational disease in the state, more than 50 percent higher than that of any other industry. These statistics did not even take into account the vast underreporting of pesticide-related illness.[10]

UNRULY ENVIRONMENTS

As occupational health experts moved onto the farm, what they found was that fields and orchards were not fully amenable to their control—far from it. The traditional methods of occupational health relied upon monitoring workers' bodies for signs of illness and limiting harmful exposures, either by modifying industrial processes and work routines or by providing protective equipment to workers. While occupational health experts acknowledged the importance of environment to health, they also assumed that the work environment was fixed, relatively predictable, and ultimately amenable to their control. A questionable assumption even for the most

[9] On these developments within the field, see Christopher Sellers, *Hazards of the Job: From Industrial Disease to Environmental Health Science* (Chapel Hill, N. C., 1997). See also Jacqueline Karnell Corn, *Response to Occupational Health Hazards: A Historical Perspective* (New York, 1992).

[10] Irma West, "Occupational Disease of Farm Workers," *Archives of Environmental Health* 9 (1964): 92–8; CDPH, *Reports of Occupational Disease Attributed to Pesticides and Agricultural Chemicals, California, 1950* (Berkeley, Calif., [1951?]).

stable of factory environments, it bore no relationship to the reality of agriculture in postwar California. For instance, toxicology assumed that the quality of toxicity was inherent to an isolated chemical substance and that data on chemical effects forged in the laboratory could be applied to any environment. In the fields of California, however, the local environment repeatedly escaped the descriptions of a delocalized laboratory science. As one investigator later lamented, in practice the amount of toxic residue on a given crop depended upon "the vicissitudes of environmental factors."[11] Place mattered in multiple ways, and the unpredictability of the natural world continually frustrated those who sought to describe and manage the agricultural environment.

From the beginning, poisonings had been clustered geographically in hot, arid regions, suggesting the importance of climatic factors. A study of parathion decay conducted in the 1970s revealed that pesticide residues in the same field could vary as much as 90-fold, depending upon the time of year the chemical was applied. The toxicity of any given residue might also be a function of weather. In 1963, more than ninety peach pickers in the northern San Joaquin Valley became severely ill, an incident that made front-page news. Yet sampling for parathion residues in the orchard suggested that levels were not high enough to produce acute effects in workers, which led investigators to suspect that a degradation product of parathion was responsible. Later work would establish the role of the "oxygen analogs" of organic phosphates. Created under conditions of intense sunlight, these compounds have a toxicity ranging from two to hundreds of times the toxicity of their predecessor chemical.[12] The California climate could produce both superior fruit and extremely toxic substances; the production of toxicity, like the yield of peaches, varied from year to year and from field to field. Once introduced into the environment, OP chemicals were subject to the uncontrolled agency of nature.

As researchers pursued the issue of worker exposure and pesticide toxicity, they found a multitude of significant environmental "variables." Important climatic factors included not only temperature, but rainfall, wind velocity, incident radiation, and humidity. Relevant as well was the type of soil: soils with high clay content were likely to bind the pesticides and slow their dissipation. Soil moisture, on the other hand, could increase the dispersion rate. Perhaps not surprisingly, then, studies of pesticide levels in different orchards located within the same area often yielded dramatically different results even when application rates had been the same. Seemingly every environmental factor researchers thought to consider had some effect on toxicity. The type of crop had a significant effect—citrus, peaches, and grapes were among the most likely to generate toxic exposures. But when researchers looked beyond the basic crop type, they found that the variety could also influence toxicity: a Minneola tangelo did not interact with a pesticide in the same way as an Orlando tangelo, nor a Temple orange in the same way as a Washington navel. As one pair of investigators

[11] William J. Popendorf and John T. Leffingwell, "Regulating OP Pesticide Residues for Farmworker Protection," *Residue Reviews* 82 (1982): 126; Jeffrey M. Paull, "The Origin and Basis of Threshold Limit Values," *American Journal of Industrial Medicine* 5 (1984): 227–38.

[12] F. A. Gunther, "Insecticide Residues in California Citrus Fruits and Products," *Residue Reviews* 28 (1968): 1–120; Thomas H. Milby, Fred Ottoboni, and Howard W. Mitchell, "Parathion Residue Poisoning among Orchard Workers," *Journal of the American Medical Association* 189 (1964): 351–6.; and Robert C. Spear, "Report of the Status of Research into the Pesticide Residue Intoxication Problem in the Central Valley of California," in *Pesticide Residue Hazards to Farm Workers: Proceedings of a Workshop Held February 9–10, 1976* (Salt Lake City, Utah, 1976), 43–62.

concluded, pesticides were so sensitive to environmental conditions that even when they had been applied in precisely the same manner, the residue levels could be highly variable. What was the weather? What types of plants were involved? Had it rained? Had it been windy or calm? Was it foggy or sunny? Was the field located in a valley? On a hill? On a north slope or a southern one? It was this environmental "sensitivity" that made mass OP poisoning events an unpredictable occurrence.[13] Toxicity was not simply a quality of a given chemical but a relationship between that chemical and the environment in which it was applied. The gridded fields, the engineered irrigation canals, the neat rows of cotton, lettuce, and orange trees lining the Central Valley all suggested a landscape highly ordered and eminently under human control. Mass outbreaks of illness among workers revealed that the modern farm was actually a chaotic, unpredictable ecology.

Like environments, bodies, too, were unruly entities that repeatedly escaped surveillance mechanisms. In contrast to the assumption of bounded and stable spaces that underlay the models of occupational health, the space that most California farmworkers occupied was, in reality, always discontinuous. Farmworkers are overwhelmingly migrants, constituted by the flows of a global labor market that recruited the most exploitable workers to pick the state's most profitable agricultural products. Unlike most modernized agriculture, which is characterized by high levels of machine mediation (think of cotton or wheat harvesting, in which a single worker steers a large machine among the rows), fruit and vegetable harvesting has never been fully amenable to mechanization. Even today it remains a labor-intensive enterprise. While the number of farmworkers declined nationally in the postwar period, it increased in California. Although some workers were permanent residents of the state, most moved in search of seasonal work at some point during the year. Others, such as the braceros—temporary workers brought into the United States from Mexico under a bilateral agreement during and after World War II—had no permanent homes in the state. Regardless of where they came from, most harvesters did not work directly for a grower; they worked for a labor boss who contracted to multiple growers. The workforce on a California farm was always changing, and workers labored in multiple environments. Migrants passed from field to field, county to county, state to state, often not knowing where they would be the following day, unable to recall all the places they had already been.[14] Occupational health professionals found they could not even begin to calculate worker exposures nor track and test worker bodies. Movement itself obscured the relationship between bodies and environments, between sick workers and modern orchards. Throughout the 1950s and 1960s, public health officials in California would complain that mobile bodies could not be adequately monitored or studied.[15]

[13] For a summary of this research, see Popendorf and Leffingwell, "Regulating OP Pesticide Residues" (cit. n. 11).

[14] According to Miriam J. Wells, in the 1990s, an estimated 78 percent of all farmwork in California was performed by hired workers. See Wells, *Strawberry Fields* (cit. n. 6), 24. For history and working conditions in California agriculture, see Stoll, *Fruits of Natural Advantage* (cit. n. 6), 124–54; Carey McWilliams, *Factories in the Field: The Story of Migratory Farm Labor in California* (Boston, 1939); Cletus Daniel, *Bitter Harvest: A History of California Farmworkers, 1870–1941* (Berkeley, Calif., 1981); Ernesto Galarza, *Farm Workers and Agri-business in California, 1947–1960* (Notre Dame, Ind., 1977); and Henry P. Anderson, *The Bracero Program in California, with Particular Reference to Health Status, Attitudes, and Practices* (Berkeley, Calif., 1961).

[15] Dr. Irma West lamented the "wasted opportunity for research." See West, "Pesticides and Other Agricultural Chemicals as a Public Health Problem with Special Reference to Occupational Disease in California," 18 July 1963, California State Library, Sacramento, Calif., 10–1.

Those bodies that could be located proved resistant to attempts to describe the effects of exposure. Though industrial toxicologists recognized in theory that the response of individual bodies to particular compounds might vary considerably, in practice they assumed the inherent similarity of all bodies, that a given exposure would generate a predictable effect.[16] In reality, workers' responses to pesticide exposure differed immensely, and attempts to quantify the effects of particular exposures seemed to yield only more variables requiring quantification. The absorption of pesticides varied in the field with work rate, work style, personal habits, and the type of clothing harvesters wore. Most disturbing was the recognition that the organophosphates had cumulative effects, which rendered exposed workers more susceptible to future exposures.[17] Moreover, researchers discovered that certain pesticide combinations had dangerous synergistic effects. Malathion, for instance, one of the OP pesticides least toxic to human beings in isolation, can become highly toxic in the presence of certain other chemicals. Yet the proliferation of new agricultural chemicals and the constant mobility of the most exposed individuals made it impossible for state professionals to record what they termed a worker's "exposure history."

An interesting phrase, "exposure history" asserts, against the normalizing tendencies of modern toxicology, the relevance of the history of an individual body. Toxicity, in other words, was not simply the result of the interaction between a given chemical and those qualities of human bodies presumed to be essential or obvious (gender or weight, for example), but of the contingent histories of the bodies in question. Had they been exposed to parathion before? How much and for how long? Where else had they worked, and under what conditions? What other chemicals had they been exposed to? Was their blood cholinesterase already depressed? By how much? The quality of "toxicity" was, in fact, a highly complex relationship among a particular chemical, the surrounding environment, and a particular body with its own history of exposures and injuries. Given the variability of toxicity among individuals, occupational health experts insisted that the preferred method for determining the existence of OP poisoning was a blood test for cholinesterase levels, a screening test suggested in 1950.[18] Yet as investigators seized upon cholinesterase as a kind of litmus test for exposure in the early 1960s, they came to realize that even "normal" cholinesterase levels varied widely among individuals. Moreover, levels of cholinesterase in the blood were only an approximation—not always a good one—for levels in the brain, which most researchers felt was the real variable of interest. While everyone agreed that OP exposure depressed cholinesterase, there was considerable disagreement over how much depression was significant. In a few cases, doctors even found that blood levels of cholinesterase could be normal, despite the presence of severe symptoms.[19]

---

[16] Paull, "Threshold Limit Values" (cit. n. 11); Robert Proctor, *Cancer Wars: How Politics Shapes What We Know and Don't Know about Cancer* (New York, 1995), 153–73.

[17] William J. Popendorf, "Exploring Citrus Harvesters' Exposure to Pesticide Contaminated Foliar Dust," *American Industrial Hygiene Association Journal* 41 (1980): 652–9; Robert C. Spear, David L. Jenkins, and Thomas H. Milby, "Pesticide Residues and Field Workers," *Environmental Science and Technology* 9 (1975): 308–13. For comment on cumulative effects of parathion exposure, see CDPH, *Occupational Disease in California Attributed to Pesticides and Agricultural Chemicals, 1959* (Berkeley, Calif., 1961), 7.

[18] D. Grob, W. L. Garlick, and A. M. Harvey, "The Toxic Effects in Man of the Anticholinesterase Insecticide Parathion (p-Nitrophenyl Diethyl Thionophosphate)," *Johns Hopkins Hospital Bulletin* 87 (1950): 106–29.

[19] "Discussion" in *Pesticide Residue Hazards* (cit. n. 12), 73–6.

The human body, like the natural environment, was unpredictable and resistant to quantification. It exhibited an agency of its own that escaped both conscious control and scientific description. As biomedical and environmental knowledge increased, so did uncertainty.

The ability, or perhaps inability, of occupational health to produce dependable knowledge about pesticide poisoning was conditioned by political and intellectual factors. In the 1950s, occupational health officials were unwilling to confront the power of California's agricultural interests directly, in part because they saw themselves as disinterested experts, with no particular politics. Yet despite the studied neutrality of CDPH reports, many field investigators slowly came to recognize that the hazards of pesticides could not be separated from the political economy of farm labor. In a 1963 report on the occupational hazards of agricultural chemicals, the agency asserted that farmworkers' limited legal protections and lack of unionization significantly increased the biological hazards of pesticides. In other words, officials within CDPH gradually acknowledged that, at some level, pesticides were a political, as well as a medical, issue.[20]

It was not only the hesitancy of the California health establishment to engage political questions that limited its ability to make pesticide poisoning visible, however, but also its reliance on modernist models of the body. Though the discipline of occupational health necessarily recognized a connection between bodies and environments, the two remained distinct entities. In the prototypical exposure model that gradually emerged, investigators focused on four critical variables: the amount of pesticide applied, the amount remaining in the orchard when workers began harvesting, the amount that entered the worker's body, and the response of the body to different amounts of exposure (see Figure 1). In such a model, the connection between body and environment is clearly indicated by the line that connects "environmental residue" to "worker dose." Nonetheless, bodies and environments are confined to their own distinct boxes, which are punctuated only by a slender arrow. This iconography is revealing. When investigators spoke of "exposure pathways," they implied that such pathways were narrow routes of entry that could be regulated or even blocked.[21] The occupational health emphasis on discrete boundaries marks the modernist desire to see bodies as both cosmopolitan and separate, or at least separable, from their environment. Such models suggested that the impact of chemicals upon health was ultimately controllable. The goal of occupational health, after all, was to engineer compatibility between fragile human bodies and the industrial work environment.

Critical to understanding the prevalence of pesticide poisoning in California, however, were the ways in which workers' bodies became completely intermixed with their work environment. Observations made in the field revealed the limits of the language of "environmental residue" and "exposure pathway." As one occupational health expert described it:

---

[20] Milby interview (cit. n. 8); CDPH, *Occupational Disease in California Attributed to Pesticides and Agricultural Chemicals, 1963* (Berkeley, Calif., [1964?]).

[21] For discussions of modern conceptions of the body, see Emily Martin, "The Body at Work: Boundaries and Collectivities in the Late Twentieth Century," in *The Social and Political Body*, ed. Theodore R. Schatzki and Wolfgang Natter (London, 1996); idem, *Flexible Bodies: The Role of Immunity in American Culture from the Days of Polio to the Age of AIDS* (Boston, 1994); and Donna Haraway, "The Politics of Postmodern Bodies: Constitutions of Self in Immune System Discourse," in *Simians, Cyborgs, and Women* (New York, 1991).

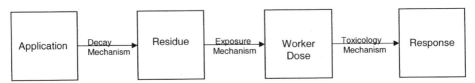

Figure 1. *Conceptual model used to understand pesticide poisioning within the discipline of occupational health. (From William J. Popendorf, "Exploring Citrus Harvesters' Exposure to Pesticide Contaminated Foliar Dust,"* American Industrial Hygiene Association Journal *41 [1980]: 652–9. Reprinted with permission of* American Industrial Hygiene Association Journal.*)*

> [Fruits] do not grow at the tips of branches. In order to cut a cluster of grapes, or pick an orange or a peach, it is . . . necessary to penetrate a leaf curtain. In the course of only an hour or two, the worker is drenched with whatever liquids may be clinging to these leaves: he has inhaled quantities of whatever dusts the picking process has rendered airborne; he has gotten these substances up his shirt sleeves, down the front of his shirt, down the back of his neck; he has gotten them in his eyes and ears; he has gotten them in his mouth and throat.[22]

This lack of mechanical mediation and the resulting bodily engagement—the mixing of bodies, leaves, sweat, and dust—has rendered fruit picking unlike the modern assembly line and particularly dangerous in the presence of organophosphates such as parathion. Absorbed rapidly through the skin, organophosphates are typically even more damaging when a person sweats or wears contaminated clothing. The hot environment of the Central Valley and the intricacy of orchard work made it impracticable for workers to wear gloves, masks, or other protective equipment. Moreover, workers might wear the same contaminated clothing for days, eat contaminated fruit, and wash with contaminated water. On several occasions, crews of workers reported being sprayed directly with chemicals as they worked. At other times, entire crews were forced to spend the night in recently sprayed orchards. Here, given the social realities of farm labor, the notion of the body as a discrete entity penetrated only by narrow "pathways" begins to break down.[23]

At a certain level, workers themselves seem to have constructed a more all-encompassing relationship between health and the environment. By the mid-1950s, some workers had connected their illnesses to the pesticide-laden landscape of rural California. When asked by a public health researcher in 1958 about their recent health problems, several braceros volunteered their concerns over pesticides: "While I was working for _____ Farms, I got sick. My mouth puffed up and swelled. I think it was because of the poison they put on the plants. It hurt a lot." Another laborer told the same interviewer: "I got sick here. My eyes hurt very much. I don't know what caused it, but it may have been something they sprayed on the trees."[24]

---

[22] Statement of Thomas H. Milby in U.S. House Committee on Education and Labor, *Occupational Safety and Health Act of 1969: Hearings on H. R. 843, H. R.. 3809, H. R. 4294, H. R. 13373,* 91st Cong., 1st sess., 1969, 1389–90.

[23] On the ability to manage and quantify respiratory exposures, for example, see Popendorf and Leffingwell, "Regulating OP Pesticide Residues" (cit. n. 11), 152; Keith T. Maddy, "Current Considerations on the Relative Importance of Conducting Additional Studies on Hazards of Field Worker Exposure to Pesticide Residues," in *Pesticide Residue Hazards* (cit. n. 12), 134–5. On working conditions, including direct spraying of workers, see CDPH, *Community Studies on Pesticides,* 15 Dec. 1970, especially 20–2.

[24] Anderson, *Bracero Program* (cit. n. 14), 286, 288.

What emerges from these scattered, translated, and necessarily partial statements is not only the braceros' recognition of the danger of "chemicals" and "spray" but also a more general sense that California, or at least the state's agricultural landscape, was itself disease-ridden. Farmworkers constructed a popular epistemology of environmental health in which changes to the agricultural environment, in this case the introduction of new pesticides, were registered in their own bodies. As one bracero stated, "I have talked to many braceros from my village who have worked in the Untied States. Many of them are in worse health when they return to Chorinzio than they were when they left. The reason for this, I believe, is that they have to breathe in too many chemicals that have been sprayed on the plants where they work."[25] These workers located disease not within their own bodies as germs or viruses but in a landscape they found foreign and physically threatening and one over which they had little or no control. In their epistemology, the environment, rather than the body, was the site of pathology.[26] Having little or no technical knowledge of the chemicals themselves, workers gradually came to associate particular illnesses with particular locales or particular crops, an assessment biomedicine would later confirm. As one farmworker reported, "My daughter gets swollen hands and feet, and welts, when picking tomatoes. Also peaches. . . . My husband gets very sick at the stomach when picking the lemons and valencia oranges. I get eye irritation with all the jobs around here: plums, grapes, peaches, tomatoes."[27] In these accounts, sick bodies were products of a larger environment and could not be neatly separated from it.

CREATING VISIBILITY: POLITICS AND EPISTEMOLOGY

From a political perspective, it is depressing, though perhaps not surprising, that the health problems of a severely marginalized group of workers drew so little notice in the 1950s and 1960s. Yet in the same period, the risks of pesticides to consumers received considerable political attention. In 1950, Representative James Delaney of New York initiated a series of hearings around the country that focused primarily on the risks of eating foods containing small quantities of DDT and other organochlorine chemicals. In 1954, Representative Arthur Miller of Nebraska successfully sponsored a bill that required the U.S. Food and Drug Administration to ensure that any pesticides remaining on food products posed no health risks to the consumer. In 1958, Congress passed the Delaney amendment to the Food, Drug, and Cosmetic Act, which established a "zero tolerance" for carcinogens in the food supply. Two years later, the publication of *Silent Spring* generated massive popular and political attention to the risks of pesticides, but again the public debate focused almost exclusively on the organochlorine compounds and the risks to consumers.[28] In the wake of *Silent Spring,*

---

[25] Ibid., 215.

[26] For a similar observation in a different context, see Michelle Murphy, "The 'Elsewhere within Here' and Environmental Illness; or, How to Build Yourself a Body in a Safe Space," *Configurations* 8 (2000): 87–120.

[27] CDPH, *Community Studies on Pesticides,* 15 Dec. 1970, 14.

[28] James Whorton, *Before Silent Spring: Pesticides and Public Health in Pre-DDT America* (Princeton, N.J., 1974); John Wargo, *Our Children's Toxic Legacy: How Science and Law Fail to Protect Us from Pesticides* (New Haven, Conn., 1998), 70–8; and Bosso, *Pesticides and Politics* (cit. n. 2), 61–78. My comments on public attention are also derived in part from reviewing entries in *The Readers Guide to Periodical Literature* for the years 1950–1968 and from the *San Francisco Chronicle Index, 1950–1980.*

environmental groups would begin lobbying for a complete ban on DDT within the United States; many of these groups tacitly accepted, and even supported, the increased use of organophosphates as a replacement for the more environmentally persistent organochlorine compounds. In contrast to the continuing attention devoted to issues of consumer health throughout the 1950s and early 1960s, federal hearings on the risks of pesticides to workers would not be held until 1969. Moreover, ongoing toxicological research was devoted almost entirely to residues on the edible portions of plants, whereas the greatest danger to workers came from the contact of their skin with contaminated leaves. No worker safety regulations would be adopted until 1971, when California established reentry intervals (mandatory waiting periods between the application of pesticides and the time at which workers would be permitted into the field to harvest), but only for the most lethal pesticide-and-crop combinations. Federal regulations, which were markedly weaker than those passed in California, would not be adopted until the mid-1970s.[29]

Yet serious poisonings among agricultural workers continued throughout the 1960s. While the most dramatic incident was that of the peach pickers in 1963, similar incidents followed, and they received little or no press. In 1966, thirty-seven workers became seriously ill in five different instances. In 1967, peach pickers in the northern San Joaquin Valley were again poisoned; this time after exposure to azinphosmethyl ethion. In 1968, nineteen orange harvesters were seriously poisoned by parathion near Lindsay. Still officials directed relatively little attention at the problem. In part this was because few workers actually died from systemic poisoning. It was also possible because pesticide illness among farmworkers was dramatically underreported. As one researcher put it, it was difficult to argue the seriousness of the problem "because you can't show bodies. . . . We just don't have the bodies, dead or otherwise."[30] Yet consumers of pesticide-treated food were not becoming obviously ill or falling dead in greater numbers. To the contrary: farmworkers suffered far greater pesticide exposures and many more illnesses than did consumers. Dr. Howard Mitchell, chief of the occupational health group at CDPH, was among those who recognized the implicit double-standard for consumer and worker health. Testifying before the state legislature in 1964, Mitchell observed that "instead of being so preoccupied with finding out whether people are being made ill from food residues from what we consider homeopathic doses, it might be interesting to note what are some measurable effects among those people who are exposed to large quantities for many years."[31] Despite Mitchell's suggestion that more money should be directed toward occupational health research, the legislative and research focus remained on consumers.

Epistemologically, concerns over consumer health always focused on ingestion. Con-

[29] On DDT and environmental groups, Dunlap, *DDT* (cit. n. 2). For a summary of the adoption of worker reentry regulations, see Victoria Elenes, "Farmworker Pesticide Exposures: Interplay of Science and Politics in the History of Regulation (1947–1988)" (master's thesis, Univ. of Wisconsin, Madison, 1991); Orville E. Paynter, "Worker Reentry Safety, III: Viewpoint and Program of the Environmental Protection Agency," *Residue Reviews* 62 (1976): 13–20.

[30] Comment of Robert C. Spear in *Pesticide Residue Hazards* (cit. n. 12), 231; Ephraim Kahn, "Pesticide Related Illness in California Farm Workers," *Journal of Occupational Medicine* 18 (1976): 693–6.

[31] Statement of Doctor Howard Mitchell in California Assembly, Interim General Research Committee, Subcommittee on Pesticides, *Hearings on Pesticides: Report of the Governor's Pesticide Review Committee*, 6 Feb. 1964, 107. Mitchell was seeking increased funding for his own division. See also Mary K. Farinholt, *The New Masked Man in Agriculture: Pesticides and the Health of Agricultural Users* (Cleveland, Ohio, [1962?]).

taminants were presumed to enter the body through a well-defined and singular pathway, the mouth. Bacteriological models and modernist constructions of the body underlay these discussions. It was easier for physicians as well as laypeople to acknowledge the orifices of the body as routes of connection to the environment—rather than the skin, for instance—for over these we have more control. Foods can be chosen. A mouth can be shut. Pesticide risks to consumers were thus rendered analogous to being infected with typhoid after drinking water from a contaminated well. Offending pollutants could be identified and cleaned up, or so legislators thought. In contrast, workers' chemical exposures were obviously not confined to a single pathway. Their illnesses pointed not to a discrete source of contamination but to an environment rendered generally toxic.

Social class and location also powerfully determined the visibility, and acceptability, of pesticide effects. The San Joaquin Valley was not suburban Los Angeles or Washington, D.C. An ethnically Mexican farmworker was not a white middle-class consumer. While many officials within CDPH increasingly acknowledged that social and political conditions were critical factors in farmworker health, they nonetheless continued to invoke racialized notions of susceptibility to explain the relatively higher incidence of poisoning among ethnically Mexican workers. CDPH officials asserted that workers' inability to understand English, "substandard" education, unfamiliarity with the danger of the substances, poor hygiene, marginal health, lack of attention to early symptoms, and reluctance to seek medical attention all put them at greater risk.[32] These discussions of bodily susceptibility to pesticide poisoning invoked and intertwined with other discourses that had raged around the body of the Mexican immigrant for decades. Nonwhite bodies have historically been seen as a source of contagion and disease in the United States, and the importation of Mexican laborers under the bracero agreement during and after World War II rekindled longstanding fears about the threats to (white) public health posed by certain racial and ethnic groups. Though public health officials rejected any form of biological racism, they portrayed Mexican cultural differences as so important and so deeply rooted as to be a kind of racial proxy. Public health officials, as well as both political supporters and opponents of farmworkers, drew on a racialized discourse of hygiene that asserted that farmworkers' lack of education made them susceptible to all kinds of disease, whether infectious diarrhea or parathion poisoning. Either way, much of the problem of pesticide poisoning seemingly lay with the workers themselves, not the work environment. As the president of the California Medical Association told the state senate, "People just must be more careful about their personal hygiene if they are going to avoid any difficulty [with pesticides]." Adopting a similar analysis, if a more sympathetic tone, a state occupational health official insisted that this group required much greater supervision in the use of pesticides.[33]

[32] West, "Occupational Disease of Farm Workers" (cit. n. 10); CDPH, *Occupational Disease in California* (cit. n. 20), 13–4.

[33] On immigrants as disease carriers, see Anderson, *Bracero Program* (cit. n. 14), 223–4; Nayan Shah, *Contagious Divides: Epidemics and Race in San Francisco's Chinatown* (Berkeley, Calif., 2001); and Alan M. Kraut, *Silent Travelers: Germs, Genes, and the "Immigrant Menace"* (New York, 1994). On racial stereotypes associated with Mexicans and Mexican Americans, including the discourse of hygiene, see David Montejano, *Anglos and Mexicans in the Making of Texas* (Austin, 1987), 220–34; Alexandra Minna Stern, "Buildings, Boundaries, and Blood: Medicalization and Nation-Building on the U.S.-Mexico Border, 1910–1930," *Hispanic American Historical Review* 79 (1999): 41–82. Quote is from statement of Ralph Teall in California Senate, Fact-Finding Committee on Agriculture, *Hearings in Regard to the Application and Use of Pesticides,* 16 June 1964, 12. Comment on need for supervision is from West, "Occupational Disease of Farm Workers" (cit. n. 10), 98.

Because many of the symptoms of organic phosphate poisoning (headaches, nausea, vomiting, cramps) could be associated with viral or bacterial infection, sickness among workers could be read as a sign of their own weak and disease-prone bodies. Growers and labor contractors frequently insisted that the sources of illness lay within the impure or substandard bodies of workers. When large groups of pickers became ill simultaneously, their symptoms were often interpreted as evidence of more typical illnesses, such as food poisoning or common diarrhea. Alternatively, when only a few pickers were affected, heat stroke was the typical explanation. A doctor from Tulare County told the U.S. Congress that "[the workers] joke about the field boss calling them all sickly, saying, 'You people are always passing the Hong Kong flu to one another on the way to work.'"[34] This construction of illness rendered the cause as a virus and also a completely foreign agent—an alien disease residing in the bodies of alien workers—with no connection to either pesticides or the local environment. For those who opposed regulation, the assumption of a body that had no relation to its environment served obvious economic interests. Modernist frameworks of disease were supported by biomedicine and effectively mobilized by growers and their supporters. This discourse of infection and diseased bodies exonerated the pesticide-ridden landscape of central California and helped render chemical exposures invisible to a broader public. Even those who advocated on behalf of farmworkers believed their poor health to be rooted in their poverty and exclusion from American society, not in the physical environment in which they worked.[35]

The problem of farmworker exposures finally emerged into public consciousness and political debates in the late 1960s, garnering public research funds and prompting limited state and federal regulation. While California occupational health specialists would be there to cite the data they had gathered over two decades, the reasons for this new visibility for the most part lay outside their own profession, in the labor and antiwar politics of the 1960s. Most critically, the United Farmworkers (UFW), which began organizing workers in the San Joaquin Valley in 1964, made visible the appalling conditions that characterized so much of California agriculture. The increased scrutiny of working conditions also brought attention to the hazards associated with agricultural pesticides, and in 1968, the union made health and safety issues a key element in its contract demands. Then in 1969, the UFW brought several farmworkers to testify before the U.S. Senate about their experiences in the field—the prevalence of spraying, the chronic headaches and nausea experienced by workers, stories of continuing seizures, nosebleeds, and persistent skin conditions. Yet recognizing that the experience of Mexican workers did not necessarily constitute "evidence" in the eyes of a white public, union leaders also mobilized the data of biomedicine, pointing out that blood tests done on children in Tulare County indicated that nearly half of those tested had abnormally low cholinesterase levels as

[34] Lee Mizrahi in House Committee, *Occupational Safety and Health Act* (cit. n. 22), 1449–50, 1453. For industry attitude, see John M. McCarthy, "Comments on Reentry Research," in *Pesticide Residue Hazards* (cit. n. 12), 223–6. On calls for research, see ibid., appendix A, 244–53. On evolution of federal policy, see Elenes, "Farmworker Pesticide Exposures" (cit. n. 29). For various diagnoses, see Griffith E. Quinby and Allen B. Lemmon, "Parathion Residues as a Cause of Poisoning in Crop Workers," *J. Amer. Med. Ass.* 166 (1958): 740–6. Statement of Milby in House Committee, *Occupational Safety and Health Act* (cit. n. 22), 1387; CDPH, *Community Studies on Pesticides,* 15 Dec. 1970, 15.

[35] See, e.g., Anderson, *Bracero Program* (cit. n. 14).

well as abnormally high levels of organochlorine pesticides, such as DDE and DDT.[36] Although the introduction of cholinesterase tests grew out of a modern notion of the body as a discrete and bounded entity, for farmworkers and their advocates, cholinesterase tests corroborated ecological understandings, old and new, that emphasized the inherent porosity of the human frame and insisted that bodies reflected the quality of their surrounding environment. Knowledge, once produced, can be put to work in many different ways. For a middle-class public growing increasingly sensitive to environmental concerns and familiar with a popular discourse of ecology, the scientific evidence of cholinesterase measurements corroborated the experience of workers and helped materialize a link between the local environment and farmworker health.

At the same time, growing concerns over chemical and biological warfare helped "denormalize" the use of OP pesticides in domestic agriculture.[37] In May 1968, 6,400 sheep suddenly died in the Skull Valley of Utah. Newspapers soon reported that the cause was an accidental release of the nerve gas VX from the U.S. Army's chemical and biological weapons installation located in nearby Dugway. Learning that VX was an OP compound significantly changed the public perception of agricultural pesticides. Farmworker advocates began referring to agricultural spraying as "chemical warfare" and to parathion as a "nerve gas." In hearings before the U.S. Senate, several witnesses pushed the connection between the notorious compounds GB and VX and the OP pesticides in wide use domestically, pointing out that parathion and GB had comparable toxicities. Protests over the use of chemical defoliants in Vietnam also resonated with concerns over the domestic use of agricultural chemicals. An increasingly militant agricultural workforce and consumer advocates argued that pesticide poisoning was indicative of an uncontrolled and dangerous modern technology that threatened all human bodies.[38] War itself helped change the meaning of certain synthetic chemicals and make visible their effects on the otherwise marginalized bodies of farmworkers.

By the end of the sixties, the UFW had been able to capitalize on this sense of vulnerability, linking producer and consumer concerns in its promulgation of an ongoing boycott against California's grape growers.[39] Arguing that consumers themselves would not be safe until the use of pesticides was controlled, that the same chemicals posed risks to both those who labored in the fields and those who ate California's produce, the UFW promoted their boycott as an issue of global environmental health as well as one of social justice—though this claim skirted the fact that the substances of

[36] On the UFW, see Jerald Brown, "The United Farmworkers Grape Strike and Boycott, 1965–1970: An Evaluation of the Culture of Poverty Theory" (Ph.D. diss., Cornell University, 1972); Galarza, *Farm Workers and Agri-business* (cit. n. 14); and Sam Kushner, *Long Road to Delano* (New York, 1975). On DDT and DDE testing, see House Committee, *Occupational Safety and Health Act* (cit. n. 22), 1448.

[37] On denormalization of dangerous working conditions, see Arthur McEvoy, "Working Environments: An Ecological Approach to Occupational Health and Safety," *Technology and Culture* 36, suppl. (1995): 145–73.

[38] Seymour M. Hersh, "The Secret Arsenal: Chemical and Biological Weapons," *New York Times Magazine,* 25 Aug. 1968; "Pesticide Jungle: The Growing Menace," *El Malcriado* (Delano, Calif.), 1 Jan. 1969, 5; and "Grower Blasts Pesticides 'Fraud,'" *El Malcriado,* 15 Feb. 1969, 15; Ronald B. Taylor, "Nerve Gas in the Orchards," *Nation,* 22 June 1970, 751–3.

[39] Samuel P. Hays, *Beauty, Health, and Permanence: Environmental Politics in the United States, 1955–1985* (New York, 1987); Robert Gottlieb, *Forcing the Spring: The Transformation of the American Environmental Movement* (Washington, D.C., 1993).

most concern to consumers were *not* those of greatest concern to farmworkers. Nevertheless, the UFW invoked environmental concerns, along with the evidence of both farmworkers and biomedicine, to articulate an ecological vision of the human body and human society, in which all persons, whether workers or consumers, were at risk. Many people found this vision compelling, and what emerged, if only incompletely, was a shared sense of physical vulnerability.[40] Popular notions of ecology and environmental health constructed the bodies of middle-class consumers and farm laborers as enmeshed in the modernized environment, as open, porous, and increasingly at risk. For a time, these concerns over environmental health crossed urban and rural boundaries as well as lines of class and race—albeit somewhat unevenly—generating what sociologist Ulrich Beck has referred to as a "generalized consciousness of affliction" and a call for alternatives to chemically intensive agriculture.[41]

CONCLUSION

In retrospect, the history of the recognition of organophosphate poisoning is less a story of conflicts about knowledge and control between experts and working people than a story of how attention to laboring bodies generated new knowledge about the larger environment and how specific constructions of bodies and environments underlay the production of certain facts. In postwar California, it was the immediate and undeniable reactions of bodies that produced new kinds of knowledge about the landscape and the changes it was undergoing. Yet the potential connection between illness and the environment remained invisible to most. The exception, of course, lay within the discipline of occupational health. However, while occupational health specialists recognized a link between the bodies of farmworkers and the environment of the Central Valley, they articulated that link very narrowly. They believed that their own discipline and skill could manage the interaction between the two, that they could "supervise" and ultimately control the material traffic between bodies and their environments. Though the political influence of California agricultural interests already limited the potential for radical change, never once did those in occupational health suggest that these pesticides should not be used or that the relationship between bodies and environments was so all encompassing that it might never succumb to quantification. Instead, as certainty continued to elude researchers and poisonings persisted, they called for more research, as if one more field study would finally allow them to describe and control how chemicals moved from a complex environment into an individual body. They believed that the techniques of public health, if politically supported, could manage the effects of the new agricultural chemicals on laboring bodies. The old dream of controlling nature emerges in many sites, in public health as well as in modern agriculture.

[40] "Chavez Blasts FDA for Condoning Poisoned Food," *El Malcriado,* 15–30 Oct. 1969, 3; "The Threat of Chemical Poisons," *El Malcriado,* 1 Jan. 1969, 5. See also Senate Subcommittee on Migratory Labor, *Hearings on Migrant and Seasonal Farmworker Powerlessness,* part 6B, 91st Cong., 1st sess., 3275–91. In 1969, the Consumers Federation of America endorsed the UFW boycott, largely on the basis of the pesticide issue. See "Consumer Group Backs Boycott," *El Malcriado,* 15 Aug.–15 Sept. 1969, 15. For other work on the UFW and pesticides, see Robert Gordon, "Poisons in the Fields: The United Farm Workers, Pesticides, and Environmental Politics," *Pacific Historical Review* 68 (1999): 51–77; Laura Pulido, *Environmentalism and Economic Justice: Two Chicano Struggles in the Southwest* (Tucson, Ariz., 1996); and Marion Moses, "Farmworkers and Pesticides," in *Confronting Environmental Racism: Voices from the Grassroots,* ed. Robert D. Bullard (Boston, 1993), 161–78.
[41] Ulrich Beck, *Risk Society: Toward a New Modernity,* trans. Mark Ritter (London, 1992), 74.

Underlying these assumptions about the manageability of pesticides was the belief that modern bodies and modern environments were separate entities and that both could be bounded, monitored, and regulated. Occupational health itself formed part of the modernist project, and blood cholinesterase tests were yet one more technique enabling the surveillance and management of fragile bodies in dangerous environments. Bodies, however, could not be monitored when embedded within a transnational capitalist system that created and demanded mobile labor. Nor could the local environment of the field be contained and managed in the way traditional industrial hygienists and the new environmental regulators ultimately hoped. Bodies as well as environments flowed outside the boxes occupational health experts constructed for them. Plants, soils, chemicals, and cholinesterase interacted with one another in unpredictable ways. In the modern orchard, this entanglement was manifest in the many acute and chronic illnesses suffered by farmworkers, what many would come to call *andando muerte,* "walking death."[42] These walking dead were themselves evidence that the boundaries between bodies and environments were never fast and that the "control of nature" might be more accurately described as its rearrangement.

[42] Statement of Cesar Chavez in Senate Subcommittee, *Farmworker Powerlessness,* 3390 (cit. n. 40).

Poisoned Food, Poisoned Uniforms, and Anthrax:

Or, How Guerillas Die in War

By Luise White*

ABSTRACT

Many people believe that Rhodesia, struggling to maintain minority rule in Africa, used chemical and biological weapons against African guerilla armies in the liberation war. Clothes and food were routinely poisoned, and Rhodesian agents, perhaps in concert with global forces of reaction, caused the largest single outbreak of anthrax in modern times. Oral interviews with traditional healers and Rhodesians' confessional memoirs of the war suggest that deaths by poisoning or disease were not so straightforward, that guerillas and healers and doctors struggled to understand not only what caused death but also what kind of death a poisoned uniform or poisoned boot was.

INTRODUCTION

When guerillas die in terrible agony, poisoned as they struggle to liberate their homeland from the rule of its white minority, what do they die of? Is a death by poisoning caused by the poison or by the motives of the poisoner? Does a man poisoned by food government agents force women to cook for him die the same horrible death as a man poisoned by a political rival among his comrades? Does a man who dies from wearing secondhand clothing doused with insecticide die a different death when the clothing comes from security forces rather than an irresponsible merchant who purchased it from farm workers? To explore these questions, this article examines two episodes of poisoning and one of biological warfare in the liberation war that wrought the nation of Zimbabwe from white-ruled Rhodesia.

  The war began with sporadic guerilla attacks shortly after Rhodesia declared its renegade independence in 1965 and by the late 1970s pitted two African guerilla armies, representing two exiled parties opposed to each other,[1] against the Rhodesian

* Department of History, 025 Keene-Flint Hall, University of Florida, Gainesville, FL 32611; lwhite@history.ufl.edu.

  The article was originally presented at a conference on health, place, and environment held at the University of Wisconsin–Madison in April 2001. I am grateful to the organizers and participants for their comments and to Stacy Langwick, Vassiliki Betty Smocovitis, and two anonymous reviewers for thoughtful readings.

[1] The differences between the older, Soviet-backed Zimbabwe African People's Union (ZAPU) and the newer, larger, avowedly pro-Chinese Zimbabwe African National Union (ZANU) were great, and their armies' conduct of guerilla struggle were different, but none of these differences seem to have influenced their ideas about death and dying.

security forces. The Rhodesian army consisted of white and black regiments, in which Africans outnumbered whites. The war on the ground, which continued until the 1979 cease-fire and negotiated settlement, was unconventional even by the standards of guerilla struggle. There were very few battles and no clearly demarcated enemy territory. Instead there were guerilla incursions, and peasants were asked to support the armed struggle. Guerilla detachments infiltrated and ambushed. At the same time, they were informed on (by sellouts, or traitors) and hunted down by Rhodesian pseudo-gangs and cost-effective helicopter units. The war's impact on all Africans was enormous.

In guerilla camps and in the countryside, many stories of biological warfare circulated. Guerillas claimed that uniforms and boots had been poisoned by Rhodesian agents desperate to kill them and that anthrax and cholera had been introduced into specific areas.[2] Almost everyone I interviewed said biological weapons had been used between 1976 and 1980, the years of most intense conflict.

In 1980 one-man one-vote elections brought majority rule to the new, legitimate nation of Zimbabwe; one of the exiled parties took power, and the other was defeated. In this article, however, I will not address the differences between the two parties, as this is not a history of political conflict but one of ontological conflict: Is a death different because of the agent causing the affliction? I argue that it is, that deaths or illnesses from poisoning are different when the agent responsible for the poisoning is different. The fact of poisoning alone does not describe the situation in which guerillas died. In this I am at odds with much of the literature about the war, a literature in which Rhodesian officials claimed African guerillas did not know they were being poisoned. I argue that guerillas insisted they were being poisoned and took precautions against it.

How did guerillas die when not in battle? Classifiable, quantifiable deaths are different from each other, just as brain death is different from cardiopulmonary death.[3] When a two-month-old is found dead in a crib, Ian Hacking notes, whether the death is determined to have been caused by Sudden Infant Death Syndrome or child abuse is a matter of separate ontologies and their respective places in medical practices.[4] More to the point, medical practices and their epistemologies of causes and risk be-

[2] Jocelyn Alexander, JoAnn McGregor, and Terence Ranger, *Violence and Memory: One Hundred Years in the 'Dark Forests' of Matabeleland* (Portsmouth, N.H., 2000), 144–5, 172–9. Such accusations are well documented in twentieth-century wars. The Japanese army accused the Chinese of polluting wells with bacteria in 1936; Chinese claimed the Japanese vaccination program caused cholera. In 1952, Chinese soldiers on the border of North Korea and China reported American aircraft dropping cardboard boxes filled with live insects and soybean stalks: Korean railway workers saw hundreds of black beetles and died from anthrax a week later. As Biafra fought a losing war for its independence from Nigeria in the mid-1960s, university students refused to eat the food they maintained loyalist cooks had poisoned. Later rumors circulated that the British food aid was poisoned; Biafra rejected the food. When Rhodesians introduced the injectable contraceptive Depo-Provera in the 1970s, Africans claimed it was intended to sterilize them, and black-ruled Zimbabwe banned it in 1981. Richard D. Loewenberg, "Rumors of Mass Poisoning in Times of Crisis," *Journal of Criminal Psychopathology* 5 (1943): 131–43; Stephen Endicott and Edward Hagerman, *The United States and Biological Warfare: Secrets from the Early Cold War and Korea* (Bloomington, Ind., 1998), 3–5; Nwokocha K. U. Nkpa, "Rumors of Mass Poisoning in Biafra," *Public Opinion Quarterly* 4 (1977): 332–46; Amy Kaler, "A Threat to the Nation and a Threat to the Men: The Banning of Depo-Provera in Zimbabwe in 1981," *Journal of Southern African Studies* 24 (1998): 347–76.

[3] Margaret Lock, *Twice Dead: Organ Transplants and the Reinvention of Death* (Berkeley, Calif., 2002).

[4] Ian Hacking, *The Social Construction of What?* (Cambridge, Mass., 1999), 143–4.

haviors do not always convey the meaning of a death, let alone its exact cause. The literature that straddles occupational and environmental health makes this point as well. In Martha Balshem's study of cancer in an industrial, largely Polish neighborhood of Philadelphia, heavy smokers disregard medical opinion and educational campaigns and insist that the high rates of cancer among them are due to pollution and the chemicals with which many adults, and cancer victims, work. At least one widow struggled for months to have her husband's death certificate altered so that his twenty-odd years of smoking cigarettes was not given as the sole cause of his death from lung cancer. She blamed his death on the explosion three years earlier in the chemical plant in which he worked.[5] Balshem's book makes a strong case for competing epistemologies, both of which explain the same medical condition with great accuracy. These epistemologies overlap, as well: both explain medical conditions and locate their prevalence in ideas about politics and history and in scientific authority. Each epistemology is a located concept, in which cigarettes, work in a chemical plant, or the food preferences of Polish communities determine—one might say overdetermine— medical conditions. But epistemologies do not classify afflictions or deaths; nowhere is this clearer than in the case of Zimbabwe's wartime poisonings. Knowing what causes death is very different from knowing how—and for which reasons—death occurred.

Only the specific experiences—and the specific practices that shaped those experiences—can explain how someone died from a poison. Very much in the way that posttraumatic stress disorder was first defined as the result of wartime experience in Vietnam, here a specific war experience determines the cause of death.[6] To push Annemarie Mol's point about atherosclerosis further than she does, wartime poisonings are performed and diagnosed by separate medical practices, all located in wartime politics.[7] The experience of war—of loyal cadres, collaborators, and agents —and the loyalties of war allow for guerillas and historians to ascertain how guerillas experienced death by poison and how they interpreted death by poison.

POISONS AND HEALTH IN ZIMBABWE

In Zimbabwe, for the most part, ancestors' spirits guide human life, and ancestors give warnings, gifts, and punishments all the time, or, as one healer told me, the living act out the wishes of the dead.[8] Ancestors make moral judgments about the living. There are no cosmological equivalents of disease, infection, or epidemic, I was repeatedly told, only punishments.[9] Not every dead Zimbabwean becomes an ancestor, however. Without the proper burial rites, spirits wander in the void, tormenting the living. Many healers insist that the spirits of the thousands of war dead still buried

[5] Martha Balshem, *Cancer in the Community: Class and Medical Authority* (Washington, D.C., 1993).

[6] Allan Young, "A Description of How Ideology Shapes Knowledge of a Mental Disorder (Posttraumatic Stress Disorder)," in *Knowledge, Power, and Practice: The Anthropology of Medicine in Everyday Life,* ed. Shirley Lindenbaum and Margaret Lock (Berkeley, Calif., 1993), 108–28.

[7] Annemarie Mol, "Missing Links, Making Links: The Performance of Some Atheroscleroses," in *Differences in Medicine: Unveiling Practices, Techniques, and Bodies,* ed. Marc Berg and Annemarie Mol (Durham, N.C., 1998), 144–65; idem, *The Body Multiple: Ontology in Medical Practice* (Durham, N.C., 2002).

[8] Edward Machingadize, interview by author and Simba Handeseni, Highfield, Harare, Zimbabwe, 7 July 1995.

[9] Author's field notes, 16, 23 July 1995; 4 August 1995; 24 July 1997.

outside the country, in common graves, are the cause of Africa's health crisis today.[10] Ancestors' spirits possess individuals who become their mediums; all mediums can heal, but not all healers are mediums. (When people become witches, as opposed to mediums, it is usually because they are very jealous or possessed by the spirit of an animal.) Mediums are critically important in Zimbabwean political life. During the war, for instance, guerillas asked mediums for ancestors' approval and help in their struggle.[11] Medical technologies are based on herbal remedies, and ingested medicines are an enormous part of healing and harming. Mediums' contact with ancestors is a key element in the diagnosis of an ailment, and that contact also helps mediums find the herbs that can remedy afflictions.

By definition, mediums do not harm on their own account. If they kill to avenge a death or to punish, they are acting on behalf of the dead. Witches, on the other hand, harm willfully and out of jealousy and anger. They often kill by poison.[12] Perhaps for that reason, poisoning has been seen as a risk of everyday life. Africans who fear for their lives could reasonably assume they will be poisoned, and even those who do not take precautions—such as drinking or eating first when serving—to avoid any blame for someone else's subsequent illnesses.[13] Yet how does someone know whether he or she has been deliberately poisoned or had a particularly bad stomachache?

By the 1960s, the spheres of traditional and western medicine had overlapped. Africans were using both healers and hospitals for diagnosis and cure. No one was bound to curative procedures because of the method used for diagnosis. If a medium told a man that he had a stomach ailment because his sister-in-law poisoned him, the man might choose to go to the hospital rather than be treated by that medium because an accusation of poisoning would create tensions at home. Many people diagnosed by practitioners of western medicine went to mediums when hospital treatments did not work.[14] When guerillas claimed their fellows were killed with poisons during the war, they generally meant a medium had divined it. Such divination was considered accurate because of the mediums' command of the local and ancestral scales: they knew more about poisons and their symptoms than anyone else, and they could identify the human agency of a poisoning.

---

[10] Author's field notes, 7, 23 July 1995; 24 July 1997. These dead raise extraordinary problems; their graves must be located, but their remains cannot be retrieved without disturbing the remains of other dead comrades. Thus the return of fallen heroes—and the courage of those who supervised their return, and the politics of reburial—has generated a substantial literature. See Mafuranhunzi Gumbo [Inus Daneel], *Guerilla Snuff* (Harare, Zimbabwe, 1995), 8–9, 26; Norma J. Kriger, "The Politics of Creating National Heroes: The Search for Political Legitimacy and National Heroes," and Jeremy Brickhill, "Making Peace with the Past: The Work of the Mafela Trust," in *Soldiers in Zimbabwe's Liberation War*, ed. N. Bhebe and T. Ranger (Harare, Zimbabwe, 1995), 118–39, 163–73; Richard Werbner, "Smoke from the Barrel of the Gun: Postwars of the Dead, Memory, and Reinscription in Zimbabwe," in *Memory and the Postcolony: African Anthropology and the Critique of Power*, ed. Richard Werbner (London, 1998), 71–102.

[11] Michael Gelfand, *The African Witch* (Edinburgh, 1967), 11–5, 93–100; Peter Fry, *Spirits of Protest: Spirit-Mediums and the Articulation of Consensus among the Zezeru of Southern Rhodesia (Zimbabwe)* (Cambridge, 1976), 22–6; David Lan, *Guns and Rain: Guerillas and Spirit Mediums in Zimbabwe* (London, 1985), 136–53; Terence Ranger, *Voices from the Rocks: Nature, Culture, and History in the Matopos Hills of Zimbabwe* (Bloomington, Ind., 1999), 239–41; author's field notes, 9, 10, 11 July 1995; 17, 24 July 1997.

[12] Lan, *Guns and Rain* (cit. n. 11), 36.

[13] Fry, *Spirits of Protest* (cit. 11), 125.

[14] For case studies of healers' diagnoses, see Gelfand, *African Witch* (cit. n. 11), 86–100.

POISONED FOOD

Guerillas relied on peasant support in the war; their food was prepared for them by the villagers the guerillas had convinced of the legitimacy of their cause. Nevertheless, many commanders had teenage auxiliaries taste the food before eating it themselves. Guerillas thought that a few new recruits were Rhodesian agents or collaborators who brought poison to put in guerillas' food or that Rhodesian security forces persuaded peasants to put something in the cadres' food.[15]

As counterinsurgency tactics, both assumptions sound quite reasonable, but one old man assured me that poisoning food was simply too big a risk for anyone to take. If guerillas were poisoned by a meal, they would retaliate against the villagers who prepared it.[16] But food was poisoned. For instance, in one small area, eleven out of thirty-two fallen guerillas had died from poisoned food.[17] I was told that Rhodesian soldiers would tell the village women the war was already lost and then give them "packets of poison" to add to the food they prepared for guerillas. Many women used the packets; some knew they contained poison, some did not.[18] The graphic descriptions of poisoned food that appear in postwar literature support this scenario. In one novel, a commander is warned by a spirit that his men are in danger; he suddenly realizes that the villager who served the guerillas had brought poisoned food. "Why would an old man with so many daughters come out to dish up the food himself?" he asks himself. Though one comrade had already eaten the food, the rest threw theirs on the ground because of the ancestor's warning. When the comrades confronted the old man, he confessed. He had two cousins in the Rhodesian army, and they gave him the poison he put in the guerillas' food.[19] In another story, a poisoned guerilla survives, and his comrades debate whether he was intentionally poisoned. One man recounts how as a child he almost died from eating mushrooms his mother had picked. "Obviously my mother didn't mean to kill us. It's difficult to decide if this case was accidental or intentional." But the woman in question turned out to have a son in the Rhodesian police, and he had given her a packet of dried, poisonous mushrooms, which she saved to cook for guerillas.[20]

Most of the men and women I spoke to in 1995 and 1997 thought Rhodesians and their agents had asked villagers to poison guerillas' food, but they also claimed guerillas sometimes poisoned each other. "[A] comrade would poison another to get his post." In fact, some people insisted that guerillas, and the mediums guiding them, were responsible for most food poisonings. Who knew which mushrooms were poisonous and which were not? Mediums noted that there were African herbalists in the Rhodesian army, men who made the poisons for food or identified noxious mushrooms. Healers maintain that all poisons have antidotes, "there is nothing that is done by someone that cannot be cured." Thus, when guerillas died, one medium insisted, it was not because poisons were so lethal but because there were no herbs with which to reverse the effects. Herbalists were away from their home areas and could not always find the right flora.[21]

---

[15] Lovemore Chabata, interviewed by Ian Johnstone, Harare, 7 Aug. 1987, National Archives of Zimbabwe (hereafter cited as NAZ)/ORAL/264.

[16] Author's field notes, Harare, 21 July 1997.

[17] Gumbo, *Guerilla Snuff* (cit. n. 10), 9.

[18] Author's field notes, Harare, 18 July 1997.

[19] Gumbo, *Guerilla Snuff* (cit. n. 10), 25–8.

[20] Shimmer Chinodya, *Harvest of Thorns* (Oxford, 1989), 79–84.

[21] Author's field notes, Harare, 7, 11 July 1995; 17, 18, 21, 24 July 1997.

Among guerillas the question was not so much what poison had been used, but who had done the poisoning and why. Were those responsible for the poisoning simply confused about varieties of mushrooms, were they villagers who had sold out to the Rhodesian regime, or were they witches? In this way, the cause of death was not chemical, but ontological: one died from specific practices, not specific toxins. The term "poisoning" was used to describe a variety of deaths; how these deaths were defined and differentiated was debated throughout the war. JoAnn McGregor's skillful investigation of a series of poisonings during the last year of the war in western Zimbabwe shows how guerillas negotiated the difference between poisonings by witches and by Rhodesian agents. These were not always exclusive categories; many people believed witches had a role in Rhodesian poisonings. There were "poisoning rings," sponsored by Rhodesians. One was headed by a woman with supernatural powers: when she was executed, black birds flew out of her corpse. Many people McGregor spoke to did not see any contradiction between witchcraft and biochemical counterinsurgency. Guerilla commanders, however, brought a new scale to their analysis of such poisonings: they began to doubt that witches caused the recent increase in poison deaths largely because of imperialist practices around the world. One commander had read about the Algerian war, in which the "French used some kind of chemical to kill guerillas." He talked about this with his comrades, who surmised that "perhaps there was a secret weapon used by the Rhodesians to kill us and . . . in such a way as to cause mistrust between us and the locals." Third world revolutions helped the commander rethink local policies. He gave instructions to his cadres to stop killing witches, who, after all, "have been with us since the beginning" and were not appropriate targets for military action.[22]

All Rhodesian and most Zimbabwean histories of poisonings do not mention witch beliefs or guerillas' internal struggles: they stick to hard science and do not allow themselves remarks on African superstition. The memoirs of a Rhodesian paid assassin, for example, report a preference for thallium, a heavy metal poison the use of which was a marker not only of scientific expertise but of racial difference. When the hit man met with an agricultural chemist who supplied poisons for various agencies of the Rhodesian military, the chemist wanted to know if the assassin planned to kill an African or white man. Bristling at such a grossly racialized question, the assassin asked, "Does it matter?" The chemist explained that with some poisons race and culture did matter. "Africans need twice as much to constitute a fatal dose than does a white. A white man moves his bowels only twice a day, while an African goes three or four times." The chemist had first learned about thallium and the different reactions when the Rhodesian government asked him to "doctor" some sacks of maize meal in a farm store that, according to military intelligence, guerillas were about to raid. The chemist complied. As expected, the guerillas raided the store and took the food, then burned the building and returned to their base in Mozambique. They died six days later—twice the time it would have taken the poison to kill a white man.[23]

---

[22] JoAnn McGregor, "Containing Violence: Poisoning and Guerilla/Civilian Relations in Memories of Zimbabwe's Liberation War," in *Trauma and Life Stories: International Perspectives,* ed. K. Lacy Rogers, Selma Leydesdorff, and Graham Dawson (London, 1999), 131–59.

[23] Stiff, *See You in November* (Alberton, South Africa, 1985), 308–9. Why did a government chemist think maize meal—the staple starch of Africans' diet—was the logical foodstuff to adulterate? Did the Rhodesian government think Africans did not have more sophisticated tastes? And were there no more sophisticated ways of poisoning food? Africans' ideas about poisoned food went far beyond those found in Rhodesian accounts. Africans reported that some cans of Leox brand canned meat was poisoned by Rhodesians; sometimes such cans were dropped on villages from the air, and sometimes they

Several academic sources have claimed that Rhodesians also poisoned food with warfarin.[24] Paul Epstein noted a significant number of deaths from warfarin poisoning among guerilla recruits training in Mozambique in the late 1970s. Only after reading the tell-all memoir of the head of Rhodesian intelligence, *Serving Secretly,* published in 1987, did Epstein realize that warfarin poisonings were the result of counterinsurgency. But warfarin is commonly used as rat poison and would have been available in almost any store in any trading center and could have been used by anyone—guerilla, villager, or Rhodesian agent.[25] How then could it be determined which deaths were the result of counterinsurgency and which were caused by insurgents' internal struggles? To return to Ian Hacking's point, it is a historical ontology, located in appropriate bodies of literature, that determines cause of death, not symptoms or autopsy.[26]

Whatever memoirs were published, whatever scales or scientific methods used, poisoning never attained the status of a legitimate wartime injury in Zimbabwe. The lingering effects of being poisoned never constituted a disease or syndrome because no scientific authority said it did.[27] Moreover, scientific authority was maintained by the exclusion of lay opinions or those of practitioners in other therapeutic systems.[28] Former guerillas were unable to claim that the conditions they attributed to wartime poisonings were indeed wartime injuries. In 1990, Zimbabwean government officials insisted that "there was no known food poison which continues to give symptoms as indicated in most claims we have, particularly nine years after the war." They stated that the symptoms were "really imaginary." In reality, officials said, they were "fresh symptoms of stress and anxiety. . . . This is probably the real war related disability."[29] In other words, the somatization of wartime trauma was not, in Zimbabwean medical opinion, the same as physiological wartime trauma.[30]

POISONED UNIFORMS AND POISONED BOOTS

Many guerillas considered the poisoned uniforms or poisoned clothing they received to have been one of the Rhodesian regime's most effective counterinsurgency strategies.[31] I was repeatedly told that many more people had died from wearing poisoned clothes than from eating poisoned food.[32] Over the years, I have heard that poisoned

---

were put on sale in shops, as were poisoned cigarettes. Josephine Nhongo-Simbanegavi, *For Better or Worse? Women and ZANLA in Zimbabwe's Liberation Struggle* (Harare, Zimbabwe, 2000), 89; author's field notes, 17 July 1997.

[24] Jeremy Brickhill, "Zimbabwe's Poisoned Legacy: Secret War in Southern Africa," *Covert Action Quarterly* 43 (1992–93): 8.

[25] Paul Epstein, "In Southern Africa, Brutality and Death," *Boston Globe,* 26 Dec. 1987, 23; Gelfand, *African Witch* (cit. n. 11), 96.

[26] Hacking, *Social Construction* (cit. n. 4), 143–5.

[27] Young, "How Ideology Shapes Knowledge" (cit. n. 6), 108–28.

[28] In this way, the lack of medical consensus about silicosis enabled doctors in the 1920s to argue that it was unclear that the condition caused disability. See David Rosner and Gerald Markowitz, *Deadly Dust: Silicosis and the Politics of Occupational Health in Twentieth-Century America* (Princeton, N.J., 1991), 86–9.

[29] Quoted in Norma J. Kriger, "War Veterans Compensation and Rehabilitation" (unpublished manuscript), 26

[30] When I spoke to a Zimbabwean psychologist about this, he noted that the World Health Organization considers somatization a third-world phenomena, but he considered somatization a culture-bound syndrome. Author's field notes, 23 July 1995.

[31] Brickhill, "Zimbabwe's Poisoned Legacy" (cit. n. 24), 4–11, 58–60.

[32] Author's field notes, 17, 18, 21, 24 July 1997.

uniforms killed anywhere from a few hundred Zimbabweans to thousands of them, though an official list of the war dead attributes only sixty-nine deaths to poisoned clothes.[33] The number of dead does not fully account for the place of poisoned uniforms in Zimbabwean guerillas' consciousness; at the end of the war, guerillas often complained about deaths from poisoned clothing to baffled Western journalists.[34] I was told that poisoned clothing was a persistent problem for guerillas between 1975 and 1977; as late as the 1990s, former guerillas attributed a stomach ailment to wearing the poisoned clothes they had been given during the war.[35]

Poisoned uniforms have come to occupy a significant place in a genre of Rhodesian postwar writing, the now-it-can-be-told confessional books, such as *Serving Secretly*, that manage to expose the evils of the Rhodesian regime while boasting of its technological triumphs. These accounts offer two distinct origins for poisoned uniforms. One is that poisoned uniforms, like so much else, were the products of collaboration: the opportunities of wartime made some Africans—some politicians and a few shopkeepers—greedy and willing to distribute poisoned clothing for Rhodesians. Such explanations give no information on how the uniforms were poisoned or with what, but do tell us that the young men given such clothing "died a slow death in the African bush," never knowing why they suffered so.[36] The other trajectory of this literature depicts the hard science of not only poisons but also racial difference and the manipulation of consumption. Thus in another confessional memoir, a Rhodesian assassin meets with the chatty agricultural chemist who extolls the science behind his exploits. The chemist explains how he used parathion in poisoned clothes, which worked its way into a body through hair follicles. As with food to be poisoned, Rhodesian security forces had brought clothing to him, several hundred pairs of underpants and t-shirts, which he soaked in the poison. These were dried, bundled, and put into farm stores "anticipated as future terrorist targets." The security forces did not tell store owners that the clothing was poisonous but did instruct them to put it on high shelves and not sell it under any circumstances. Many of these stores were raided, and guerillas took the clothes, wore them, and died.[37]

Certainly Rhodesian surveillance of guerillas was often surveillance of guerilla dress. Rhodesians claimed that guerillas abandoned their uniforms for civilian clothes all the time, which they got either through individual contact men or by breaking into stores, especially white-owned ones. Guerillas preferred blue denim trousers and jackets as well as khaki felt hats with leopard skin bands.[38] *The Rhodesian Front War*, a book by a former Rhodesian intelligence agent, blames the Selous Scouts, a secret, pseudogang unit whose exploits have become a minor publishing industry since 1983, for poisoning guerillas' clothes. According to its author, Selous Scouts purchased

[33] Government of Zimbabwe, *Fallen Heroes of Zimbabwe* (Harare, 1983), 6–68, passim. The figure comes from Henrik Ellert, *The Rhodesian Front War: Counter-Insurgency and Guerilla Warfare, 1962–1980* (Gwelo, 1989), 267 n. Author's field notes, 18, 21, 23 July 1997; 1 Aug. 2001.

[34] See, e.g., transcript of interview with Commander Gedi Ndlovu, Assembly Point Mike, 18 Feb. 1980, NAZ/IDAF/MS 591/4.

[35] Author's field notes, 18, 21, 23, 24 July 1995.

[36] Ranger, *Voices from the Rocks* (cit. n. 11), 235; Ken Flower, *Serving Secretly: Rhodesia's CIO Chief on Record* (Alberton, South Africa, 1987), 137.

[37] Stiff, *See You* (cit. n. 23), 307–10.

[38] Ellert, *Rhodesian Front War* (cit. n. 33), 142–4; Supt. Isemonger, British South African Police HQ, Salisbury, Terrorist Tactics, 28 June 1977, Rhodesian Army Association Papers, British Empire and Commonwealth Museum, Bristol, 2001/086/010/869.

large consignments of these denim items and soaked them in a huge vat of poison in a secret laboratory in their barracks. The poisons used were two organophosphate pesticides, "antichlorinesterase" and parathion. When the guerillas wore the clothing for seven days, they experienced fevers and bleeding from the nose and mouth, which supposedly "caused the guerillas to suspect that malaria or some other malady was responsible."[39]

Did Africans really not know what was making them so sick? Given their general familiarity with malaria, for instance, it is unlikely that they would confuse the symptoms listed above with those of malaria, which is not hemorrhagic. For similar reasons, guerillas would have known that clothing could transmit poisonous substances. Pesticide poisoning was a minor but persistent occupational health problem among farmworkers in Rhodesia in the 1970s: men who carried knapsack spray units and those who handled and distributed pesticides were at the greatest risk.[40] Farmworkers knew the symptoms of exposure to organophosphates, and it is likely that others knew them as well, at least from hearsay. Guerillas probably knew the symptoms of poisoned clothing and assumed it was a cause of death in most cases. But did every African believe every item of poisoned clothing was an act of counterinsurgency, coming from Rhodesian sources? One wartime study indicates that clothing poisoned with pesticides was sometimes sold by itinerant vendors. In addition, many of the men who became sickest were those who had suffered organophosphate poisoning on the job.[41] Cholinesterase remains in the blood long after all symptoms of poisoning disappear. New exposure brings the symptoms back rapidly.[42] As noted above, Zimbabwe's official figures claim that almost seventy guerillas died from poisons on their clothing.[43] Whether the fatal clothing came from Rhodesian intelligence, pilfering farmworkers, or itinerant traders, or all of the above, is another question.

Accounts by Zimbabweans, in the struggle and out of it, were more concerned with the characteristics of the poison than with the clothes so adulterated. The clothing preferences researched by Rhodesian intelligence did not seem noteworthy to guerillas or villagers. Some of them believed their clothing was poisoned because they recognized the symptoms of an easily obtained herbal poison: whoever wore the clothes would become sick and die, in great pain, if they did not get treatment. Some clothes were so poisonous that a person only had to wear them for a "moment," and he or she would die a day or two later. According to other people, at times whites gave poisoned clothing to collaborators who passed it on to guerillas as a "donation." Rhodesians also gave villagers poisoned powder to be applied to clothes. This clothing caused sores and rashes; no one knew if it caused deaths. Guerillas responded by policing the sources of clothing, refusing to take donations of used clothes or asking the donor to wear them first.[44] Guerillas used two means for determining whether someone had

[39] Ellert, *Rhodesian Front War* (cit. n. 33), 142–4; see also David Martin and Phyllis Johnson, *The Struggle for Zimbabwe* (New York, 1981), 276–7. I could not find the term "antichlorinesterase" in any of the literature cited below, although several articles referred to cholinesterase.

[40] M. E. MacKintosh, D. E. Crozier, R. Guild et al., "A Survey of Risk Factors Associated with Organo-Phosphate and Carbamate Pesticide Poisoning," *Central African Journal of Medicine* 24 (1978): 41–4.

[41] R. O. Laing, "Relapses in Organo-Phosphate Poisoning," *Central African Journal of Medicine* 25 (1979): 225–6.

[42] MacKintosh et al., "Survey of Risk Factors" (cit. n. 40).

[43] Ellert, *Rhodesian Front War* (cit. n. 33), 144.

[44] McGregor, "Containing Violence" (cit. n. 22), 183.

died of poisoned clothing, without concern for the distinct domains in which certain poisons made sense. The cause of death was confirmed by the divination of mediums and by sending the clothing in question to a laboratory for "forensic analysis."[45] Clearly guerillas were concerned not with different therapeutic systems but with diagnosing whether poisoning had occurred and what kind of death a comrade had suffered.

Being poisoned and wearing clothes are very different things. What does it mean to fear that the clothing you wear might kill you? What does it mean to conceive of a counterinsurgency campaign that kills through undergarments? Such meanings had been circulating around the region for more than a century, among both Africans and Europeans. Africans' appropriations of western clothing had caused whites in the region great alarm. Africans reported intense police harassment for "dressing smart" and beatings by whites for wearing hats or shoes—or any other aspect of western dress that seemed to a white person to be evidence of great audacity and pride.[46] At the same time, until 1918 South Africa's medical establishment blamed tuberculosis on Africans' inability to dress and treat their clothing exactly as Europeans did. For instance, Africans let their clothes dry on their persons, either after washing or a rainstorm, which more than anything else, white doctors claimed, led to pleurisy, bronchitis, and the beginnings of tuberculosis.[47]

This is not to say that poisoned uniforms are simply a wartime version of the stories of dangerous clothes that Africans, and whites in Africa, told daily. There is a broader context animated by these stories, a context in which Africans' ideas of the power of clothing is matched by whites' ideas about the power of African dress preferences. These fictions of consumption and difference were debated in Rhodesian manufacturing circles as the guerilla war was fought. Local firms sought to define an African market, to determine what Africans preferred to wear. (Such research was not confined to clothing. Manufacturers wanted to find out, for instance, how different African hair was so that new and profitable shampoos could be sold to Africans.)[48] The science of poisoning was joined with the science of market research to determine, for example, which hat and which hatband should be poisoned. Such widespread and shared ideas about the power of western clothing do not, of course, make poisoned uniforms any more or less true than the evidence at hand indicates, but they do provide a deeper local context that explains why poisoned clothing, real or imagined, was so fearsome. For Africans, the idea of being poisoned by jeans or a hat would seem a natural outgrowth of whites' outrage at Africans' appropriations of western fashion. For whites, the idea of poisoned clothes would seem a natural way to exploit Africans' attachment to fashionable wear.

Guerillas seem to have had ideas about how to deal with poisoned clothing. A sixteen-year-old guerilla volunteer described his arrival at a camp in Mozambique in 1976. He found

---

[45] Ibid., 134.

[46] For a selection of references, see Richard Werbner, *Tears of the Dead: The Social Biography of an African Family* (Washington, D.C., 1991), 53; Timothy Burke, *Lifebuoy Men, Lux Women: Commodification, Consumption, and Cleanliness in Modern Zimbabwe* (Durham, N. C., 1996), 102–4. For an overview see Luise White, "Work, Clothes, and Talk in Eastern Africa: An Essay about Masculinity and Migrancy," in *African Historians and African Voices: Essays Presented to Honor Professor B. A. Ogot*, ed. E. S. Atieno Odhiambo (Basel, Switzerland, 2000), 69–74.

[47] Randall M. Packard, *White Plague, Black Labor: Tuberculosis and the Political Economy of Heath and Disease in South Africa* (Berkeley, Calif., 1989), 49–50.

[48] Burke, *Lifebuoy Men* (cit. n. 46), 163–5, 196–202.

strict security measures there. . . . First, when you arrived you put off your clothes. And they'd take your clothes somewhere . . . and what you'd do was you jumped, jumping until you were fifty meters away, 500 meters, jumping without clothes. . . . Because they thought you might have poisoned clothing, they'd experienced that before. . . . So that was a security measure, to try to keep such things from the people.[49]

I do not want to turn this fragment into a medical argument, in which I claim it shows that guerillas understood something about the toxicology of how they were poisoned (although they probably did) and thus may have sought to increase the rate at which organophosphates—parathion especially—were absorbed into the recruit's bloodstream. Instead I want to suggest that this provides a way to scale down my focus and look at how guerillas thought about poisons and what actions they took to protect themselves. In the final years of the war, guerillas seemed less interested in the cures herbalists could offer and began to demand drugs and syringes from missions or from Africans who worked in rural hospitals, although many guerilla commanders insisted that Rhodesians had withdrawn "antipoisons" from rural hospitals.[50]

But what is counterinsurgency, and what is a local ontology? Africans spoke of poisoned boots, although no such item appears in any of the Rhodesian confessional literature. Some people said white soldiers would provide comrades with boots; these boots had been poisoned with a powder, and guerillas would develop sores on their feet and could not walk. Many died from these sores. Most people I spoke to did not know what the poison was or how it was put into the boots, but they knew it would make the comrades feet rot. A medium, however, explained that boots were poisoned the same way clothes were: the boot was put in water and the inside of the boot contained the poison, which caused sores on the feet. The only cure was to sit "without walking" for several weeks.[51]

ANTHRAX

The largest recorded outbreak of anthrax in humans occurred between November 1978 and October 1980 in Zimbabwe (during this period it was also Rhodesia and Rhodesia-Zimbabwe). There were almost 11,000 human cases, and 182 human deaths, and perhaps 10,000 cattle deaths. The onset of the disease was rapid: the number of reported cases of anthrax in all of Rhodesia went from 2 in 1978 to 4,002 in 1979. More important, perhaps, was that the epidemic spread over a larger area than most anthrax epidemics do and lasted longer.[52] All this has led many medical authorities to maintain that this anthrax epidemic was the result of Rhodesian biological warfare. Whether or not this particular disease was part of Rhodesia's wartime arsenal, it was part of its

---

[49] Chabata, interview (cit. n. 15).

[50] Ian Linden, *Church and State in Rhodesia, 1959–1979* (Munich, 1979), 272–3; Nhongo-Simbanegavi, *For Better or Worse* (cit. n. 23), 89; author's field notes 17, 18, 24 July 1997.

[51] Author's field notes, 17, 18, 24 July 1997.

[52] J. C. A. Davies, "Transmission of Anthrax," *Central African Journal of Medicine* 26 (1980): 47; idem, "A Major Epidemic of Anthrax in Zimbabwe: The Spread of the Epidemic in Matabeleland, Midlands, and Mashonaland Provinces during the Period November 1978 to October 1980," *Central African Journal of Medicine* 28 (1982): 291–8; Meryl Nass, "Anthrax Epizootic in Zimbabwe, 1978–80: Due to Deliberate Spread?" *Physicians for Social Responsibility* 2 (1992): 198–209; idem, "Zimbabwe's Anthrax Epizootic," *Covert Action Quarterly* 43 (1992–93): 12–8, 61. For cattle deaths, see Alexander, McGregor, and Ranger, *Violence and Memory* (cit. n. 2), 145 n.

wartime political imagination, so it is possible to look at anthrax in scientific writing, white writing, and mediums' accounts.

Meryl Nass, an American physician working in Zimbabwe in the late 1980s, listed five questions about the spread of anthrax; the answers suggest that this epidemic was deliberate. First: Did insect vectors spread the disease? Rhodesian doctors had argued against this when the outbreak first occurred: insects do not transmit a volume of blood great enough to spread anthrax. Second: Was the disease spread by the consumption of infected cattle? If infected cattle were transported and butchered in neighboring districts, this would have spread the disease beyond its original site, but no country—including Rhodesia before the epidemic—had ever reported anthrax resulting from contact with, or consumption of, infected meat. Third: Given that Rhodesia had one of the world's lowest anthrax rates in the 1960s, what could explain the precipitous increase in human and animal anthrax starting in 1978? Fourth: Did the breakdown in veterinary services in rural Rhodesia during the war and guerillas' demands that Africans boycott cattle dips (pens where cattle were doused with insecticide to control tick-borne diseases) spread anthrax? Nass notes that cattle usually become infected with anthrax from eating spores, not contact with insects. Nevertheless, the antidipping campaign made most Africans unwilling to ask veterinary services to treat their sick cattle. Fifth: How did anthrax spread to previously uninvolved areas? The data on the spread of anthrax, most of it from research in the United States, was crystal clear: the disease might take on epizootic proportions in areas with a history of anthrax outbreaks, but areas with no history of anthrax tended to have only sporadic and mild epidemics of the disease.[53] All of this, Nass argued, indicated that anthrax was a Rhodesian biological weapon. Her reasoning was global in its scale. The United States, England, and Japan had developed anthrax weapons during World War II; such weapons could have been available to Rhodesian scientists at any time. According to Nass, the deployment of biological and chemical weapons and the use of poisoned clothing was part of Rhodesia's intensified militarization of the war in 1979, which also included aerial bombing of guerilla camps in neighboring countries and isolating rural Africans in protected villages.[54]

Such an analysis makes a great deal of sense. The transitional government in Rhodesia-Zimbabwe allowed militarists a new authority in the war, and they may well have tried biological weapons on a massive scale. Any country with as large and well-funded a veterinary department as Rhodesia would have had anthrax available in laboratories, even if it had not been developed as a weapon. What is not at all clear is why the Rhodesians would have targeted the central and western parts of the country when the eastern regions were those most regularly and thoroughly infiltrated by guerillas. Of course, there is no reason to assume that biological weapons are the prerogative of good military intelligence.

A different assertion about the origins of the anthrax epidemic was made decades after the war. An anonymous Rhodesian told a researcher that one guerilla army had used anthrax against the other guerilla army, possibly with Rhodesian help. This explanation would explain why the epidemic was never investigated in Zimbabwe.[55] It also parallels the explanations about the anthrax that appear in lay white writing, in

[53] Nass, "Epizootic in Zimbabwe," 200–4; see also Davies, "Transmission of Anthrax," 47. (Both cit. n. 52.)

[54] Nass, "Epizootic in Zimbabwe" (cit. n. 52), 203–7.

[55] Ian Martinez, "The History of the Use of Bacteriological and Chemical Agents during Zimbabwe's Liberation War, 1965–80, by Rhodesian Forces," *Third World Quarterly* 23 (2002): 1173

which the disease was either caused by Africans themselves or by bold Rhodesian assassins. Many whites, in private and in print, blamed anthrax on guerillas' antidipping campaigns, which they claimed made cattle susceptible to anthrax and other diseases.[56] Other white writings locate anthrax in fanciful science and clever biological weapons. In one wartime novel, the incomprehensible plot of which takes place in Rhodesia, South Africa, Biafra, and a fictionalized island off the coast of Nigeria, one assassin, representing currency running interests, says to another, "Your lot would want to spread anthrax over his tooth brush or slip some fungus into his socks. We still kill like gentlemen."[57]

The former guerillas in western Zimbabwe interviewed by Jocelyn Alexander, JoAnn McGregor, and Terence Ranger shared some of these ideas about the subtlety of biological weapons used by Rhodesia. Many Africans recalled seeing airplanes drop "white powder" on pastures and cleared land; they believed this caused anthrax and other diseases. A few people said they saw Rhodesian soldiers sprinkle "small pills" into dams; these "killed the fish and poisoned cattle, which then became bearers of anthrax." Others thought the disease was spread by ticks, now out of control because African cattle were not dipped or resulted from the pollution caused by gunfire and smoke. Guerilla commanders had more sophisticated ideas about Rhodesian counterinsurgency, however. Several thought that cattle on white-owned, commercial farms had been deliberately infected with anthrax so that if the cattle were stolen by guerillas the disease would spread to Africans. One man imagined what Rhodesians thought about Africans: "they suspected we could kill the animals and give the meat to freedom fighters" so they infected the cattle with anthrax to give the disease to guerillas.[58]

The spirit mediums I spoke to in 1997 did not have any doubts about what, and who, caused anthrax during the war: they did. They insisted that this disease was brought by mediums after guerillas failed to perform proper rituals—or any rituals at all—or do what mediums asked of them; the mediums brought anthrax to punish certain areas. I asked how it was possible to make a disease spread over such a wide area. "Mediums don't need to be near someone to cure them or to harm them," I was told. Different mediums have different powers, "there are grades of mediums like grades of tobacco." During the war mediums gave comrades warnings about what they should do; "guerillas were warned," one medium said. I asked if the anthrax the mediums brought was to kill people or cattle. "Cattle," one medium replied, "but some greedy people ate these cattle and they got anthrax and they died."[59]

CONCLUSIONS

Zimbabwean guerillas and healers and Rhodesian assassins all struggled to find out what, other than combat, killed guerillas and to accurately classify those deaths. Each group located the various causes of death in poisons and diseases, to be sure, but more

[56] Paul Moorcraft and Peter McLaughlin, *Chimurenga* (Marshalltown, South Africa, 1982), 179; David Caute, *Under the Skin: The Death of White Rhodesia* (Harmondsworth, United Kingdom, 1983), 209; Doris Lessing, *African Laughter: Four Visits to Zimbabwe* (New York, 1992), 94.

[57] Jack Watson, *Conspire to Kill* (Salisbury, Rhodesia, 1976), 178.

[58] Alexander, McGregor, and Ranger, *Violence and Memory* (cit. n. 2), 145–6. The idea that anthrax was spread through the pollution of warfare is not so farfetched. One possible source of the epidemic is spores released from the earth by land mine explosions.

[59] Author's field notes, 18, 24 July 1997.

important in their own sense of being—the dreadful expertise of disloyal herbalists, the ambition of guerillas, and the racial science of Rhodesian officials. What killed guerillas was indeed anthrax and parathion, but what made anthrax and parathion and other poisons so lethal—and powerful enough to be talked and written about for more than two decades—was that they inscribed politics and history on to the afflictions and additives that caused guerilla deaths.

This transformed death. Workers' exposure to parathion on white-owned farms caused one kind of death, while guerillas fighting against white rule risked another kind of death from parathion-drenched clothing supplied by Rhodesian agents. The multiple ontologies of death were shaped by overlapping (and interacting) therapeutic systems and the various writings by proponents of those systems. Thus spirit mediums and modern laboratories might determine that someone was poisoned by parathion, but it was the practices of mediums and the now-it-can-be-told writings of Rhodesian officials that made it clear that a specific guerilla was poisoned in specific circumstances. No single system could explain how guerillas died, and no single—or overlapping—system could refute the cause of death. When North American physicians and Zimbabwean spirit mediums disagree about what brought anthrax to Zimbabwe in 1979, they disagree not only about the epidemiology of the disease but also about the history of the war. Guerilla deaths were the result of war, not simply of poison or disease; they were political. The circumstances of poisoning or exposure caused guerilla deaths, and the only way to understand how guerillas died from poisonings and anthrax was to understand, and take sides in, the specific history of the war.

Oral History, Subjectivity, and Environmental Reality:

Occupational Health Histories in Twentieth-Century Scotland

By Ronnie Johnston and Arthur McIvor[*]

ABSTRACT

This essay uses oral histories of dust disease in twentieth-century Scotland to illustrate the ways in which such history can illuminate how the working environment and work cultures affect workers' bodies and how workers come to terms with the ill-health caused by their employment. It emphasizes the agency of the interpreter but argues further that oral histories of dust disease in twentieth-century Scotland are simultaneously influenced by, and evidence for, material conditions. The essay explores the notion that the bodies, not just the voices of interviewees, are material testament to health-corroding work practices, cultures, and habitat. The focus is the problems caused by the inhalation of coal and asbestos dust.

SCOTTISH INDUSTRY AND THE CLYDESIDE REGION

By the early twentieth century, Scotland had developed into a major industrial nation, and Clydeside, the area in the west of Scotland centered around the city of Glasgow, was one of Europe's most important manufacturing regions. At peak, there were forty-two shipbuilding and ship-repairing yards in Scotland, thirty-two of them in Clydeside. In addition, Glasgow was (and remains) one of the unhealthiest cities in Europe. Prior to the introduction of clean air acts and smoke-free zones in the late 1950s, Glasgow was dirty, grimy, and heavily polluted. In 1950, it had one of the highest levels of urban overcrowding in the world and, linked to this, the highest tuberculosis mortality rate among European cities. Mortality levels for bronchitis and emphysema were more than double the U.K. average, while the city had the highest heart disease and cancer death rates in Europe in the second half of the twentieth century. At the end of the twentieth century, 67 percent of Glasgow's post code areas were still classed as "highly deprived" compared with only 5 percent in Edinburgh, where average life

[*] Ronnie Johnston: School of Law and Social Sciences, Glasgow Caledonian University, Cowcaddens Road, Glasgow, Scotland, G40BA; ronnie.johnston@gcal.ac.uk. Arthur McIvor: Department of History, University of Strathclyde, McCance Building, 16 Richmond Street, Glasgow, Scotland, G11XQ; a.mcivor@strath.ac.uk.

We would like to thank Nuffield Foundation, the Thriplow Trust, and the University of Strathclyde for funding our oral history research. More important, our warmest thanks go out to all our respondents who gave up their time to relate their experiences to us.

expectancy was five years longer. Currently, the three most deprived areas in the United Kingdom are located in Glasgow (Shettleston, Springburn, and Maryhill).

Harsh environmental factors were not just confined to Glasgow. Many of the smaller towns of the Clydeside industrial conurbation (a radius of roughly forty kilometers around Glasgow)—such as Clydebank, Greenock, Motherwell, and Paisley—had similar economic legacies as well as histories of poor-quality working-class housing. Moreover, because two major coalfields extended throughout the Scottish central belt—one in the county of Lanarkshire (southeast of Glasgow) and the other in Ayrshire (to the southwest)—the region was also peppered with a network of mining communities, and these, too, have followed a similar rising and falling graph of economic success and contraction. Thus heavy industries and extraction dominated the Scottish economy in the first half of the twentieth century, and many of these witnessed the generation of considerable quantities of dust during labor processes. This was notably so in coal mining, quarrying, iron and steel making, pottery, construction, heavy engineering, and shipbuilding. The first asbestos factories appeared in Clydeside as early as the 1870s. This was a region where asbestos was heavily used across a wide range of industries and processes, from locomotive manufacture to shipbuilding and repair.[1]

COAL AND ASBESTOS

As far as the environmental and bodily damage caused by coal and asbestos in the United Kingdom is concerned, there is a contentious debate within the literature regarding the extent to which the key "gatekeepers"—government, employers, management, and the medical profession—were culpable. Some scholars defend the industry and the regulators, arguing that, given the existing state of medical knowledge, the "value" to society of asbestos and coal, and the lack of "countervailing forces" (such as effective trade union pressure), little more could have been done.[2] At the other end of the scale are those who posit that industry was well aware of the dangers and sacrificed workers' health for profit. Critics argue that rapacious mine owners, asbestos producers, manufacturers, and secondary users—such as the subcontracting insulation companies—opposed protective legislation, suppressed vital medical evidence, ignored regulations, refused to accept liability for their actions, and fought tooth and nail to minimize compensation to victims.[3] The ongoing controversy has been focused on the surviving documentary evidence. While we engage with this debate, our main concern here is to

[1] For recent work on Britain, see Arthur McIvor, *A History of Work in Britain, 1880–1950* (New York, 2001), 111–47; Geoffrey Tweedale, *Magic Mineral to Killer Dust: Turner and Newall and the Asbestos Hazard* (Oxford, 2000); and Ronald Johnston and Arthur McIvor, *Lethal Work* (East Linton, Scotland, 2000). The literature in the United States is more extensive: see, e.g., Christopher Sellers, *Hazards of the Job: From Industrial Disease to Environmental Health Science* (Chapel Hill, N.C., 1997); David Rosner and Gerald Markowitz, *Deadly Dust: Silicosis and the Politics of Occupational Disease in 20th Century America* (Princeton, N.J., 1991); Alan Derickson, *Workers' Health, Workers' Democracy* (Ithaca, N.Y., 1988); and idem, *Black Lung* (Ithaca, N.Y., 1998). See also Penny Summerfield, *Reconstructing Women's Wartime Lives* (Manchester, 1998); Alistair Thomson, *Anzac Memories: Living with the Legend* (Melbourne, 1994); Alessandro Portelli, "The Peculiarities of Oral History," *History Workshop* 12 (1981): 96–107; R. Behar, "Rage and Redemption: Reading the Life Story of a Mexican Marketing Woman," *Feminist Studies* 16 (1990): 223–58; and D. Weber, "Raiz Fuere: Oral History and Mexicana Farmworkers," *Oral History Review* 16 (1989): 361–72.

[2] See Peter Bartrip, *The Way from Dusty Death: Turner and Newall and the Asbestos Industry* (London, 2001); and idem, "Too Little, Too Late? The Home Office and the Asbestos Industry Regulations, 1931," *Medical History* 42 (1998): 421–38.

[3] See Tweedale, *Magic Mineral* (cit. n. 1).

open up a new dimension, exploring workers' perceptions through participant testimony and the evidence these oral sources offer in relation to bodily damage caused by work.

The main illnesses caused by asbestos exposure are asbestosis, pleural plaques, pleural thickening, lung cancer, and mesothelioma. The first, very limited, legal regulations to control the hazard of asbestos dust in the United Kingdom appeared in 1931. These did not, however, include any provision for workers who cut and fitted asbestos products—such as the insulation engineers (known as laggers) and building trade joiners. Deaths of such workers escalated in the post–World War II years.[4] The following map illustrates quite clearly how the asbestos death figures in Glasgow radiate out from an epicenter stretching along the River Clyde in the west of the city, where the main engineering, insulation, and shipbuilding and ship-repair outlets and the workers' neighborhoods were clustered.

In contrast to the amount of attention given to the asbestos disaster, the history of coal dust disease in the United Kingdom has been neglected, despite the fact—as Peter Bartrip correctly points out—that coal dust was a much more widespread problem.[5] The west Scotland coalfields were vital to the country's industrial development, and from very early on many of the larger coal mines, especially those in Lanarkshire, pioneered machine coal-cutting and conveying techniques.[6] Beginning in the 1920s, however, employment in Scottish coal mining contracted sharply, with numbers falling from almost 150,000 in 1920 to 82,000 in 1951 to 27,000 in 1971. The last deep mine in Scotland (Longannet) closed after flooding in 2001.

The most serious coal dust disease was pneumoconiosis. Before 1950, this disease was most common among Welsh miners because of the dominance of anthracite coal seams in South Wales. Until around 1920, approximately 150 to 200 Scottish miners died annually from "pneumo." Thereafter, with increasing mechanization, the number of certified cases north of the border grew substantially.[7] In 1928, the condition was partially recognized as a disease subject to state compensation, but it was not until 1943 that the Coal Mining Industry (Pneumoconiosis) Compensation Scheme created comprehensive cover for disabled miners. The 1943 regulations were the product of medical research showing that coal dust—as distinct from stone and other dusts—could cause pneumoconiosis. Important in this process was the "discovery" of pneumo among the coal heavers who loaded the mineral onto ships. In the first decade of certification, more than 25,000 miners in the United Kingdom were found to be pneumoconiotic, more than 3,000 of them in Scotland. The level of risk varied greatly, however, from coalfield to coalfield and even from pit to pit. One survey in the 1950s found 5 percent of miners affected at one Scottish pit (Frances) and 23 percent at another (Northfield). Although beginning in the mid-1950s preventative measures reduced the number of new cases diagnosed, the legacy of the past reaped a high death and disability toll. Between 1951 and 1971, more than 17,000 U.K. miners were officially certified as having died of pneumo. In 1999, the news that 13 miners at the Longannet pit had been diagnosed with

    [4] *Glasgow Herald,* 24 Feb. 1967, 24.

    [5] Bartrip, *Way from Dusty Death* (cit. n. 2), 26. We are currently working on this topic—see our forthcoming book: Arthur McIvor and Ronald Johnston, *Miners' Lung: A History of Respiratory Disease in British Coal Mining* (Burlington, Vt. [2005]).

    [6] Alan Campbell, *The Scottish Miners, 1874–1939,* vol. 1, *Industry, Work, and Community* (Burlington, Vt., 2000), 110.

    [7] T. Christopher Smout, *A Century of the Scottish People, 1830–1950* (London, 1986) 103–4; National Coal Board Scottish Division, *A Short History of the Scottish Coal-Mining Industry* (Edinburgh, 1958), 81.

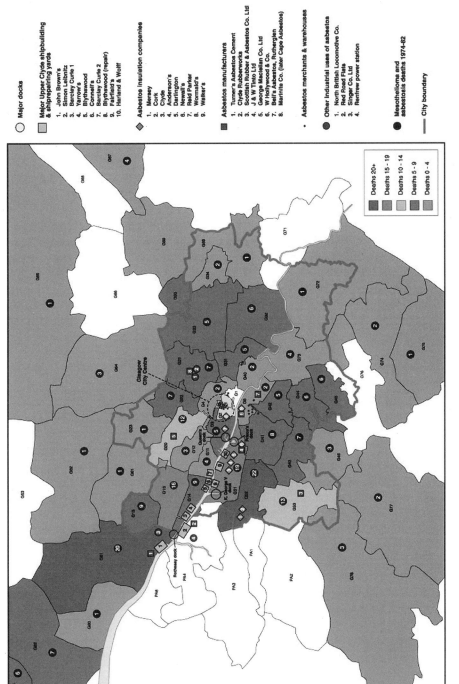

Figure 1. Asbestos exposure sites and mortality in post-war Glasgow

pneumoconiosis rocked the Scottish mining community. These were the first cases in
Scotland for more than twenty years. Meanwhile, other respiratory diseases affecting
miners were recognized. In 1998, a High Court judge awarded compensation to 6
British coal miners who had contracted emphysema and bronchitis while working for
British Coal—the final name of the National Coal Board—before denationalization. As
a consequence of this landmark legal decision, emphysema and bronchitis joined pneumoconiosis on the U.K. list of prescribed coal miners' lung diseases, and 48,000 British
miners filed claims for damages against the now-defunct nationalized industry—the
largest industrial compensation case to date in the world.

BRINGING THE WORKERS IN:
ORAL TESTIMONY OF WORKPLACE EXPOSURE TO ASBESTOS AND COAL DUST

Recently, Arthur McEvoy has argued that the key to understanding the relationship
between industrial accidents and industrial technology is to view the workplace as an
ecological system in which the worker's body forms the biological core.[8] The same
argument would apply to industrial disease. Our ongoing research project on dust disease in Scotland incorporates a systematic oral history program in which twenty
miners and twenty-six men who worked with asbestos have so far been interviewed,
together with thirteen occupational health professionals—hygienists, nurses, technicians, and academics. The choice of this methodology was partly the product of our
dissatisfaction with the existing literature, which tended to either ignore the sufferers'
perspective, to depict workers as relatively passive victims in these public health
crises, or both. William Ashworth's brief survey of occupational health in mining,
Alan Campbell's otherwise exhaustive pre–World War II history of Scottish miners,
and Geoffrey Tweedale's and Peter Bartrip's company-oriented analyses of the asbestos tragedy in the United Kingdom would be examples.[9] Our technique here is
influenced by older traditions of "social science"–based oral history and the approaches of the "new oral history"—such as those of Penny Summerfield and Alistair
Thomson, in which the authors stress the diversity of respondents' discourses and reflect on the duality of oral evidence as informative and interpretative.[10] While emphasizing the agency of the interpreter, we argue that oral interviews tell us much more
about the material world in which individuals existed than some commentators would
concede.

What came across very strongly in many of the oral testimonies was the sheer
harshness and degradation of the working environment. Workers frequently recalled
the long hours, the dirt and dust, the debilitating toil, the overbearing supervision, and
the unsanitary conditions. The shipyard insulation engineers, or laggers, were frequently required to work alongside laborers from other finishing and maintenance
trades. Several men described asbestos cuttings as "falling like snow" from above
them; others described this as having been more like a "mist" or "fog." One worker

---

[8] Arthur F. McEvoy, "Working Environments: An Ecological Approach to Industrial Health and
Safety," in *Accidents in History: Injuries, Fatalities, and Social Relations,* ed. Roger Cooter and Bill
Luckin (Amsterdam, 1997), 62.

[9] William Ashworth, *The History of the British Coal Industry,* vol. 5, *1946–1982* (Oxford, 1986);
Campbell, *Scottish Miners* (cit. n. 6); Tweedale, *Magic Mineral* (cit. n. 1); and Bartrip, *Way from Dusty
Death* (cit. n. 2).

[10] Thomson, *Anzac Memories;* and Summerfield, *Reconstructing Women's Wartime Lives.* (Both cit.
n. 1.)

remembered how the air had been "thick with this asbestos dust. And it passed through my mind then, I thought 'this *cannot* be good for us.'"[11] Coal miners usually accepted their dusty environments as a normal part of their notoriously dangerous job. However, the greatly increased dust levels brought about by mechanization had stuck in several miners' minds. For example, an eighty-year-old retired miner, reflecting on the nature of work in the early 1960s in the Ayrshire pits, remembered the dust levels as being so dense that he could not see the light of the coal-cutting machine operator.[12] This was a common observation. Another miner who worked in the Lanarkshire coalface remembered, "You couldnae see your neebor. You couldnae see from me to you 'cause you'd nae dust suppression at that time."[13]

What is evident from this kind of testimony is the entanglement of material reality with the rudiments of a "heroic" discourse (something akin to what Thomson found with his interviews of Anzac soldiers).[14] Many of the interviews contained vivid commentary on the dusty and dangerous nature of the work, commentary embedded in a story of sustained struggle against all-powerful employers and managers, with little, if any, effective assistance from the state, medical professionals, or even the trade unions. The testimony, then, is double-edged: informative and interpretative. What emerges requires careful and sensitive treatment, and it can challenge conventional wisdom constructed upon surviving documentary sources.

As soon as nationalization was achieved in 1947, the National Coal Board (NCB) began experimenting with dust monitoring and suppression, such as fitting water sprays to coal-cutting machinery in an effort to keep dust levels down. From this period on, the official government *Mines Inspector Reports* (published annually) contain several glowing reports about the success of these measures. Oral interview evidence, however, presents a different perspective and suggests that what the inspectors saw in their visits was not always the norm. For example, one retired Ayrshire miner recalled that because of the discomfort brought about by having to work in wet surroundings, the water jets would frequently be switched off, and the machinemen would "cut dry."[15] Another man remembered that although the idea of using water jets to suppress the dust was a good one, on many occasions the water pressure was far too low, and the jets tended to become clogged with dust.[16]

Several of our respondents gave us important insights into NCB attempts to monitor high dust levels. For example, a fifty-four-year-old retired miner said:

> They had a wee thing like this that men carried for measuring the level of the dust right. Now the men would come in and if it was awfully stoorie [dusty] and there was a danger of the face shutting, they flung their jacket over the top of it, ken? That's the kind of tricks. . . . They put their jacket over the top of these machines so that there was hardly any stoor going into it.[17]

Oral history, then, can provide an important counter to documentary evidence, in-

[11] Scottish Occupational Health Oral History Project (hereafter cited as SOHOHP), Oral History Centre, Strathclyde University, Scotland, Archive Deposit n. 016, Interview A8.
[12] SOHOHP, Interview C7.
[13] SOHOHP, Interview C9.
[14] Thomson, *Anzac Memories* (cit. n. 1).
[15] SOHOHP, Interview C2.
[16] SOHOHP, Interview C10.
[17] SOHOHP, Interview C4.

cluding the "official" discourses that tended toward a more sanguine and naïve portrayal of work conditions. This is also true in regard to the efforts workers took to protect their bodies from harm. Here again, the *Inspector Reports* suggest that, in some cases, miners refused to protect themselves against coal dust. In 1952, for example, an inspector noted that "trivial excuses are used by some workmen to explain, or attempt to explain, their failure to use the [safety] appliances when they have been provided." Three years later inspectors noted, "There are still some men who refuse to operate dust suppression apparatus fitted to coal cutters."[18] There was also some annoyance that the miners were not wearing masks. This "blaming the victim" discourse was prevalent in both the employers' trade journals—such as *Colliery Guardian* and *Textile Mercury*—and the government regulatory agencies' reports.

Several of the miners we interviewed remembered that older forms of protection—such as chewing tobacco—were preferred for some time over the masks provided by the NCB. Masks were also discarded because they were ineffective or because they interfered with productivity.[19] The testimony also revealed that, in any case, the masks were too flimsy to do much good, as each one only consisted of "a thin piece of gauze held on with two clips. . . . You cannae breath with them on. You couldnae walk with them."[20]

Mining in any form is a dangerous profession, and the body's battle against the underground work environment is one that has striking global similarities.[21] To a certain degree among the coal miners we interviewed there was a general acceptance of the risks involved in the job. Some of the men told us how, for the sake of speed and bigger pay packets, they would ignore health and safety regulations.[22] Some interviewees said management condoned such practices for the sake of productivity, and it was only when the mines inspector paid a visit that conditions would be temporarily improved. It was also the case that the miners saw themselves as a unified work group in a distinct industry. By contrast, most of our respondents who had worked with asbestos came from a fairly wide cross-section of engineering, shipbuilding, and construction. Consequently, although they had seen their jobs as dangerous in some way or another, these workers, unlike miners and their view of coal dust, had not considered asbestos dust as a significant hazard during their working lives. Therefore, their reconstructions of how the working environment damaged their bodies differ in some respects from those of the miners.

Up until the late 1960s in the United Kingdom, there was a widespread lack of understanding among those working with asbestos regarding the *extreme* health risks inherent in the material. As Tweedale's work has shown, from the 1930s to the late 1960s, information on the hazards of contact with asbestos was in many cases withheld from the industry, asbestos users, and the workforce. This contributed to a laxity on the shopfloor in handling such toxic material. According to some oral testimony, young workers playing with asbestos cuttings was a common sight.[23] Criminal negligence and managerial mal-

---

[18] *Annual Reports of the Scottish Mines Inspector* (London, 1955), 19.

[19] SOHOHP, Interview C10.

[20] SOHOHP, Interview C8. See also Interview C1.

[21] See, e.g., A. Derickson, *Black Lung* (Ithaca, N.Y., 1998); and Rosner and Markowitz, *Deadly Dust* (cit. n. 1). For asbestos mining specifically, see, e.g., Jock McCulloch, *Asbestos Blues: Labour, Capital and the State in South Africa* (Oxford, 2002), 39.

[22] For an oral history-based assessment of the impact of productivity on industrial accidents in manufacturing, see Theo Nichols, *The Sociology of Industrial Injury* (London, 1997), 35–61.

[23] SOHOHP, Interview A14.

Table 1: Employment in Main Industries on Clydeside, 1951

Industry	Clydeside (000s)	Male Employment (000s)	Male to Total Employment (%)	Female Employment (000s)
Transportation/ Communication	94.5	78.4	83.0	16.1
Mechanical engineering	81.7	69.1	84.6	12.6
Construction	71.8	68.4	95.3	3.2
Shipbuilding	58.2	55.5	95.4	2.7
Textiles	55.7	18.9	33.9	36.8
Public administration	53.6	44.1	82.2	9.5
Food/Drink	47.1	26.4	56.1	20.7
Metals	44.2	41.6	94.1	2.6
Mining	37.5	36.9	98.4	0.6
Clothing	30.9	6.7	21.7	24.2
Vehicles	29.4	27.3	92.9	2.1
Paper	21.8	12.6	57.8	9.2
Chemicals	21.1	15.4	73.0	5.7
Misc. metals	19.3	15.0	77.7	4.3
Timber	18.4	14.7	79.9	3.7
Electrical engineering	11.4	6.7	58.8	4.7
Instrument engineering	6.6	4.1	62.1	2.5
Not classified	0.8	0.5	—	0.3
Total	1,093.3	748.7	68.5	344.6

SOURCE: C. H. Lee, *British Regional Employment Statistics* (1979)
NOTE: Clydeside comprises the counties of Renfrewshire, Lanarkshire, Dumbartonshire, Ayrshire, and Argyllshire.

practice were also reported by workers in the building industry or those involved with asbestos-stripping operations in Scotland. These ranged from management's not informing workers that they were in contact with asbestos to bringing in cheap casual labor to deal with asbestos disposal and expecting workers—as noted above with coal mining—to cut corners when handling the material. However, even when the dangers were realized, many of our respondents—both in asbestos working and coal mining—accepted a high degree of risk and bodily damage at work. Part of the reason—as discussed earlier—was the need to maximize production in an economy in which insecurity of work was commonplace. The prevalence of stoic fatalism was also a factor. In conjunction with this was the fact that, in many cases, to be seen backing away from doing dangerous or dirty work would have been considered to be unmanly. This leads us on to the effects of machismo work practices and attitudes—an area that is neglected in the literature and one that is especially accessible through oral history interviewing.

MASCULINITY IN THE CLYDESIDE HEAVY INDUSTRIES

The issue of masculinity at work is a difficult and complex one. The sharply segmented and gendered nature of work in Clydeside (see Table 1) incubated identities and, by assigning the most toxic and dangerous laboring jobs to men, ensured that occupational injury, disease, and mortality rates were substantially higher among men than women—even accounting for underrecording of female incidence. This constitutes one of the most significant of the health differences between the genders.[24]

As many of our respondents recalled, an intensely machismo, or "hard man," culture prevailed in the shipyards, factories, coal mines, and building sites in west Scotland. Famous shipyard shop steward and communist Jimmy Reid immortalized this culture in 1971 when he said: "We not only build ships on the Clyde, we build men." This was an environment in which workers tolerated high levels of risk because of peer pressure and managerial expectations. Not to conform could mean being pilloried in public, losing face, and having your sexuality questioned. This machismo posturing could contribute to an undermining of health and well-being. For younger workers this might be expressed in reckless bravado. For the older generation there was more likely to be a long acculturated disregard for personal health, especially when the health implications were uncertain and long term—as with occupational disease and smoking.

In part, this may help explain the failure of the Scottish trade unions to prioritize occupational health before the 1970s, placing wages and the protection of jobs higher up on the agenda. When trade unions did become involved, they sometimes appeared to workers as interfering unnecessarily in their lives. One Transport and General Workers' Union activist recalled members telling him he was "always moaning and groaning about something" and to "get to fuck" when he warned workers at a building contract in Glasgow about the dangers of asbestos dust.[25]

The role of the trade unions, however, is complex. Clearly, they operated at a number of levels and were capable of both retrogressive and progressive impulses. On the asbestos issue, for example, the Scottish Trades Union Congress remained remarkably silent until well into the 1970s. By contrast, some individual unions—such as the cotton textile unions in northern England—were much more aggressive and proactive in campaigning to protect workers against the dangers of dust inhalation. Here, however, the main initiatives occurred at the local and regional levels. One recent study has shown that the national executive of the National Union of Miners (NUM) continued to pursue a policy of prioritizing compensation over prevention and used respiratory disease as a lever in negotiations to force up miners' wages throughout the period 1945–1970.[26] This continued a long tradition of collectively bargained agreements providing higher wages for handling dirty, dusty, and toxic materials—what one commentator called "the fetishism of the wage packet."[27] Thus while there were significant initiatives—including in mining and cotton—overall, the trade unions were

---

[24] See, e.g., Karen Messing, *One-Eyed Science: Occupational Health and Women Workers* (Philadelphia, 1998).

[25] SOHOHP, Interview A22.

[26] Susan Morrison, "Pneumoconiosis in British Mining, c1930–1970" (unpublished honors diss., Univ. of Strathclyde, 2001).

[27] Paul Willis, "Shop Floor Culture, Masculinity, and the Wage Form," in *Working Class Culture,* ed. John Clarke, Charles Critcher, and Richard Johnson (London, 1979), 196.

remarkably ineffective as a countervailing force in the area of occupational health in Scotland, even in this period, the peak of their power.

The prevailing workplace culture in Clydeside shipyards, engineering shops, mines, and building sites provided an environment conducive to the incubation of life-threatening occupational disease. Moreover, the contraction of mining, shipbuilding, and the heavy industries in postwar Scotland may also have perpetuated and fossilized such attitudes, making workers—especially those less skilled—even more willing to tolerate dangerous work practices for the sake of employment. At the same time, as we might expect, attitudes toward work and occupational hazards varied widely among a diverse section of workers. However, a kind of stoic fatalism permeates much of the oral testimony, combined with a machismo work culture that frowned upon "grumbling" to management. "You didnae think of the future, we just thought from day to day," one respondent noted.[28] Alan Campbell has stressed that a masculine discourse pervaded the Scottish coalfields in the interwar period, and several of the coal miners to whom we spoke recalled that being seen to be tough was, for a great many men, an accepted part of the job.[29] This included standing up to management: "A deputy would say . . . , 'Why is this conveyor stopped? . . . Why is this fucking belt no going? . . . ' If you were a *weak* man, you would have did what the boss said."[30]

Clydeside workers developed something of a machismo reputation, embracing, as Robert Connell has claimed for Australian workers, "the processes that consume their bodies, as their way of 'doing' masculinity, and claiming some self-respect in the damaging world of wage labour."[31] Behind the myths, caricature, and exaggeration a cult of toughness characterized the Clydeside heavy industries, and young male workers adapted to this and absorbed it through peer pressure. Life histories could be couched in heroic terms. Common elements in the oral narratives were criticisms of trade union inactivity and of the ineffectiveness of state intervention; respondents found composure in accounts stressing their long struggles against exploitative managers, employers, and foremen in dirty, hazardous, and dusty work environments. Masculinity, then, was fostered and celebrated in physically demanding and often-dangerous labor, with men gaining much self-esteem from hard work in an alienated and sometimes brutal environment. A retired west Scotland steel maker looked back with pride at the reputation his trade had gained by the 1930s:

> About one steel worker in every ten could stand up to them [the work conditions] successfully, which was one reason why the furnacemen were looked up to in the world of heavy industry. That they got the biggest pay packets was another reason. They also had the biggest thirsts and that too was a prideful possession in that part of the world. . . . A legend grew up about the steel smelter.[32]

Danger, then, went with the territory: you had to be seen "to be a man."

There was a rationale operating here, designed ultimately to maximize the earnings of the household. Men would say, for example, that they were working for the family or to give the kids "a better chance." However, this had the potential to lead to bodily

---

[28] SOHOHP, Interview C3.
[29] Campbell, *Scottish Miners* (cit. n. 6), 238-40.
[30] SOHOHP, Interview C1 (our italics).
[31] Robert William Connell, *The Men and the Boys* (Oxford, 2000), 188.
[32] Ronald Fraser, *Work: Twenty Personal Accounts* (London, 1969), 56–7.

damage—fatigue, injury, disease, disability—representing a sacrifice of sorts by the workingmen. In turn, such risk-taking provided another important justification for male power and male dominance within society. It gave entitlement to a privileged position. Paul Willis has argued persuasively that this bodily forfeit "also gives a kind of emotional power—a hold over other members of the family who do not make the sacrifice."[33]

Therefore, workers endured (and even embraced) dangerous work conditions for a number of reasons, including the high wages and the manly status conferred. For many others, the alternatives were less palatable. With deindustrialization and contracting job opportunities in Clydeside, it was more difficult—as in the 1930s—to walk out of a job and find other work. Here we see an intertwining of capitalist exploitation, managerial chauvinism, and dominant workplace masculine values. The wife of an asbestos worker told us that even when her husband became aware that his job was health threatening, "he was frightened to walk off the job because he was married with a family." She then went on to say, "What could we do? We were trying to bring up two kids." Her husband—now suffering from an asbestos-related disease—expressed the same sentiments: "When you have two of a family to bring up it was better than walking the streets. I never was idle in my life."[34] However, some of our respondents looked back ruefully and with evident regret at the risks they had taken in the past:

> The filth that we worked in right fae 14 years of age, and being a man with no education— the only thing you had was the muscle in your arm and what experience you got with metal, and a very willingness to work. I would go in and say to people, "Yes I'll do that in that time." And whatever it took to do that, I would do it. Silly now, looking back through the years, you know.[35]

This, of course, should alert us to the fact that masculinity, or machismo, is not a static concept, but one that is constantly changing as society evolves and, equally as important, changing throughout the course of men's lifetimes.[36] Our focus here has been upon the dominant masculinity in Clydeside workplaces up to the 1970s—that of the "hard man" in the heavy industries. There also coexisted a range of attitudes, ranging from the "rough" masculinity of some laborers and colliers through to more "respectable" modes of masculinity, exhibited by many—though not all—craftsmen and some of those who had risen to white-collar jobs. Clearly, moreover, a new generation of workers brought new attitudes to their employment: less tolerant of dusty work atmospheres and embracing something of the greater health consciousness of the post-1970s period.[37] Nonetheless, we would hypothesize that the persistence of the "hard man" mode of masculinity in the heavy industries of Clydeside up to the 1970s played a significant part in the pervasiveness of patriarchal attitudes within such working-class communities. As late as 1977, a safety officer at one Clydeside shipyard expressed his frustration thus: "Somehow we have to persuade people to take a safe at-

---

[33] Cited in Daniel Wight, *Workers Not Wasters: Masculine Respectability, Consumption, and Unemployment in Central Scotland: A Community Study* (Edinburgh, 1999), 107.

[34] SOHOHP, Interview A26.

[35] SOHOHP, Interview A9.

[36] See Michael Roper and John Tosh, eds., *Manful Assertions: Masculinities in Britain Since 1800* (London, 1991).

[37] Wight, *Workers Not Wasters* (cit. n. 33), 110–1.

titude to their work. It is easier said than done . . . in a traditional industry like ship-building where men are set in their ways."[38] Such attitudes, and the prevailing work practices that went along with them, contributed to the heavy price that male workers paid in the dusty heavy industries of Clydeside in terms of bodily damage and mortality.

THE WORK ENVIRONMENT AND BODILY DAMAGE

The materiality of workers' bodies should be seen as an important constant when trying to gain an understanding of the working environments the workers describe in their oral testimonies. Consequently, whereas some scholars have argued that illness and disability are socially constructed, we would suggest that this argument can only be taken so far because bodily damage is there for all to see. In this final section we want to draw on oral testimony to briefly illustrate the effects the environment had on workers' bodies, including their sense of manliness.

Inhaling dust at work affected people in many ways. The most obvious effect perhaps is premature death. In the early 1970s, coal miners in the United Kingdom had fifteen times the average mortality rate from occupational lung disease, were three times more likely to die of bronchitis, and were four times more likely to die of pneumoconiosis complicated by tuberculosis.[39] There was also physical impairment, deterioration, and damage of varying degrees. The experience of pain is difficult to convey but a critical part of any evaluation of the impact of occupational disease. One volunteer for the victims' pressure group Clydeside Action on Asbestos talked of "intolerable suffering," while Lillian Nicholson noted of her asbestotic husband: "During the last two years he was very ill. Right up to his last breath he was in agony."[40] Frequently, individuals and their families were drawn into a closer relationship with medical institutions, doctors, and nurses, as initial diagnosis was followed by treatment—where possible—palliative care, or both. More important, loss of employment could have tangible and disastrous economic effects on individuals and dependents. Disability-induced impoverishment was particularly acute in Clydeside, where income replacement strategies were limited due to deindustrialization and mass unemployment. Erosion of physical prowess, breathlessness, and lung damage often led to a more prescribed lifestyle—in many cases involving a painful journey to what we would now term "social exclusion," or relative poverty: "excluded from ordinary living patterns, customs and activities."[41] In the mid-1950s, Andrew Meiklejohn, Glasgow's foremost academic specialist in pneumoconiosis, commented on the difficulties faced by an individual losing work as a consequence of being diagnosed by the Medical Panels in Scotland as disabled: "he becomes the victim of a vicious circle of unfavourable influences and circumstances, largely beyond his control, and in which he can get little or no help."[42] One of our respondents conjured up an evocative image to describe this, saying that they were all now "industrial lepers."

[38] Safety officer for Scott Lithgows shipyard, cited in Martin Bellamy, *The Shipbuilders* (Edinburgh, 2001), 74.

[39] David Douglass and Joel Kreiger, *A Miner's Life* (London, 1983), 64–5.

[40] *Aberdeen Evening Express,* 9 July 1992.

[41] Stephen Hutton, "Testing Townsend," in *Quality of Life,* ed. Sally Baldwin and Christine Godfrey (London, 1994), 180.

[42] Andrew Meiklejohn, "Silicosis," in *Industrial Medicine and Hygiene,* ed. E. R. A. Merewether (London, 1956), 32.

Many of those affected by asbestos and coal dust disease fell into the category of the socially excluded for two main reasons: they could not afford to pursue the same life patterns as they once could, and their poor health prevented them from taking part in these accepted social norms. Social exclusion could involve loss of self-esteem, loss of companionship, and a more prescribed lifestyle. A retired engineer suffering from asbestos-related disease commented that he had gone "from one extreme to another. . . . I led a very full social life . . . used to go to parties, used to have friends round to the house. I no longer do that. I have shut myself off from life completely."[43]

Many miners, the archetypal tough workers, also suffered the indignity of not being able to breathe easily and enjoy their retirement. Here's what one former leading hand[44] remembers about the onset of his coal dust–induced breathlessness:

> Being a leading man I never ever asked any man to do anything that I couldnae do myself, and I was embarrassed walking in the tailgate in the 1970s, and I tried to get in before the men got in 'cause if they hear me panting they'd be saying, "He's done," which I presume I was, but I was embarrassed.[45]

Another miner expressed similar sentiments when he said: "Well, obviously I'm no fit to work, eh, and that. I worked a' my life. . . . It was a big blow to me to be told that I'd never work again. Eh, your pride's dented, ken?"[46] The loss of self-worth, then, is clearly evident in such testimonies. Respiratory disease such as this eroded masculinity by removing some of the foundations of male power and authority. Victims of dust disease invariably underwent a transition from an active, independent life in a male-dominated work environment to dependence, inactivity, and confinement in the traditionally female-dominated sphere of the home.

The physical and psychological ramifications could affect masculine roles, attitudes, and behavior in many ways, including sexual behavior. A sixty-five year-old miner commented that his lung condition meant he was "too fucked to have sex."[47] In another case, an asbestotic lagger was attending a clinic to deal with a dysfunctional erection problem.[48] The loss of libido was an important consequence of industrial disability for some people. Coupled with this was a loss of self-regard brought about by not being able to perform as a man, and not just in the sexual sense. One miner forced to take early retirement from the pits because of his failing health expressed this quite poignantly:

> I mean when you're out and your wife's to come out and say to you "Come on I'll get that. . . ." Wee jobs outside eh, that you're no fit to do, and your son or whatever eh will say to you, "Right, come on. . . ." It definitely hurts your pride.[49]

Clearly, within such a work-dominated masculine culture the removal of employment combined with disability could have deep social, physical, and psychological ef-

---

[43] SOHOHP, Interview A12.
[44] Sometimes called a "charge hand," the leading man was a person in charge of a work squad. "Foreman" was not used in the Scottish coal mines.
[45] SOHOHP, Interview C16. See also Interview C15.
[46] SOHOHP, Interview C2.
[47] SOHOHP, Interview C10.
[48] SOHOHP, Interview A14.
[49] SOHOHP, Interview C2.

fects. Joanna Bourke has documented problems of impotence among unemployed men in the 1930s, while Annemarie Hughes has identified a close correlation between loss of employment in the interwar depression in west Scotland, a culture of "consolation"—in pub and football—and wife-beating.[50] We, too, found in some cases masculine frustration boiled over into rage and violence, as was the case with one ex-lagger whose frustration over his disability led to his beating his wife. Adriana Petryna comments elsewhere in this volume on similar effects upon victims of the Chernobyl disaster. The ramifications on the male psyche—especially, perhaps, that of the archetypal Glasgow "hard man"—could be even more damaging when loss of the provider role was combined with physical deterioration as a consequence of industrial injury or disease. The adjustment necessary in this forced transition from independent provider in a male-dominated work environment to socially excluded dependent, often confined within the home, was frequently painful and traumatic. To many, this represented the reversal of all that encapsulated being a man.

CONCLUSION

Although the validity of oral evidence has been questioned, we have argued here that workers' testimonies provide us with critical insights into *both* the perceptions and the realities of health-threatening labor processes, enabling a more evocative and fuller picture to emerge of the interactions between dust and the body in the workplace. The oral testimony indicates quite clearly the significant gulf that existed between regulatory frameworks purported to establish minimum standards and actual workplace practice. Legislation and reform initiatives were capable of being subverted or blunted at the point of production, and this is one important reason why occupational health hazards persisted. The reasons are complex and multifaceted and cannot be adequately understood within the simplistic paradigm of exploitative capitalism versus innocent victim. Undoubtedly, as we have argued elsewhere, workers in the Scottish central belt worked within a profit-orientated system in which their health and well-being took second place to capital accumulation.[51] However, workers themselves, as well as their organizations, were agencies in this process. Through their actions, and those of their trade unions and local pressure groups, workers and activists struggled to minimize the dust hazard and attain decent levels of compensation for victims.

At the same time, a pervasive and persistent work culture in Clydeside celebrated hard graft and induced a competitive spirit and a machismo ethos, which manifested itself in taking risks, ignoring protective masks, and tampering with dust suppression equipment. More work needs to be undertaken to elucidate this gray area at the intersection of capitalist and patriarchal social relations. What appears evident in Scotland is that in a brutal, insecure, exploitative relationship, industrial workers experienced some empowerment through bodily sacrifice at work—notably in gender relations within the private spheres of family and home. Concurrently, the possibilities for maneuver by workers were severely constrained by prevailing power relations and economic and social circumstances, including the peer pressure of the machismo work environment, the imperatives of an unpredictable labor market, and in many cases, the

---

[50] Joanna Bourke, *Working Class Cultures in Britain, 1890–1960* (London, 1994), 133; Annemarie Hughes, "*A Rough Kind of Feminism*": *The Formation of Working Class Women's Political Identities, Clydeside, c1919–1936* (unpublished Ph.D thesis, Univ. of Strathclyde, 2001), 230–1, 316.

[51] See Johnston and McIvor, *Lethal Work* (cit. n.1).

basic needs of having to provide for a family. With deindustrialization and contracting job opportunities in Clydeside, it was more difficult to walk out and find less dangerous work—this was especially the case in mining communities dedicated to one
industry. In such circumstances, very real structural constraints curtailed agency.

Moreover, with few exceptions, employers and managers promoted this competitive, productionist, masculine work culture and failed to use their power and influence
to inculcate a more safety-conscious and healthy work environment. Clydeside workers in the heavy industries were thus caught within an intensely machismo work culture and an aggressive managerial ethos dominated by the profit motive, in a period of
deindustrialization and economic insecurity. This set of unfavorable circumstances
ensured that workplace exposure to damaging levels of dust persisted in this region
over the period 1930–1975, despite ameliorative pressures, medical research, state intervention, and reform.

Dust inhaled at work affected people's lives in numerous ways and at a variety of
levels: premature death, bodily damage, and deterioration, trauma, social exclusion,
and loss of masculinity. Oral testimony has elucidated the experience of victims of
pneumoconiosis, asbestosis, and mesothelioma. Again, though, we would stress that
workers and their organizations were agencies in this process, not just passive victims.
The role of the victims' pressure group Clyde Action on Asbestos and the miners'
unions in community welfare provide tangible examples of positive action. Similarly,
Giovanna Di Chiro has examined the philosophy and active campaigning of the environmental justice activist Teresa Leal in a quite different community on the U.S.-
Mexican border.[52] In Scotland, coping strategies were probably more effective, especially in close-knit and well-unionized mining communities, in which extended
families, the union, and welfare facilities acted to ameliorate social isolation and provide a more effective support network.

We have also argued here that many of those disabled through exposure to dust
found the very basis of their masculinity undermined by their dependent status and
corrosion of their physical capacities. Hence we posit that the Clydeside heavy industries, which incubated a dominant machismo work culture and exalted the "hard
man" from the 1930s to the 1970s, also had the capacity to undermine masculinity as
disabled and unemployed miners and shipyard and metal workers found it impossible
to fulfill the "breadwinner" role. Whether this tentative hypothesis will stand up to
closer scrutiny remains to be seen.

Finally, our contention here has been that a failure to move much beyond the documentary evidence has resulted in occupational health studies in the United Kingdom
that fail to adequately understand the dynamics of workplace practices and culture,
especially the interactions of the latter upon the body. However, one of the criticisms
of the kind of oral history methodology used here is that it can be far too subjective in
nature. Postmodern historians would argue that the true value of oral history testimony is that it can open up a range of different discourses and, in so doing, reveal
more about the people being interviewed than the actual subject under inquiry. When
interviewing workers it is important to realize that they are not only reflecting on their
past but also trying to make some sense of how the past has contributed to their present circumstances. Their testimonies, then, are a constant and complex interplay of the

[52] Giovanna Di Chiro, "'Living Is for Everyone': Border Crossings for Community, Environment,
and Health" (this volume).

past upon the present and the present upon the past, in which feelings of anger, the need to attribute responsibility, combined with elements of guilt that the victims may have in some way contributed to their own and others' injuries, intermingle and sometimes clash with the need to recount to a researcher as accurately as possible what happened to them many years ago. Historians of a more linguistic bent would argue that these men were steeped in a victim's discourse, in which their main agenda was to apportion blame and to reestablish self-respect. Certainly, the notion of discourse analysis is a fascinating one, and we would agree that this way of treating oral history—if done properly—can reveal a multilayering of experiences and perceptions and tell us a lot about the nature of memory. Moreover, we have argued here that a critical interrogation of oral testimonies can elucidate other layers of experience—getting the historian within the choking dust clouds of the workplace and the hidden, private world of disabled, breathless workers and their families. Not only are oral history respondents telling us what happened in the past in the best way they can, but they are also frequently doing so from the point of view of the indisputable materiality of their damaged bodies.

Biological Citizenship:

The Science and Politics of Chernobyl-Exposed Populations

*By Adriana Petryna**

ABSTRACT

In the transition out of socialism to market capitalism, bodies, populations, and categories of citizenship have been reordered. The rational-technical management of groups affected by the Chernobyl disaster in Ukraine is a window into this contested process. Chernobyl exemplifies a moment when scientific knowability collapsed and new maps and categories of entitlement emerged. Older models of welfare rely on precise definitions situating citizens and their attributes on a cross-mesh of known categories upon which claims rights are based. Here one observes how ambiguities related to categorizing suffering created a political field in which a state, forms of citizenship, and informal economies were remade.

INTRODUCTION

> "Common sense is what is left over when all the more articulated
> sorts of symbol systems have exhausted their tasks."
> —Clifford Geertz, *Local Knowledge*[1]

This essay explores the forms of scientific cooperation and political management that emerged after the Chernobyl nuclear disaster of 1986. It is about how such managements are interconnected with global flows of technology and their integration into state-building processes, new market strategies, and governance and citizenship in post-Soviet Ukraine. Together with such dynamics, the essay considers, through ethnographic example, how local claims of disease and health are refracted through such institutions, how the sociopolitical contexts in which scientific knowledge is made can influence particular courses of health and disease and outcomes of these conditions. The aim here is to articulate the circumstances through which communities of "at-risk" populations come into being; to show how norms of citizenship are related to such circumstances; and to show how such norms propagate through everyday scientific understandings and practices related to institutions of medicine and law in Ukraine. A set of working relations informs or is at stake in the propagation of

---

* Graduate Faculty of Political and Social Science, New School University, 65 Fifth Avenue, New York, NY 10003; petrynaa@newschool.edu.

I thank the editors and reviewers of this volume for their helpful comments and suggestions.

[1] Clifford Geertz, *Local Knowledge: Further Essays in Interpretive Anthropology* (New York, 1983), 92.

individual claims of being at risk. They involve the sciences of global institutions and experts, national sciences and laws, local bureaucratic contingencies, and familial dynamics of suffering. These relations are indeed "working" in the sense that they affect perceptions of the seriousness and scale of the disaster, claims to its continuing harm, and the scientific, economic, and political modes through which such harm is addressed. How do different systems of modeling risk from Chernobyl affect people's capacities to reason politically? How might the choice of illness, rather than health, become a form of "common sense" expressive of these models? These questions are explored in a context in which science is inextricably connected to state-building processes, and market developments are quite productively intertwined, generating new institutions and social arrangements through which citizenship, experience, and ethics are being altered.

My book, *Life Exposed: Biological Citizens after Chernobyl,* elucidates how scientific knowledge and Chernobyl-related suffering were tooled to access social equity in a harsh market transition. More generally, it showed that in this new state, science and politics were engaged in a constant process of exchange and mutual stabilization.[2] This essay builds on that material by showing how contested attempts to intervene and to quantify radiation risk shaped the nature of the postdamage legal and political regime. Viewed longitudinally, the Chernobyl aftermath exemplifies a process wherein scientific knowability collapses and new categories of entitlement emerge. Ambiguities related to categorizing suffering create a political field in which a state, forms of citizenship, and informal economies of health care and entitlement are remade. This appropriation of suffering at all levels is one aspect of how images of suffering are becoming increasingly objectified in their legal, economic, and political dimensions.[3] This essay is specifically concerned with how these objectifications become a form of common sense and are enacted by sufferers in ways that can intensify the political stakes of suffering and promote protection, as well as new kinds of vulnerability, in domestic, scientific, and bureaucratic arenas.

THE EVENT

The Chernobyl nuclear reactor's Unit Four exploded in Ukraine on April 26, 1986. The damages from this disaster have been manifold, including immediate injury in the form of radiation burns and death to plant workers, damaged human immunities and high rates of thyroid cancer among resettled populations, and substantial soil and waterway contamination. Soviet reports attributed the cause of the disaster to a failed experiment. According to one official report, "The purpose of the experiment was to test the possibility of using the mechanical energy of the rotor in a turbo-generator cut off from steam supply to sustain the amounts of power requirements during a power failure."[4] Many of the reactor's safety systems were shut off for the duration of the experiment. A huge power surge occurred as technicians decreased power and shut off

---

[2] Adriana Petryna, *Life Exposed: Biological Citizens After Chernobyl* (Princeton, 2002).

[3] Arthur Kleinman and Joan Kleinman, "The Appeal of Experience; The Dismay of Images: Cultural Appropriations of Suffering in Our Times," *Daedalus* 125 (1999): 1–24. See also Veena Das, *Critical Events: An Anthropological Perspective on Contemporary India* (Oxford, 1995). I use pseudonyms for the majority of people interviewed for this essay. Names that appear in scientific and legal print are in some cases actual.

[4] See Soviet State Committee on the Utilization of Atomic Energy, *Report to the IAEA* (Vienna, 1986), 16.

the steam. The unit exploded once at 1:23 A.M. and then again. Due to particular wind-pressure gradients that day and in the following weeks, the radioactive plume moved to an estimated height of eight kilometers. Subsequent attempts to extinguish the flames of the burning graphite core proved only partly successful. By most accounts, they even exacerbated the danger of the situation. For example, an attempt was made to suffocate the flames with tons of boron carbide, dolomite, sand, clay, and lead dropped from helicopters. As a result, the core's temperature increased. The cloud of radiation rose dramatically and moved across Belarus, Ukraine, Russia, Western Europe, and other areas of the Northern Hemisphere.[5]

An official announcement of the disaster came almost three weeks after the event. In that time, roughly 13,000 children in contaminated areas took in a dose of radiation to the thyroid that was more than two times the highest allowable dose for nuclear workers for a year.[6] A massive onset of thyroid cancers in adults and children began appearing four years later. Had nonradioactive iodine pills been made available within the first week of the disaster, the onset of this disease could have been significantly reduced. Soviet administrators contradicted assessments of the scale of the plume made by English and American meteorological groups. The Soviets claimed the biomedical aspects of Chernobyl were under control. Dr. Angelina Guskova of the Institute of Biophysics in Moscow initially selected 237 victims to be airlifted to her institute's acute radiation sickness ward. Acute radiation syndrome (ARS) was diagnosed among 134 of them. The official death toll was set at 31 persons, most of them fire fighters or plant workers.

The disaster continued, especially among the groups of workers who were recruited or went voluntarily to work at the disaster site. Among the hundreds of thousands of paid and unpaid laborers,[7] work ranged from bulldozing polluted soil and dumping it in so-called radiation dumpsites (*mohyl'nyky*), to raking and shoveling pieces of the reactor core—radioactive graphite—that had dispersed over a vast area, to constructing fences around the reactor, to cutting down highly contaminated surrounding forests. By far the most dangerous work involved the adjacent reactor's roof. In one-minute intervals, workers (mainly military recruits) ran onto the roof, hurled radioactive debris over parapets into containers below with their shovels, and then left. Many of these volunteers called themselves "bio-robots"; their biologies were exploited "and then thrown out." Based on extensive interviews, some laborers felt trapped and unable to leave the disaster area; this sentiment was particularly felt by unpaid military recruits and local collective farmworkers recruited to do the most menial and dangerous of tasks. Some said they went gladly, believing their tripled salary more than compensated for their risk. However, it cannot be definitively said that money truly compensated them for the suffering that was to come.

Five months after the disaster, a so-called sarcophagus (now simply called the Shelter) was built to contain the 216 tons of uranium and plutonium in the ruined reactor. At present, the power plant is decommissioned. Some fifteen thousand people conduct maintenance work or service the Zone of Exclusion. Most of the exclusion zone is located in Ukraine. The zone circumscribes the disaster site and covers thirty kilometers in diameter. Zone entry is limited to the plant's workers.

---

[5] See Alexander Sich, "The Denial Syndrome (Efforts to Smother the Burning Nuclear Core at the Chernobyl Power Plant in 1986 Were Insufficient)," *Bulletin of Atomic Scientists* 52 (1996): 38–40.

[6] See Yurii Shcherbak, "Ten Years of the Chernobyl Era," *Scientific American,* April 1996, 46.

[7] Estimates vary from 600,000 to 800,000. These workers came from all over the Soviet Union. The labor pool, however, drew heavily from the Russian and Ukrainian populations.

Ukraine inherited the power plant and most of the Zone of Exclusion when independence was declared in 1991. The government announced new and ambitious standards of safety. It focused its resources on stabilizing the crumbling Shelter, implementing norms of worker safety, decreasing the possibility of future fallout risk, and decommissioning all units of the Chernobyl plant. These acts were important from a foreign policy standpoint. Showing that it could adhere to strict safety standards, Ukraine became the recipient of European and American technical assistance, loans, and trading partnerships. The legacy of Chernobyl has been used as a means of signaling Ukraine's domestic and international legitimacy and staking territorial claims; and as a venue of governance and state building, social welfare, and corruption.

Some maintenance workers lived in government-constructed housing units in Kyiv, the country's capital, sixty miles south of the disaster area. They work in the zone for two weeks and then return home for two weeks. I met one such worker in 1992, the first time I traveled to the country. He identified himself as a "sufferer," a legal classification instituted in 1991 for Chernobyl-affected individuals. He complained about how little his compensation (about five U.S. dollars a month) was in relation to rising food prices.[8] The man was in absolute despair, trapped because he had nowhere else to work. He said he had attempted to find employment elsewhere, but nobody would hire him on account of his bad health and work history. The man linked his suffering to first a precarious and dangerous Soviet management of the aftermath, and then a complex medical and legal apparatus he felt unable to navigate. He then showed me a work injury, a flap of skin that had puckered and formed a kind of ring just above his ankle. Direct contact with a source of ionizing radiation had apparently caused it. His sense of violation and loss were clear when he referred to himself as a "living dead," whose memory of who he was in a former life "is gone."

In 2000, I interviewed the director of the Shelter complex. What I learned was that almost a decade after independence, worker protections, in spite of some improvements, were still deficient. The director told me that norms of radiation safety were inoperative. In a place of tremendous economic desperation, people *competed* for work in the Zone of Exclusion, where salaries were relatively high and steadily paid. Prospective workers engaged in a troubling cost-benefit assessment that went something like this: if I work in the Zone, I lose my health. But I can send my son to law school. "Taking this risk is their individual problem. No one else is responsible for it," the director told me. He compared Ukraine's mode of enforcing safety standards with European modes and told me that the "value" of a dose exposure remained untallied in Ukraine. In Europe, such values are calculated on the basis of the rem-expenditures workers incur; international safety standards limit the amounts. Despite the existence of these international limitations, the director's comment suggests that norms of worker exposures are in fact being decided locally and within the constraints of a national economy. In effect, he was revealing to me the extent to which workers' lives are undervalued by being overexposed (for much less pay). Yet however undervalued his workers' lives may be, they are still driven to work by a situation in which

---

[8] The karbovanets (Krb) was Ukraine's legal tender from 1992 to 1996. Exchange rates per US$1.00 plunged between 1992 and 1993. In March 1992, the exchange rate was Krb640:$1. By March 1993, that rate had fallen to Krb12,610:$1. Subsequent rates were as follows: 1994—Krb104,200:US$1; 1995—179,900:$1; 1996—188,700:$1. The hryvnia (Hrn) replaced the karbovanets at Hrn1: Krb100,000 in September 1996. The exchange rates were as follows: 1997—Hrn1.84:US$1; 1998—2.04:$1; 1999—4.13:$1; 2000—5.44:$1.

economic forces are overwhelming. In such an environment, physical risks escalate and risky work is seen as acceptable and even normal.

"As a result of all the compounding uncertainties in the factors involved," wrote Frank von Hippel, "our estimates of the long-term health consequences of the Chernobyl accident are uncertain even as to the order of magnitude."[9] Indeed, available models of assessment could not account for the scope of the disaster. As the short history of the disaster indicates, rational-technical responses and political administrations (both in the Soviet and Ukrainian periods) have been compounding factors in the medical and welfare tragedy that now affects more than 3.5 million people in Ukraine alone. Contested scientific assessments of the disaster's extent and medical impact, the decision to postpone public communication, and the economic impetus to work in the exclusion zone have made Chernobyl a *tekhnohenna katastrofa* (a technogenic catastrophe). This is a term that was used among my informants, including people fighting for disability status, local physicians, and scientists. It suggests that not only radiation exposure but also political managements have produced new biological uncertainties.

Ulrich Beck noted that Chernobyl was an "anthropological shock" in Western Europe. The shock came from the fact that everyday knowledge proved useless in the face of this catastrophe, as did expert knowledge.[10] This "collapse" of knowledge also occurred, but in another way, in the other Europe. Chernobyl was associated with the collapse of Soviet life in general. Knowledge about risk, how to deliver it, how to value it, became something of a political resource. In this disaster's wake a state, a society, and knowledge and experience of health have been reconfigured.

In exploring this aftermath, I use a methodological approach that involves moving back and forth between vulnerable persons and the everyday bureaucracies and procedures by which they express their desires, claims, and needs for protection and security. Such an ethnographic mode of engagement is in itself meant to question the possibility of a linear account or an all-or-none moral or political solution to this complex reality. Instead, its dynamics are approached from a prismatic point of view to gain a broader perspective on the interests and values involved in particular claims and sites.

EXPERIMENTAL MODELS AND ETHNOGRAPHIC METHODS

Between 1992 and 1997, I conducted archival and field research in Ukraine, Russia, and the United States. In Ukraine, I worked with resettled families and radiation-exposed workers. I also carried out archival research in the country's new Chernobyl Ministry, the Health Ministry, and Parliamentary Commissions on Human Rights. I conducted interviews with key scientific and political actors in Kyiv and Moscow, comparing scientific standards informing concepts of biological risk and safety in the Soviet and post-Soviet administrations of the aftermath. The very nature of the problem, that is, understanding the everyday lived aspects of the Chernobyl aftermath, led me to a number of different sites and challenges. One of those challenges involved understanding how scientific knowledge about radiation risk was being circulated, assimilated, or rejected at the various levels (international, national, and local) in which interventions were being made.

---

[9] Frank von Hippel, *Citizen Scientist* (New York, 1991), 235.
[10] Ulrich Beck, "The Anthropological Shock: Chernobyl and the Contours of a Risk Society," *Berkeley Journal of Sociology* 32 (1987): 153–65.

I examined claims about the scale of the disaster made by scientific experts affiliated with the International Atomic Energy Agency. I compared expert knowledge with that of basic scientists in U.S. radiation laboratories and learned about how radiobiologists went about evaluating radiobiological effects at the cellular and subcellular levels.

As a consequence, I could better situate expert claims and their measures in the context of their laboratory production and testing. I soon discovered that there was a "black box" separating knowledge about the effects of low-dose radiation at the animal (laboratory) level and human (field) level. The dose-effect curves for high doses of radiation were one to one and fairly straightforward. The same could not be said for ongoing exposures at low doses (a typical condition after Chernobyl). On the one hand, experts promoted their authority, based in part on their mastery of what composed appropriate evidence of Chernobyl-related injury. On the other hand, there was considerable disagreement at the laboratory level over what the terms for interpreting radiation-induced biological risk in human populations are. International experts' projections about the health effects of Chernobyl often contradicted people's lived sense of those effects. For Ukrainian scientists, the lack of consensus at the basic science level meant that the criteria of evaluation of injury were, in essence, contestable.

Ukraine became a most compelling place to examine the relations between risk, rational-technical power, and the emergence of new populations. Indeed, a new political, economic, and moral arena had been thrown open owing to the absence of consistent evaluative criteria. During the period of my field research, the country saw the growth of a population claiming radiation exposure qualified them for some form of social protection. Social protections included cash subsidies, family allowances, free medical care and education, and pension benefits for sufferers and the disabled. This new population, named *poterpili* (sufferers), numbered 3.5 million and constituted 7 percent of the population. A political economy of Chernobyl-related illnesses with new kinds of social categories and hierarchies of entitlement was emerging. An individual classified as "disabled" received the best entitlement package as compared with a mere "sufferer." Nonsufferers, that is, people outside the Chernobyl compensation system, had even less or no chance of receiving state social benefits. Scientific know-how became essential to the negotiation of everyday life and the maintenance of one's status in the Chernobyl system. One had to know one's dose and be able to relate it to one's symptoms and work experiences in the Zone of Exclusion. The effectiveness of this knowledge determined the place one could occupy and how long one could occupy it in the system of management of Chernobyl populations.

Today, approximately 8.9 percent of Ukraine is considered contaminated. On average, 5 percent of its state budget is spent on Chernobyl-related expenses. This includes costs related to the environmental cleanup and technical support of the destroyed reactor. The majority of funds (65 percent), however, are spent on social compensations and financial maintenance of the Chernobyl public health and scientific apparatus. Belarus was much more heavily affected than Ukraine. Nearly 23 percent of its territory is contaminated. Contrastively, Belarus expends much less than its southern neighbor does on affected populations; it has curbed its sum of Chernobyl claimants—as has Russia.[11] Dr. Guskova, who oversees the Russian compensation

[11] In Russia, the number of people considered affected and compensable has been kept to a minimum and remains fairly stable (about 350,000, including 300,000 Zone of Exclusion laborers and 50,000 resettled).

system for workers of nuclear installations, including Chernobyl, is a well-known critic of Ukraine's compensation system. She told me that Ukrainians were inflating their numbers of exposed persons, that their so-called invalids "didn't want to recover." She saw the illnesses of this group as a "struggle for power and material resources related to the disaster."

In response to her former colleague's indictment, Dr. Angelina Ceanu, a neurophysiologist and physician to Chernobyl victims in Kyiv, told me, "It is inconceivable that an organism of any kind is passive to its own destruction." Her response was based on evidence from experiments conducted by the Soviet radiobiologist V. L. Komarov. In one experiment conducted in the late 1950s, he observed that sleeping rats, without provocation, woke up when exposed to small amounts of ionizing radiation. From these examples one can begin to appreciate how competing scientific models (animal vs. human; psychometric vs. biological; laboratory vs. field-based), financial agendas, and distinct moral attitudes regarding the need for scientific work in this arena were not simply at odds with each other. Their confrontation opened up a novel social arena consisting of contested claims around radiation illness. Indeed, a number of civic organizations lobbying for the right to compensation for such illnesses evolved with the biomedical and political institutions promoting "safe living" in Ukraine. These so-called *fondy* (funds) were conduits of international charity and represented the concerns of exclusion zone workers and resettled persons living in Ukraine. These funds enjoyed tax-exempt status and with their numbers (more than 500 in 1996) established an informal economy of a variety of imported goods, including vehicles, drugs, and frozen and dry foodstuffs. In short, the Chernobyl aftermath became a prism of the troubled political-economic and social circumstances that typified the Ukrainian transition to a market economy. The production of scientific know-how, markets, and state formations were mutually embedded, generating new inequalities and opportunities in the redefinition of citizenship and ethics.

This work is based on multiple lengthy research visits to various state, scientific, and domestic contexts during 1992–1995, fieldwork conducted during 1996–1997, and a follow-up visit in 2000. The Radiation Research Center, also known as Klinika, became a primary focus of the field research. The center was established in 1986 to monitor the health of zone laborers; shortly afterward it began providing similar services for resettled persons. Its national-level Medical-Labor Committee (Ekspertiza) comprises scientists, physicians, and administrators who have the authority to diagnose illnesses as Chernobyl-related (there are twelve regional committees). Patients with illnesses diagnosed as such receive a document, the so-called Chernobyl tie, which qualifies bearers to receive compensation privileges as a result of their Chernobyl-related illnesses. By 1996, the center had become the site of intense scientific and legal disputes. I observed physicians, nurses, and patients as they negotiated over who should receive the tie. I looked into current research, particularly in the center's neurological division. I also carried out interviews with sixty middle-age male and female patients and reviewed their medical histories, their illness progressions, and their experiences in attempting to qualify for disability status. A significant aspect of my research focused on the daily lives of the clinic's male patients and their families. I was concerned with how their belonging to a political economy of illness displaced their self-perceptions and roles as breadwinners and paternal figures. I traced changing experiences of *lichnost'*, a Russian-Soviet model of personhood evidenced in a person's work ethics and level of commitment to a collective of laborers,[12] the effects such changes had on domestic life, and the techniques household members used to have their illnesses count in the rational-technical domain in which their futures came to be addressed.

These anthropological concerns illustrate the extent to which definitions of health and illness are embedded within spheres of politics and economics and are almost always connected with dimensions that go beyond the immediate body, such as interpersonal and domestic relationships. Arthur Kleinman has elucidated the "social course" of illness.[13] Other anthropologists, such as Veena Das and Nancy Scheper-Hughes, have been concerned with constructions of health as they indicate discrepancies in power, social position, and inequality, particularly as lived by marginal groups and individuals. Recent ethnographies of science have portrayed how, more and more, biomedical technologies play a key role in that constructedness. PET scans, genetically based diagnostics, and sonograms image biological facts and are therefore inseparable from the objects they recognize and remake as disease.[14] Social problems, health problems, and the technologies that image them are also linked. Anthropologist Paul Farmer has shown how patterns of "structural violence" affect the construction and expansion of populations at risk for diseases. Deteriorating health care, limited treatments, and inequalities are worsened by structural adjustment programs and have led to epidemics of preventable infectious diseases such as multidrug-resistant tuberculosis. Indeed, "social forces and processes come to be embodied as biological events."[15] In Ukraine, efforts to remediate the health effects of Chernobyl have themselves contributed to social and biological indeterminacy and novel formations of power. Radiation exposures and their unaccountability, bureaucratic interventions by the state and failures to intervene, the growth of clinical regimes, and harsh market changes intensified the course of illness and suffering. Thus in the Chernobyl aftermath, illness and health are engendered and made sense of within the technical and political domain in which they come to be addressed.

CONSTRUCTED UNKNOWNS

In what follows, I address some of the scientific elements that played a key role in measuring and delineating the scope of the disaster and defining remediation and compensation strategies. In this context, matters such as atmospheric dispersion maps, international scientific cooperations, and local scientific responses, as well as people's involvement in bureaucratic and testing procedures, led up to what can be called a "technical and political course of illness." Examples of people's engagement with, and influence on, such courses will then be discussed.

Most scientists today would agree that given the state of technology at the time of the disaster, specialists "did not know how to make an objective assessment of what had

---

[12] Oleg Kharkhordin, *The Collective and the Individual in Russia: A Study of Practices* (Berkeley, Calif., 1999).

[13] Arthur Kleinman, *Social Origins of Distress and Disease* (New Haven, Conn., 1986).

[14] Emily Martin, *Flexible Bodies: Tracking Immunity in American Culture from the Days of Polio to the Age of AIDS* (Boston, 1994); Rayna Rapp, *Testing Women, Testing the Fetus: The Social Impact of Amniocentesis on America* (New York, 1999); Joseph Dumit, *Picturing Personhood: Brain Scans and Biomedical Identity* (Princeton, N.J., 2004).

[15] Paul Farmer, *Infections and Inequalities: The Modern Plagues* (Berkeley, Calif., 1999), 5.

happened."[16] Tom Sullivan, who until recently directed the Atmospheric Release Advisory Capability (ARAC) group at Lawrence Livermore Laboratory in Livermore, California, agrees with this general appraisal.[17] Prior to the Chernobyl disaster, Sullivan's ARAC team had generated atmospheric dispersion models of the size and movement of nuclear plumes resulting from American and Chinese aboveground nuclear weapons tests and the Three Mile Island accident. "A 200 by 200 kilometer area had been sufficient to model prior radiation releases," he told me. "We did the imaging near the Chernobyl plant using this 200 kilometer square grid, but the grid was so saturated, I mean, you couldn't even make sense of it because every place had these enormously high radiation values. . . . *Our codes were not prepared for an event of this magnitude.*"[18]

Soviet scientists, too, were unprepared, but they did not admit their ignorance. In an August 1986 meeting with the International Atomic Energy Agency (IAEA), they presented a crude analysis of the distribution of radiation in the Zone of Exclusion and in the Soviet Union: "assessments were made of the actual and future radiation doses received by the populations of towns, villages, and other inhabited places. As a result of these and other measures, *it proved possible to keep exposures within the established limits.*"[19]

The issue at stake is the state's capacity to produce and use scientific knowledge and nonknowledge to maintain political order. Historian Loren Graham, for example, has written about how "false" sciences such as Lysenkoism, which denied the existence of the gene and advocated labor-intensive methods of accelerating crop yields, have been instrumental in shaping work psychology and social life in the socialist project.[20] The fact is that limited Soviet maps of Chernobyl helped to justify limited forms of dosimetric surveillance and resettlement actions. Nonknowledge became essential to the deployment of authoritative knowledge. High doses absorbed by at least 200,000 workers during 1986–1987 were insufficiently documented. According to one biochemist, many of the cleanup workers "received 6–8 times the lethal dose of radiation."[21] "They are alive," he told me. "They know that they didn't die. *But they don't know how they survived.*" His statement speaks to the extent to which not only knowledge but also ignorance were constructed and used as state tools for maintaining public order. As science historian Robert Proctor tells us in his informative book on how politics shapes cancer science, ignorance "is not just a natural consequence of the ever shifting boundary between the known and the unknown." It is a "political consequence" of decisions concerning how to approach what could and should be done to mitigate danger or disease.[22]

---

[16] *One Decade After Chernobyl* (Vienna, 1996).

[17] ARAC is a national emergency response service for real-time assessment of incidents involving nuclear, chemical, biological, or natural hazardous material.

[18] Sullivan's team offered technical assistance through a Swedish intermediary, but the offer was refused by Soviet administrators.

[19] Soviet State Committee on the Utilization of Atomic Energy, *The Accident at Chernobyl Nuclear Power Plant and Its Consequences.* Information complied for the IAEA Expert's Meeting, Aug. 25–29, 1986, Vienna; Zhores Medvedev, *The Legacy of Chernobyl* (New York, 1990).

[20] Loren Graham, *What Have We Learned about Science and Technology from the Russian Experience?* (Stanford, Calif., 1998).

[21] Symptoms of acute radiation sickness begin at 200 rem. At 400 rem, bone marrow failure sets in. Lethal dose (LD100) is a dose exposure that causes 100 percent of the death of cells or the human. LD50/30 is a dose exposure that causes 50 percent of the death of cells or the human within thirty days.

[22] Robert Proctor, *Cancer Wars: How Politics Shapes What We Know and Don't Know about Cancer* (New York, 1995), 7.

Chernobyl also became a venue for unprecedented international scientific cooper-ation and human research. President Mikhail Gorbachev personally invited a team of American oncologists led by leukemia specialist Robert Gale (UCLA) to conduct experimental bone marrow transplantations upon individuals whose exposures were beyond the lethal limit and for whom these transplantations were deemed appro-priate. Additionally, 400 workers selected by Dr. Guskova and others received a genetically-engineered hematopoietic growth factor molecule (rhGM-CSF), thought to regenerate stem cell growth. Though the results of the transplantations and trial proved unsuccessful, the medical work on this cohort (and the objective indices cre-ated around them) helped consolidate an image of a biomedical crisis that was being successfully controlled by cutting-edge scientific applications. In an effort to allevi-ate the public's fear, Dr. Gale appeared on television and walked barefoot in the zone with one of his children.

As this internationalization of science ensued, however, the physical management of contamination at the accident site was internalized—to the sphere of Soviet state control. One policy statement released by the Soviet Health Ministry at the height of these cooperations, for example, directed medical examiners in the Zone of Exclusion to "classify workers who have received a maximum dose as having "vegetovascular dystonia," that is, a kind of panic disorder, and a novel psychosocial disorder called "radiophobia" (or the fear of the biological influence of radiation). These categories were used to filter out the majority of disability claims.[23] Substantial challenges to this Soviet management came from certain labor sectors in subsequent years. At the end of 1989 only 130 additional persons were granted disability; by 1990, 2,753 more cases had been considered, of which 50 percent were authorized on a neurological ba-sis. Levels of political influence of specific labor sectors are reflected in the order they received disability: coal miners, then Ministry of Internal Affairs workers (the police), and then Transport Ministry workers. These various labor groups would soon realize that in the Ukrainian management of Chernobyl, forms of political leveraging had to be coupled with medical-scientific know-how.

Arguably, the new Ukrainian accounting of the Chernobyl unknown was part and parcel of the government's strategies for "knowledge-based" governance and social mobilization. In 1991 and in its first set of laws, the new parliament denounced the Soviet management of Chernobyl as "an act of genocide." The new nation-state viewed the disaster as (among other things) a key means for instituting domestic and international authority. Legislators assailed the Soviet standard for determining bio-logical risk to populations. The Soviets had established a high of 35 rem (a unit of ab-sorbed dose), spread over an individual's lifetime (understood as a standard seventy-year span), as the threshold of allowable radiation dose intakes. This threshold limited the scale of resettlement actions. Ukrainian law lowered the Soviet threshold dose to 7 rem, comparable to what an average American would be exposed to in his or her life-time. In effect these lowered measures for safe living increased the size of the labor forces going to the exclusion zone (since workers had to work shorter amounts of time if they were to avoid exceeding the stricter dose standards). The measures also ex-panded territories considered contaminated. A significant new sector of the population

[23] In my interviews, I heard instances of workers mimicking symptoms of ARS (vomiting, for ex-ample). This shows the level of desperation on the part of some of them to receive permission to leave the zone.

would want to claim itself as part of a state-protected post-Soviet polity. A biophysicist responsible for conducting retrospective dose assays on resettlers told me: "Long lines of resettlers extended from our laboratory doors. It wasn't enough that they were evacuated to 'clean' areas. People got entangled in the category of victim, by law. They had unpredictable futures, and *each of them wanted to know their dose.*"

Statistics from the Ukrainian Ministry of Health gave evidence of the sharp increase in 1991 of zone workers, resettled persons and inhabitants of contaminated territories registering their disability, and the annual patterns of enrollment of this new population for which the state committed itself to care. The statistics also show that the sharpest increase in the clinical registration of illnesses occurred under the category "symptoms and other inadequately known states," Class 16 in the International Classification of Disease, ICD 10 (see Figure 1). These states typically include afflictions such as personality changes, premature senility, and psychosis.

Ukrainian claims to a sudden expansion of Chernobyl health effects became a target of international skepticism. Ukrainian scientists were often rebuked for their "failure to use modern epidemiological methods and criteria of causality and a reliable data system." As a World Bank consultant noted, "Right now virtually any disease is attributed to Chernobyl, and no effort is being made either to prove or disprove these claims that would satisfy standard epidemiological criteria of causality."[24] For the government, however, one can argue that these new statistics became a kind of "moral science,"[25] a resolute display of its intention to make visible the effects of the Soviet mismanagement of the disaster and to guarantee its own social legitimacy while keeping world attention on the Chernobyl risk.

In this daily bureaucratic instantiation of Chernobyl, tensions among zone workers, resettled individuals and families, scientists, physicians, legislators, and civil servants intensified. Together, these groups became invested in a new social and moral contract between state and civil society, a contract guaranteeing them the right to know their levels of risk and to use legal means to obtain medical care and monitoring. The sufferers and their administrators were also supported by the nonsuffering citizens, who paid a 12 percent tax on their salaries to support compensations. The hybrid quality of this postsocialist state and social contract comes into view. On the one hand, the Ukrainian government rejected Western neoliberal prescriptions to *downsize* its social welfare domain; on the other hand, it presented itself as informed by the principles of a modern risk society. On the one hand, these Chernobyl laws allowed for unprecedented civic organizing; on the other hand, they became distinct venues of corruption through which informal practices of providing or selling access to state privileges and protections (*blat*) expanded.[26]

Ethnographic accounts have illustrated that postsocialism's future cannot be based in predictive models or treated as unproblematic flows toward free markets. Michael Burawoy and Katherine Verdery point to the links between the socialist and postsocialist worlds as well as growing dependencies between postsocialist state formations and global economics. Such dependencies "have radically shifted the rules of the game, the parameters of action within which actors pursue their daily routines

---

[24] World Bank, *Managing the Legacy of Chernobyl* (Washington, D.C., 1994), 7:6.
[25] Ian Hacking, *Taming of Chance* (Cambridge, 1990).
[26] For an elaboration of the concept of *blat,* see Alena Ledeneva, *Russia's Economy of Favours: Blat, Networking, and Informal Exchange* (Cambridge, 1998).

Figure 1. Symptoms and Other Inadequately Known States (per 10,000)

1982	1983	1984	1985	1986	1987	1988	1989	1990	1991	1992
1.3	1.7	1.7	1.9	2.3	2.7	5.9	34.7	108.3	127.4	141.3

SOURCE: Ministry of Statistics, Kyiv, Ukraine.

and practices."[27] Ethnographic methods are critical for elucidating such interrelated processes at local levels. This is particularly true with regard to assessing the decisions people make based on limited choices available to them and the informal aspects of power that inform those decisions.

Shifts in aggregate human conditions and the circumstances of citizenship are also at stake in these changing political and economic worlds. The principles of a "classical citizenship" endow citizens with natural and legal rights protected as matters of birthright.[28] Regardless of nationality, such protections were granted to all Ukrainian inhabitants when the country declared independence. Yet birthright remains an insufficient guarantor of protection as the lives of inhabitants of some Ukrainian areas cannot be fully, or even partly, protected owing to long-term environmental challenges. For these inhabitants, the very concept of citizenship is charged with the superadded burden of survival. The acquisition and mastery of certain democratic forms related to openness, freedom of expression, and the right to information are primary goals to be sure. Yet populations are also negotiating for the even more basic goal of protection (i.e., economic and social inclusion) using the constituent matters of life. Such negotiations expose certain patterns that are traceable elsewhere: the role of science in legitimating democratic institutions, increasingly limited access to health care and welfare as the capitalist trends take over, and the uneasy correlation of human rights with biological self-preservation.

BIOLOGICAL CITIZENSHIP

In Ukraine, where democratization is linked to a harsh market transition, the injured biology of a population has become the basis for social membership and for staking claims to citizenship. Government-operated radiation research clinics and nongovernmental organizations mediate an informal economy of illness and claims to a "biological citizenship"—a demand for, but limited access to, a form of social welfare based on medical, scientific, and legal criteria that recognize injury and compensate for it. These demands are being expressed in the context of losses of primary resources such as employment and state protections against inflation and a deterioration in legal-political categories. Struggles over limited medical resources and the factors that constitute a legitimate claim to citizenship are part of postsocialism's uncharted terrain. Against a stark and overwhelming order of insecurity, there are questions to be asked about how the value of another's life is being judged in this new political

---

[27] Michael Burawoy and Katherine Verdery, *Uncertain Transition: Ethnographies of Change in the Postsocialist World* (Lanham, Md., 1999), 2.

[28] Dominique Schnapper, "The European Debate on Citizenship," *Daedalus* 126 (1997): 201

economy, about the ability of scientific knowledge to politically empower those seeking to set that value relatively high, and about the kinds of rationalities and biomedical practices emerging with respect to novel social, economic, and somatic indeter-
minacies. The indeterminacy of scientific knowledge about the afflictions people face
and about the nature of nuclear catastrophe materializes here as both a curse and a
source of leverage. Ambiguities related to the interpretation of radiation-related injury, together with their inextricable relations to the social and political uncertainties generated by Soviet interventions and current political-economic vulnerability,
make the scope of the afflicted population in Ukraine and its claims to injury at once
plausible, ironic, and catastrophic.

One instance of how these scientific and political dynamics operated in the everyday: the country's eminent expert on matters related to the disaster, Symon Lavrov,
was well-regarded internationally for having developed computerized fallout models
and calculating population-wide doses in the post-Soviet period. He told me, however, that "when a crying mother comes to my laboratory and asks me, Professor
Lavrov, 'tell me what's wrong with my child?' I assign her a dose and say nothing
more. I double it, as much as I can." The offer of a higher dose increased the likelihood that the mother would be able to secure social protection on account of her potentially sick child. Lavrov and the grieving mother were two of the many figures
whose efforts I documented. The point is the following: the mother could offer her
child a dose, a protective tie with the state, which is founded on a probability of sickness, a biological tie. What she could offer, perhaps the most precious thing she could
offer her child in that context, is a specific knowledge, history, and category. The
child's "exposure" and the knowledge that would make that exposure an empirical
fact were not things to be repressed or denied (as had been tried in the Soviet model)
but rather things to be made into a resource and then distributed through informal
means.

Specific cases illustrate how these economic and state processes, combined with the
technical dynamics already described, have laid the groundwork for such "counterpolitics."[29] Citizens have come to depend on obtainable technologies and legal procedures to gain political recognition and admission to some form of welfare inclusion.
Aware that they had fewer chances for finding employment and health in the new market economy, these citizens accounted for elements in their lives (measures, numbers,
symptoms) that could be linked to a state, scientific, and bureaucratic history of mismanagement and risk. The tighter the connection that could be drawn, the greater the
chance of securing economic and social entitlement. This dimension of illness as
counterpolitics suggests that sufferers are aware of the way politics shapes what they
know and do not know about their illnesses and that they are put in a role of having to
use these politics to curb further deteriorations of their health, which they see as resulting, in part, from a collapsing state health system and loss of adequate legal protections.

Probability in relation to radiation-related disease became a central resource for local scientific research. This play with probability was being projected back into nature, so to speak, through an intricate local science. Young neuropsychiatrists made
the best of the inescapability of their political circumstances (they could not get visas

---

[29] Colin Gordon, "Government Rationality: An Introduction," in *The Foucault Effect: Studies in
Governmentality,* ed. G. Burchell, C. Gordon, and P. Miller (Chicago, 1991), 5.

to leave the country) as they integrated international medical taxonomies into Soviet ones and developed classifications of mental and nervous disorders that in expert literatures were considered far too low to make any significant biological contribution. For example, neuropsychiatrists were involved in a project designed to find and assess cases of mental retardation in children exposed in utero in the first year after the disaster. In the case of one such child, a limping nine-year-old boy, researchers and parents pooled their knowledge to reconstruct the child's disorder as having a radiation origin. Even though the boy's radiation dose was low, he was given the status of sufferer because of his mother's occupation-related exposure (she was an emergency doctor who elected to work in the zone until late in her pregnancy) and also because a PET scan did reveal a cerebral lesion that was never hypothesized as being related to anything other than radiation. (It could have been birth trauma.) As researchers constructed a human research cohort, they were also constructing a destiny for the newly designated human research subjects. It was precisely the destiny the parents were intent on offering to this child—a biological citizenship.

These radiation-related claims and practices constituted a form of work in this market transition. A clinical administrator concurred that claims to radiation illness among the Ukrainian population amounted to a form of "market compensation." He told me, "If people could improve their family budgets, there would be a lot less illness. People are now oriented towards one thing. They believe that only through the constitution of illnesses, and particularly difficult illnesses, incurable ones, can they improve their family budgets." Administrators such as he informed me that they should not to be "blamed too much" for fueling an informal economy of diagnoses and entitlements. Complicities could be found at every level, and the moral conflicts they entailed were publicly discussed. Another administrator who authenticated compensation claims told me illnesses had become a form of currency. "There are a lot of people out of work," he said. "People don't have enough money to eat. The state doesn't give medicines for free anymore. Drug stores are commercialized." He likened his work to that of a bank. "The diagnosis we write is money."

The story of Anton and Halia (age forty-two in 1997) shows the ways such complicity functioned in the most personal arenas. The new institutions, procedures, and actors that were at work at the state level, at the research clinic, and at the level of civic organizations were making their way into the couple's *kvartyra* (apartment). Anton's identity as a worker, his sense of masculinity, and his role as a father and breadwinner were being violently dislocated and altered in the process. In 1986, the state recruited Anton to work for six months in the Zone of Exclusion, transporting bags of lead oxide, sand, and gravel to the reactor site. The bags were airlifted and deposited using helicopters. He had no idea how much radiation he absorbed during those six months. From 1991 on, Anton routinely passed through the clinical system, monitored like any "prospective" invalid. His symptoms mounted over time. He had chronic headaches, lost his short-term memory, exhibited antisocial behavior, developed a speech disorder, and experienced seizures and impotence, as well as many other problems. Despite the growing number and intensity of his symptoms, his diagnosis did not "progress" from an initial listing as a "psycho-social" case.

When I met Anton and his wife, Halia, they were trying to manage on a small pension he received as a sufferer. Anton saw himself as bankrupt, morally as well as economically: "The state took my life away. Ripped me off, gone. What is there to be happy about? An honorable man cannot survive now. For what? For what? We had a life. We had butter. We had milk. I can't buy an iron. Before I could buy fifty irons. The money was there. My wife's salary is less than the cost of one iron." He told me that he did not know "how to trade goods" or to sell petty goods on the market. His meager pension left Anton with few options. He found himself confronting the shameful option of breadwinning with his illness in the Chernobyl compensation system or facing poverty. Over time, and in a concerted effort to remove Anton's psychosocial label, the couple befriended a leader of a disabled workers' activist group in a clinic. Through him they met a neurologist who knew the director of the local medical-labor committee. The couple hoped this individual would provide official support for Anton's claim of Chernobyl-related disability.

The economic motives for these actions were clear. Yet it was difficult for me to see this man giving up everything he knew or thought about himself to prove that his diffused symptoms had an organic basis. Neurology was a key gateway to disability; neurological disorders were most ambiguous but most possible to prove using diagnostic technologies, self-inducements, and bodily display. At each step, Anton was mentally breaking down; he fell into a pattern of abusive behavior. His legal-medical gamble—this gaining of life in the new market economy through illness—reflected the practices of an entire citizenry lacking money or the means of generating it. This approach has become common sense, in Clifford Geertz's words, or that which is "left over when all [the] more articulated sorts of symbol systems have exhausted their tasks."[30]

When I returned in 2000 to Kyiv to conduct further research, I discovered that current democratic politicians, many of whom drafted the original compensation laws as sovereignty-minded nationalists, now saw the Chernobyl compensation system as a dire mistake that has "accidentally" reproduced a socialist-like population. Funds and activist groups were now supported by socialist and communist leaderships, who lobbied for continued aid in an increasingly divided parliament. Meanwhile, international agencies such as the World Bank cited the Chernobyl social apparatus as a "dead weight" to Ukraine's less-than-ideal transition to a market economy. Bank officials were so ill-disposed toward the system that they made its quick extinction a condition of future loan contracting. The disappearance of this exposed population from the state's radar seems ever more likely. Once "protected" by a safety-conscious state, this exposed population is being left alone to their symptoms and social disarray.

Opinions about how the state should address the fate of these Chernobyl victims also serve as a kind of barometer of the country's changing moral fabric. Rural inhabitants who normally received the least in terms of socialist redistribution tended to be sympathetic to the victims' struggles. Among inhabitants of Kyiv and other urban centers, there is a growing consensus that the invalids are "parasites of the state, damaging the economy, not paying taxes." Many youths who had been evacuated from the zone do not want to be associated with groups of sufferers as this association makes it more difficult for them to find employment.

Chernobyl was a key political event, generating many effects, some of which have yet to be known; its truths have been made only partly known through estimates derived from experimental science. The immediate postindependence discourse in Ukraine centered on the "truth" of Chernobyl. Ukrainians tried to put their suffering in perspective vis-à-vis the repressive model of science and state: the number of

---

[30] Geertz, *Local Knowledge* (cit. n. 1).

people who died, how the government deceived citizens about the scale of the disaster, how the maps of contamination were misrepresentative, and so on. As harsh market realities entered everyday life, this model of organizing suffering quickly gave way to a different kind of scientific and political negotiation, one which had directly to do with the maintenance, and indeed the remaking, of a postsocialist state and population.

If, at the level of the modern state, spheres of scientific production and politics are in a constant process of exchange and mutual stabilization, then what I have suggested here is that stabilization proves to be a much more difficult task. At stake in the Chernobyl aftermath is a distinctive postsocialist field of power-in-the-making that is using science and scientific categories to establish the state's reach. Scientists and victims are also establishing their own modes of knowledge related to injury as a means of negotiating public accountability, political power, and further state protections in the form of financial compensation and medical care. Biology becomes a resource in a multidimensional sense—versatile material through which the state and new populations can be made to appear. This postsocialist field of power has specific physical, experiential, political, economic, and spatial aspects. It is about knowledge and constructed ignorance, visibility and invisibility, inclusion and exclusion, probabilities and facts, and the parceling out of protection and welfare that do not fit predictive models. It is also about how individuals and populations become part of new cooperative regimes in scientific research and in local state-sponsored forms of human subjects protection. In this context, suffering is wholly appropriated and objectified in its legal, economic, and political dimensions. At the same time, these objectifications constitute a common sense that is enacted by sufferers themselves in ways that can promote protection as well as intensify new kinds of vulnerability in domestic, scientific, and bureaucratic spheres.

Uncertain Exposures and the Privilege of Imperception:

Activist Scientists and Race at the U.S. Environmental Protection Agency

By Michelle Murphy[*]

ABSTRACT

This paper locates the EPA national headquarters within the racialized local geography of southwest Washington, D.C. By focusing on the formation of a scientist union and the union's struggle to make visible an episode of chemical exposure in its own offices, the paper connects the work of racialized privilege with the difficulty of proving chemical exposures in the 1980s.

INTRODUCTION

> These things happen to people in poor areas. Society is set up in such a way that it is the poor and the uneducated who suffer the main impact of natural and man-made disasters. People in low-lying areas get the floods, people in shanties get the hurricanes and tornadoes. . . . I'm not just a college professor, I'm a head of a department. I don't see myself fleeing an airborne toxic event.
> —Jack Gladney in Don DeLillo's novel *White Noise* (1985)[1]

What happens when chemical exposures do not obey systems of privilege? Society is set up to protect the privileged from toxic events, or so the neurotic protagonist of the novel *White Noise,* Jack Gladney, insists. In the novel, Gladney's affluence produces both an anxiety and a blindness about his own vulnerability to errant plumes and accidental spills. When a toxic cloud from a train accident floats over his suburban neighborhood, Gladney finds himself on the run despite his worldly advantages. Even systems of privilege can disappoint.

"White noise" is a technical term describing a steady complex unobtrusive sound, such as the drone of a fan, that drowns out, or makes imperceptible, other surrounding sounds. As a metaphor, white noise suggests that imperception can be produced.

---

[*] History Department and Institute for Women's Studies and Gender Studies, 100 St. George St., Rm. 2074, Toronto, Ontario, Canada M5S 3G3; michelle.murphy@utoronto.ca.
I would like to thank NFFE Local 2050 and Lance Wallace for their assistance with my research into the EPA. Thank you to the participants in the "Environment, Health and Place in Global Perspective" workshop, Gregg Mitman, Matthew Price, and Evelynn Hammonds for their help in the development of this paper.
[1] Don DeLillo, *White Noise* (New York, 1985), 114–7.

Perception and subjectivity in modernity, as the historian Jonathan Crary argues, is characterized by historically specific modes of paying attention.[2] Attention, moreover, always involves disengaging from a broader field of stimuli for the sake of focusing on, isolating, and rendering intelligible a more narrowly-delineated set of phenomena. In other words, focusing on a single signal entails a learned inattention to other noise. Thus modern (and postmodern) subjects apprehended the persistent environmental bombardments that surrounded them through strategic *suspensions of perception*.[3] These suspensions of perception, moreover, resulted in not just passive disengagement but also production of historically specific terrains of invisibility, or what I call regimes of imperceptibility.

This essay explores imperceptibility and its relation to chemical exposures and race through a case study of activism by government scientists in the U.S. Environmental Protection Agency (EPA), the federal agency charged with investigating chemical exposures and setting national standards. Even more narrowly, this paper will focus on one idiosyncratic event at the EPA: the political activism of EPA scientists organized around an incident of chemical exposure at the agency's own Washington, D.C., headquarters in the 1980s. Through workplace activism and unionization, these scientists sought to resist the production of uncertainty and imperceptibility generated at the nexus of state, corporate, scientific, and juridical practices. At the same time that the EPA was developing national standards and embroiled in questions of scientific uncertainty, it was also being shaped by local racialized geographies of the particular neighborhood in Washington, D.C., in which it stood.

In the late-twentieth-century United States, both critics of and apologists for racism typically saw it as an issue concerning the disadvantaging of people with marked racialized identities, often called "visible minorities," emphasizing the role of perception in defining difference. Scholarly attention to relationships between racism and science, including environmental issues, has followed this same pattern. Historians of science have tended to take up questions of race only when examining acts of racism or when "race" has been the subject of science. Much less attention has been paid to the inverse subject of racialized disadvantage—the work of racialized privilege.[4] Furthermore, virtually no attention has been paid to the work of racialization in scientific practices not explicitly about race. One of the reasons this gap exists is that racialized privilege itself has often operated through its invisibility to those who possessed it and thus was rarely named as such.[5] It is difficult to research the work of race when historical actors did not mark it themselves.

In contrast to the early twentieth century, when those who benefited from and upheld white supremacy explicitly and frequently named and invoked it, the desegre-

[2] Jonathan Crary, *Suspensions of Perception: Attention, Spectacle, and Modern Culture* (Cambridge, Mass., 1999).

[3] Ibid.

[4] Important contributions to the study of white privilege in science include Warwick Anderson, "The Trespass Speaks: White Masculinity and Colonial Breakdown," *American Historical Review* 102 (1997): 1343–70; and Jill Morawski, "White Experimenters, White Blood, and Other White Conditions: Locating the Psychologist's Race," in *Off White: Readings on Race, Power, and Society,* ed. Michelle Fine, Lois Weis, Linda Powell et al. (New York, 1997), 13–28.

[5] On white privilege see Richard Delgado and Jean Stefancic, eds., *Critical White Studies: Looking Beyond the Mirror* (Philadelphia, 1997); Brigit Brander Rasmussen, Erick Klinenberg, Irene Nexica et al., eds., *The Making and Unmaking of Whiteness* (Durham, N.C., 2001); George Lipsitz, *The Possessive Investment in Whiteness: How White People Profit from Identity Politics* (Philadelphia, 1998).

gating cold war era produced a liberalism that provided a newly-articulated and powerful refashioning of "race" as a social, rather than biological, phenomenon. Racism
was increasingly defined as an individually held psychological prejudice that prevented the ideal colorless meritocracy. Race was thus an artifact of color vision. In
turn, whiteness could be held as an unraced identity; its very colorlessness fostered a
belief among those who enjoyed it in the possibility of a better, "color-blind" society.[6]
By the 1980s, the colorless location of "whiteness" and the confinement of racism to
the realm of the psychological encouraged white U.S. citizens to suspend their awareness of persistent racialized distributions of privilege and to look only for expressions
of racialized disadvantage. White privilege operated through this regularized suspension of perception—in other words, through a regime of imperception. Instead of
government-sanctioned signs over water fountains and doorways, in the late twentieth
century white privilege was generated, like white noise, precisely by "seeming not to
be anything in particular."[7]

This paper seeks to explore the inverse of how U.S. communities of color theorized
race in their development of environmental justice—that is, how the practice and activism of predominantly white state environmental scientists in the 1980s was shaped
by the racialized location of their work and lives. I use the term "racialization" to underscore that "race" was not a possession of persons prior to social arrangements of
power but rather produced by those arrangements. Likewise, individuals did not own
privilege. They enacted and generated it both intentionally and unintentionally by
virtue of pervasive racialization. Instead of asking whether individual scientists held
racist views, this paper tries to understand how scientists' various political and scientific positions were shaped by the racialized world in which they lived. Thus this paper
takes up the difficult task of connecting two different, yet coexisting, regimes of imperceptibility in late-twentieth-century America: first, the uncertainty of chemical exposures, and second, the unmarked location of racialized privilege.

RACING AND PLACING THE EPA

The physical condition and location of the EPA's headquarters, in southwest Washington's Waterside Mall, was symbolic of both the inequalities within the capital and
the agency's neglect and low standing under the Reagan administration. An ugly,
beige, concrete and glass complex, Waterside Mall was the direct result of one of the
federal government's biggest urban renewal programs of the 1960s. Previously, the
southwest neighborhood had been notorious for its alleyway slums, in which the
city's poorest black residents were crowded together by a segregated housing market.[8]
The bulldozing and redevelopment of the area displaced more than 10,000 African
American "alley dwellers."[9] In place of low-income housing came a shopping mall
called Waterside, flanked by two twelve-story towers of upscale apartments. When, in
an era of white flight from city centers, the apartment towers failed to attract renters,

[6] Howard Winant, "White Racial Projects," in Rasmussen et al., *Making and Unmaking of Whiteness*
(cit. n. 5), 97–112.

[7] Richard Dyer, "White," *Screen* 29 (1998): 44.

[8] James Borchert, *Alley Life in Washington: Family, Community, Religion, and Folklife in the City,
1850–1970* (Urbana, Ill., 1980).

[9] Howard Gillette Jr., *Between Justice and Beauty: Race, Planning, and the Failure of Urban Policy
in Washington, D.C.* (Baltimore, 1995).

the real estate developer leased his building to the federal government, which in turned assigned it in 1971 to the newly-founded EPA.

The presence of the EPA headquarters in this downtown area was part of the more general schizophrenic character of Washington, D.C., a finely segregated southern city profoundly shaped in terms of "black" and "white." Yet serving as the nation's capital, it was also a meeting place for the North and the South, as well as for the national and the local.[10] In the EPA's southwest neighborhood, located within walking distance of the National Mall, grand government agencies and luxury apartments sat uneasily next to public housing, racialized unemployment, and homelessness. Waterside was also just a ten-minute drive from the site of one of the most violent riots that followed Martin Luther King's 1968 assassination. It was a neighborhood to which many of the poor residents of southwest alleys had been displaced; some of whom lost their neighborhood, yet again, in the riot's flames.[11] Nonetheless, segregation was imperfect; in 1989 African Americans from a wide spectrum of classes made up 61 percent of the southwest neighborhood's residents, while European Americans made up 35 percent.[12] Racialized spaces were minutely distributed within neighborhoods, buildings, and workplaces as much as between them.

The racialized geography of Washington, D.C., exemplified the persistence of geographic distributions of privilege that, as scholar George Lipsitz has argued, characterized the cold war era of government-mandated *desegregation*.[13] For example, large corporations accommodated state desegregation orders by channeling recently hired African Americans into racially segregated departments that relied on devalued technologies and skills.[14] Suburbanization and middle-class "white flight" during desegregation moved 4 million whites out of city centers, while the number of whites living in suburbs increased by 22 million from 1966 to 1977.[15] Access to mortgages, loans, the provisions of municipal services, and other distributions of government-sponsored privileges became correspondingly concentrated in the suburbs.[16] By 1993, the results of white flight meant that 80 percent of the nation's suburban whites lived in places with a black population under 1 percent.[17] In the District, 90.3 percent of European Americans lived in the suburbs, helping to make African Americans the vast majority of metropolitan residents. African Americans constituted 71.1 percent of the city's residents but only 8.2 percent of its suburban residents.[18]

[10] Steven Diner, "Washington: The Black Majority: Race and Politics in the Nation's Capital," in *Snowbelt Cities: Metropolitan Politics in the Northeast and Midwest since World War II*, ed. Richard M. Bernard (Bloomington, Ind., 1990), 247–65; Beverly W. Jones, "Before Montgomery and Greensboro: The Desegregation Movement in the District of Columbia, 1950–1953," *Phylon* 43 (1982): 144–54.

[11] Ben Gilbert, *Ten Blocks from the White House: Anatomy of the Washington Riots of 1968* (New York, 1968); Nelson Kofie, *Race, Class, and the Struggle for Neighborhood in Washington, D.C.* (New York, 1999).

[12] U.S. Bureau of the Census, *U.S. Census 1990,* demographic data for zip code 20024.

[13] Lipsitz, *Possessive Investment* (cit. n. 5).

[14] See, e.g., Venus Green, *Race on the Line: Gender, Labor, and Technology in the Bell System, 1880–1980* (Durham, N.C., 2001).

[15] Lipsitz, *Possessive Investment* (cit. n. 5), 7.

[16] In addition to Lipsitz, see Martha Mahoney, "Residential Segregation and White Privilege," and Karen Brodkin Sacks, "The GI Bill: Whites Only Need Apply," in Delgado and Stefancic, *Critical White Studies* (cit. n. 5).

[17] See sources in note 16.

[18] Robert Manning, "Multicultural Washington, D.C.: The Changing Social and Economic Landscape of a Post-Industrial Metropolis," *Ethnic and Racial Studies* 21 (1998): 337.

Such patterns of desegregation in Washington, D.C., also extended to workplaces of the federal government, which ever since passage of the Civil Rights Act of 1964 had hired large numbers of local African Americans to work within its bureaucracies. Within the EPA headquarters and Waterside Mall, if not in their home neighborhoods, predominantly white scientists and professionals had daily interactions with African Americans: in the shopping mall with African American customers, sales clerks, and cleaning staff and in the EPA itself with African American secretaries, administration assistants, and security personnel—and a few African American professionals. The passage from the shopping mall into the headquarters marked a boundary between local and national spaces as well as between majority African American and majority European American spaces. While the EPA has not published longitudinal data, in 2001 black workers made up more than half of the clerical staff at the agency's headquarters but only 8.2 percent of the professional class workforce, of which whites composed 81.5 percent.[19] It is also useful to look at racialization of "grade levels," the system by which seniority and pay is ranked in federal bureaucracies. White workers made up 87.9 percent of the highest rank, GS15, and only 21.8 percent of the lowest ranks, GS1–4; black workers composed the bulk of workers in the ranks below GS 9 and only a sliver, 6.8 percent, of the highest rank.[20]

While the EPA was physically situated in a particular neighborhood, its charge was to establish regulations and standards that would encompass the whole nation. By the 1980s, most people saw the agency as failing in that mission. A multitude of reasons explained why so many EPA investigators were thwarted in their efforts to make strong claims about the health effects of chemical exposures. For one, exposures themselves were often transient and complicated. For another, sometimes the failure was a product of the difficulty and uncertainty plaguing good faith efforts to stand up to the narrow scrutiny of juridical standards that asked scientists to find causality in an individual chemical signal separated from the white noise of the built environment. At other times, the failure was a product of the EPA's methods and instruments, originally designed to detect straightforward exposures in factory settings and not chronic or transient exposures in neighborhoods and offices. Thus a terrain of imperceptibility was hardwired into the very instruments investigators used. However, failure could also result from their positions as government scientists whose ability to communicate findings or design studies was strictly circumscribed by politically appointed administrators whose ideology often rejected the notion that the state should regulate capital. EPA scientists were awkwardly positioned as civil servants accountable to the citizens, the state, and corporations on the one hand, and as scientific spokespersons for "nature" and truth on the other.

Since the 1980 televised struggle at Love Canal, New York, which culminated sensationally in two EPA agents' being taken hostage by community women to force the government to help residents, EPA scientists have become the regular villains in all too common dramas around chemical exposures.[21] This plot line has agency investigators

---

[19] U.S. Environmental Protection Agency, *Affirmative Program Plan for Women and Minorities: FY 2002 Plan Update & FY 2001 Accomplishment Report* (Washington, D.C., April 2002), 49.

[20] Ibid, 52.

[21] Love Canal, a working-class neighborhood near Niagara, New York, was the site of one of the earliest and most-documented instances of grassroots toxic waste activism concerning the health effects of toxic waste disposal. See Allan Mazur, *A Hazardous Inquiry: The Rashomon Effect at Love Canal* (Cambridge, Mass., 1998).

arriving at the instigation of local community activism (as they did at Love Canal), and then failing to come up with evidence useful to or commensurate with residents' accounts. As sociologist Celene Krauss argues, the white working-class communities, especially women, who protested against toxic waste in the 1970s and early 1980s, saw their problems as tied to the failure of government-sponsored protections.[22] Lois Gibbs, an influential toxic waste activist who got her start at Love Canal, described her investment in the government this way:

> I grew up in a blue-collar community, it was very patriotic, into democracy. . . . I believed in government. . . . I believed that if you had a complaint, you went to the right person in the government. If there was a way to solve the problem, they would be glad to do it.[23]

It was when the state, often the EPA, failed to provide expected aid that women such as Gibbs became politicized as activists. The ill repute with which EPA scientists contended was reflected in the advice activists gave one another about government investigations:

> The government studies are also bogus. We just have to start out knowing that the Centers for Disease Control, the EPA, or any of these regulatory agencies are not telling the truth. When they come your way, tell them to go away. Tell them, "We don't need your studies." You don't need their studies, because then you are countering more than you were before they got there.[24]

For many grassroots activists, as well as environmental journalists, the agency was simply not trustworthy. In just a short span of time, the EPA had gone from optimistic offshoot of Earth Day to obstructer of environmental justice in many eyes.

Over the 1980s, this already unsatisfying situation turned worse; EPA scientists went from frustrated to obstructed. When Republican candidate Ronald Reagan was elected president in 1980, he was forthright about his anti-environmentalist, pro-industry, deregulation politics. The EPA—just ten years old when Reagan began his first term—had been founded in a reformist moment, when the expansion of the state was greeted with liberal optimism, guided by a "progressive" and technocratic conviction that objective scientific expertise would solve problems of the social and natural orders. Many of the scientists hired then believed their science could improve the nation if not the world.[25] The 1980s, however, brought a backlash against state regulation in the name of economic progress, and EPA scientists saw their positions as trustworthy and privileged experts expire. Reagan proposed a 60 percent slash in the agency's budget and a 40 percent cut in staff.[26] Though the Democrat-controlled Congress put up some resistance, most cuts went through, and the administrative staff was overhauled.

---

[22] Celene Krauss, "Challenging Power: Toxic Waste Protests and the Politicization of White, Working-Class Women," in *Community Activism and Feminist Politics: Organizing across Race, Class, and Gender,* ed. Nancy Naples (New York, 1998), 129–50.

[23] Lois Gibbs, *Love Canal: My Story* (Albany, N.Y., 1982), 12.

[24] Quote from Patty Fraser in Robbin Lee Zeff, Marsha Love, and Karen Stults, eds., *Empowering Ourselves: Women and Toxic Organizing* (Arlington, Va., 1989), 13.

[25] For the history of how university industrial hygienists fashioned themselves in this way, see Christopher Sellers, *Hazards of the Job: From Industrial Disease to Environmental Health Science* (Chapel Hill, N. C., 1997).

[26] Robert Proctor, "The Reagan Effect," in *Cancer Wars: How Politics Shapes What We Know and Don't Know About Cancer* (New York, 1995).

Most detrimental was the appointment of Ann Gorsuch (1981–1983) as head of the EPA. Gorsuch filled the agency's upper administrative ranks with professionals who had made their livings defending industry against regulation. Rather than acting as the EPA's conservative steward, she set out to declaw the agency, stripping it of its regulatory capacity in practice if not in rule.[27] Gorsuch, nicknamed the "Ice Queen" within the agency, tacked up a brightly colored "hit list" in her office, a flagrant posting of career staff targeted for dismissal.[28] Scientists who resisted pressure to repress damning data or acted as whistleblowers could find themselves fired, harassed, or transferred to new positions in which their only tasks would be answering phones or filing papers. Reagan made Gorsuch's job easier by signing a series of executive orders that prevented the EPA from collecting information about a chemical for possible regulation without the Office of Management and Budget's (OMB) sanction. OMB only gave its sanction if the cost-benefit economic analysis it ran proved to be economical. The regulatory process could now be stopped before it even began, placing economic considerations squarely before those of science. Gorsuch was followed by William Ruckelshaus (1983–1985), who left his position at the timber company Weyerhaeuser, a frequent target of environmentalist groups. Next came Lee Thomas (1985–1989), who afterward became senior vice-president at the pulp and paper company Georgia-Pacific. At its nadir in the 1980s, the regulatory agency was being run, with little pretense at neutrality, by representatives of the companies it was supposed to regulate, a pattern of movement between industry and the agency that one critical EPA scientist labeled the "revolving door."[29]

Yet even under such difficult circumstances, EPA had instances when it successfully enforced a regulation or fined a company. However, such successes actually added to an unevenness of enforcement that exacerbated distributions of privilege and disadvantage.[30] The agency's tendency to levy its heaviest fines against those polluters near middle-class neighborhoods compounded widespread corporate strategies of locating garbage incinerators, dumps, and toxic waste sites near working-class or underemployed neighborhoods, neighborhoods constituted through racialized and geographic arrangements of power.[31] In the early 1980s, African American civil rights activists associated with the Washington-based United Church of Christ's Commission for Racial Justice gave this arrangement a name—"environmental racism." This term was defined as

> racial prejudice plus power. Racism is the intentional or unintentional use of power to isolate, separate and exploit others. . . . Racism confers certain privileges on and defends the

[27] On the history of the EPA during the 1980s, see Jonathon Lash, Katherine Gillman, and David Sheridan, *A Season of Spoils: The Story of the Reagan Administration's Attack on the Environment* (New York, 1990); Marc Land, Marc Roberts, and Stephen Thomas, *The Environmental Protection Agency: Asking the Wrong Questions* (New York, 1990).

[28] William Ruckelshaus, interview by Michael Gain, EPA History Office, Washington, D.C., Jan. 1993.

[29] William Sanjour, "EPA's Revolving Door," *Sierra Magazine* (Sept./Oct. 1992): 77.

[30] M. Lavelle and M. Coyle, "A Special Investigation; Unequal Protection: The Racial Divide in Environmental Law," *National Law Journal,* 21 Sept. 1992, S1–S16.

[31] See, e.g., General Accounting Office (GAO), *Siting of Hazardous Waste Landfills and Their Correlation with Racial and Economic Status of Surrounding Communities* (Washington, D.C., 1983), GAO/RCED-83–168; Bryant Bunyan and Paul Mohai, "Environmental Injustice: Weighing Race and Class as Factors in the Distribution of Environmental Hazards," *University of Colorado Law Review* 63 (1992): 921–32. Robert D. Bullard, ed., *Confronting Environmental Racism: Voices from the Grassroots* (Boston, 1993).

> dominant group, which in turn sustains and perpetuates racism. . . . Racism is more than just personal attitude; it is the institutionalized form of that attitude.[32]

Analysts saw the concentration of pollution and hazard in African American and other disenfranchised groups' neighborhoods as a continuation of government-sanctioned unequal distributions of services such as housing, education, and health care.

Unionizing EPA scientists knew about disenfranchised communities' efforts to represent environmental problems through civil rights discourse in terms of the racialized and unequal distributions of hazards. The inaugural incident of environmental justice activism in Warren County, North Carolina, for example, even included a few EPA scientists as rally speakers. Moreover, over the 1980s community environmental activists had ceaselessly lobbied the agency to incorporate environmental justice analyses of racial disbursements of hazardous waste into its mandate.[33] Environmental justice critics of the EPA used civil rights legislation against inequality. Thus they employed a different strategy than had toxic waste activists, who used popular epidemiology to demonstrate chemical exposure and trigger a state-sponsored scientific investigation or pursue toxic torts. These two forms of activism depended on different stances toward the state. Critics of environmental racism portrayed the state as historically complicit in the production of racialized inequalities, while toxic waste activists tended to see the state as failing to secure protections that had been expectations in the past. What the movements shared, however, was a distrust of the EPA.

Caught between the activists' criticism and an antiregulation administration, a small group of EPA scientists, many with "backgrounds in environmental, political and labor activism," took the unusual and impressive step of organizing a union of "toxicologists, chemists, biologists, attorneys and other environmental professionals" in the name of scientific ethics.[34]

THWARTED SCIENCE

Chartered in 1983, the National Federation of Federal Employees (NFFE), Local 2050, represented approximately 1,200 EPA professionals.[35] The leadership was primarily composed of left-leaning white male senior scientists, many of whom had been with the agency since its inception, were dedicated environmentalists, and had risked their jobs as outspoken critics of EPA positions. The union argued that as government scientists they had "a duty and a right to perform our work in an ethical environment, and to see that our work is not distorted, misrepresented, stolen or lied about in devising false cover for Agency policies."[36] Their professional ethics, they argued, were being corrupted through the influence of "economically powerful industries that are

[32] United Church of Christ Commission on Racial Justice, *Toxic Wastes and Race in the United States* (New York, 1987), ix–x.

[33] On the relationship between environmental justice and the EPA, see Stephen Sandweiss, "The Social Construction of Environmental Justice," in *Environmental Injustices, Political Struggles: Race, Class, and the Environment,* ed. David Camacho (Durham, N.C., 1998), 31–57.

[34] National Treasury Employees Union (NTEU), Chapter 280, "The Official History of NTEU Chapter 280: 17 years of Public Service at the EPA," April, 9, 2001, http://www.nteu280.org/history.htm (accessed 14 June 2001).

[35] In February 1998, members voted to change affiliation from NFFE to National Treasury Employees Union, Chapter 280.

[36] NTEU, Chapter 280, "The Official History" (cit. n. 34).

doing things harmful to the environment."[37] They claimed a lost "right" to a neutral work environment. The union's criticism was very much like that of toxic waste activists.

Through the union, EPA scientists fashioned themselves as champions of objectivity and spokespersons for nature.[38] The particular version of objectivity they upheld was that of traditional modern Western science, what Donna Haraway has called the "view from nowhere," which relied on scientists acting as "modest witnesses" who kept the details of their persons separate from the practice of their science.[39] This particular construction of objectivity dovetailed with the way white privilege functioned in postwar America—both relied on holding an unmarked and neutral location. Though a "view from nowhere" was produced by and supported racialized privilege, it was not necessarily a conservative ideology. It could also be deployed in a liberal frame to argue for the desegregation of science by asserting that the race, sex, class, or religion of a scientist was irrelevant to the scientific method. In positioning themselves as defenders of objective science, Local 2050 complained not of a disruption to the neutrality of their identities but rather of a violation by EPA administration to the neutrality of their workplace. As civil servants working for the nation's citizens, they had the "right" to a disinterested workplace in which they could execute their duty.

The union blamed the corruption of science in the EPA on the influence of large companies and industry-sponsored organizations on everything from the agency's promotions, to its research agendas, to the wording of its reports and brochures. The presence of corporate interests in environmental science even extended to experimental design and practice at the EPA. As practitioners of toxicological studies, the scientists knew that tinkering with humidity levels, changing strains of mice, modifying forms, or using stationary, rather than body-mounted, air samplers could determine whether a chemical exposure was detected or remained invisible.[40] They also knew that corporate scientists were expert at these same manipulations.

EPA administrators could be both flagrant and subtle in obstructing their own scientists. At their most flagrant, administrators would prevent scientists from going public with findings critical of powerful corporations. Anyone who went ahead risked ruining his or her career. At their subtlest, administrators could counter almost any positive finding by an EPA scientist by pointing to a nearly identical corporate-sponsored experiment that produced a negative or more ambivalent result. Uncertainty justified the need for the proliferation of yet more studies and the continuing tinkering with protocols. Ceaseless agency calls for more studies allowed the production of ever more ambiguities and thus the generation of uncertainty ad infinitum, helping to make regulation next to impossible.

This purposeful production of uncertainty was indeed the subtlest means of distorting the founding goals of the agency and serving the antiregulation agendas. The

---

[37] NTEU, Chapter 280, "Why We Need a Code of Professional Ethics," August, 25, 1999, http://www.nteu280.org/issues/NTEU-%20Professional%20Ethics.htm.

[38] Ibid.

[39] On the "view from nowhere," see Donna Haraway, "Situated Knowledges: The Science Question in Feminism and the Privilege of Partial Perspective," in *Simians, Cyborgs, and Women: The Reinvention of Nature* (New York, 1991), 183–202.

[40] For a detailed analysis of how this worked at the EPA in the case of carpet, see the five-part series by activist Cindy Duehring, beginning with "Carpet, Part One: EPA Stalls and Industry Hedges While Consumers Remain at Risk," *Informed Consent* 1 (1993): 6–11, 30–3.

union fought to counter such tactics, as illustrated by its resistance to the production of uncertainty in its struggle with the EPA policy on the potential toxicity of new carpet. The union went so far as to commission separate experiments at one of the few independent toxicology labs in the United States. That lab found that some samples caused severe neuromuscular reactions in mice.[41] Yet, ultimately, because the carpet industry and an official EPA lab—when using a slightly modified experimental apparatus—could not replicate the independent lab's findings, the union could not successfully subvert the development of the EPA's industry-friendly carpet "green tag" program.

In the late twentieth century, the extreme difficulty of making visible the health consequences of chemicals became, I believe, the single most significant characteristic of "chemical exposures" as a scientific artifact. Yet rather than locating the problem in the way the EPA attempted to resolve questions of chemical exposure—that is, exclusively on the narrow terrain of laboratory toxicology—the union largely sought to hold on to the terms of their scientific practice by critiquing the conditions under which it occurred.

The formation of Local 2050, however, was motivated solely by the scientists' defense of state scientific practice against corporate interests. Though exceptional in many ways, their workplace activism was also part of a late-twentieth-century swell in unionization among government and office workers, one that went against the tide of a waning and beleaguered industrial labor movement. Through unionization, scientists were recognizing their devalued status. No longer glorified as the influential and disinterested arbiters of disputes between citizens and corporations about the consequences of industrial pollution, EPA scientists implicitly aligned themselves with the downtrodden proletarianized service sector rather than with the upper administration's managerial class. Thus the way they expressed their activism and its protection of objectivity was partially predicated on the conflicted assumption of a subjugated position in an era of undermined white privilege. Yet their ideology was the inverse of epistemological claims made in terms of identity politics (including feminist arguments), which typically argued that subjugated viewpoints provided better access to the truth.[42] Instead, members of Local 2050 saw their oppressive circumstances as disturbing the neutral ground they held as necessary for the production of good science.

ENVIRONMENTAL ANXIETY

As a union concerned with workplace conditions, Local 2050 soon became focused not just on the political but also on the environmental conditions in the agency's headquarters. There was, in fact, a nationwide surge of distress over the unexpected presence of chemical exposures inside nonindustrial, ordinary spaces such as office buildings. EPA scientists, whose very livelihoods connected the politics of environmental

[41] This study was eventually published as Rosalind C. Anderson, "Toxic Emissions from Carpets," *Journal of Nutritional and Environmental Medicine* 5 (1995): 375–86.

[42] See, e.g., Nancy Harstock, "The Feminist Standpoint: Developing the Ground for Specifically Feminist Historical Materialism," in *Discovering Reality: Feminist Perspectives on Epistemology, Metaphysics, Methodology, and Philosophy of Science,* ed. Sandra Harding and Merrill Hintikka (Dordrecht, 1983), 283–310; Sandra Harding, "From Feminist Empiricism to Feminist Standpoint Epistemologies," in *The Science Question in Feminism* (Ithaca, N.Y., 1986), 136–62; George Lukacs, "Reification and the Standpoint of the Proletariat," in *History and Class Consciousness* (Boston, 1971).

exposure and suspended perceptions, were not immune to this distress nor the way privilege shaped its articulation.

Pervasive white privilege was imperfect and insecure in a cold war period that was as much about emancipation and civil rights as about conservative containment.[43] Whites were losing the geographic and workplace monopolies they enjoyed before desegregation, while the intensification of globalized capital flows in the 1980s saw U.S. industrial jobs moved abroad and middle-class managerial jobs downsized. Anxiety triggered by new forms of insecurity abounded (brilliantly satirized by DeLillo), helping to create a middle-class "risk society" worried over errant chemical exposures that violated expected protections.[44] Chemical exposures were not confined to factories. They could come from consumer products, the construction materials of office buildings, passing luxury vehicles, perfect green lawns, designer pharmaceuticals, shiny blemish-free food, or plush carpets. Objects that were the very hallmarks of suburban privilege could let loose exposures that violated racialized protection.

The inability to absolutely contain chemical exposures through privilege prompted many middle-class Americans to ask nervously, Is it happening here? On the one hand, the regularity and expectedness of distributions of racialized privilege made it possible to displace unearned privilege on to the naturalness of place—bad things happen to people in low-lying areas or in scrubby parts of the country. Better yet, chemical accidents happened to people who lived in faraway places such as Bhopal.[45] On the other hand, illnesses such as breast cancer and leukemia indicated that no one was absolutely safe. While the environmental justice movement highlighted raced and classed distributions of toxic exposure, white middle-class America expressed its own version of environmental politics by seeking to isolate and prohibit all dangers, no matter how small, from the home, road, playground, and workplace. Thus not only were exposures themselves racialized and classed, but so, too, were environmental anxieties, shaped by an insecure yet pervasive white privilege.

The EPA headquarters was ripe to be identified as a "sick building" and environmental hazard. Under Reagan, conditions at the Waterside Mall represented the adverse workplace circumstances.[46] Approximately 5,000 agency staff members now crowded inside. A jumble of hallways led to a crazy quilt of tiny individual offices, most without windows, cut out of what had originally been apartments, creating a "warren of people crammed into rooms." In typical energy-efficient construction, those windows that did exist were unopenable. The building's interior was filthy and neglected: roaches and mice infested the offices, burnt-out light fixtures left corridors "dark for days," and toilets were often out of order. The air was stale, and vent grills were clogged with debris, grit, and fibrous matter that caused a fine black powder to settle on surfaces. In the words of its inhabitants, the building was "oppressive" and "a dull, dirty, and depressing place to work." One EPA manager drew a comparison to

[43] A. Yvette Huginnie, "Containment and Emancipation: Race, Class, and Gender in the Cold War West," in *The Cold War American West, 1945–89,* ed. Kevin Fernlund, (Albuquerque, 1998), 51–70.

[44] Ulrich Beck, *Risk Society: Towards a New Modernity,* trans. Mark Ritter (London, 1992).

[45] *White Noise* was published around the time of the Bhopal chemical disaster so coverage of Bhopal in the mainstream press and reviews of the novel occurred simultaneously.

[46] This description of Waterside Mall is compiled from my own observations during a visit in 1996, press coverage of the indoor pollution episode, the Indoor Air Quality and Work Environment Survey, and most importantly the approximately 1,200 essays EPA employees wrote on the back of the survey, administered in February 1989. The essay responses were compiled in an unpublished manuscript: Lance Wallace, "Preliminary Analysis of the Essay Question" (1991).

conditions of disadvantage outside the headquarters' door, writing, "I understand how poor housing project occupants feel"; another employee dramatically likened the conditions to ones in "Third World public hospitals."[47]

The administration tried to give the building a quick facelift in October 1987 by installing new carpet. Immediately, some EPA staff, including scientists, began to complain of tearing eyes, irritated throats, burning lungs, shortness of breath, crippling headaches, and dizziness.[48] As the carpet installation pushed its way through the building, the trickle of complaints became a torrent. The EPA's Emergency Response "SWAT" Team, usually held in reserve for toxic spills, was called in. The facilities management director reported the results at a staff meeting: 68 different airborne chemicals had been detected, but all were at concentrations "no more higher [sic] than your living room."[49] EPA scientists found themselves facing the same regime of imperceptibility that they had participated in through their fieldwork.

Local 2050 became consumed with the issue of the building's "indoor air quality," members channeling their challenge to corrupted science into proving the existence of harmful chemical exposures in the union's own workplace. Union leaders were fashioning themselves as defenders of the victims of exposure among their own. With their expert technical skills, EPA scientists had a unique insider opportunity to demonstrate how the detection of harmful exposures was purposefully avoided or, in other words, how suspensions of perception had been strategically generated at the agency.

Some of the sickest and most outraged employees formed the Committee of Poisoned Employees (COPE). With NFFE Local 2050 and the American Federation of Government Employees (AFGE), Local 3331—which represented clerical and other workers in the headquarters, many of them women and persons of color—organized a protest outside the building in May 1988. Approximately sixty employees assembled there, carrying signs festooned with upside-down EPA logos or declaring "EPA is a Superfund Site." Inspired by toxic waste activist practices of popular epidemiology, the EPA employees handed out a survey. Out of necessity, but not without a sense of drama, some of the sickest employees had begun wearing gas masks to work. Placards reading "Canaries in a Coal Mine" underlined the belief that if chemical exposures could be found in white-collar workplaces, they might occur anywhere. Office buildings were "ordinary" unmarked places, that is, places where privilege was expected to operate and where systematic disadvantage was expected to be rare. Moreover, carpet was a ubiquitous artifact of contemporary living and might be found in almost any kind of building. A chemical exposure through carpet at the EPA headquarters, of all places, simultaneously signaled the subjugation of agency workers under Reagan and the lack of immunity that privilege was providing more generally.

[47] Ibid., 6–9.

[48] These symptoms were recorded in the health survey handed out by Local 2050 at the 25 May 1988 protest. NFFE Papers, NFFE Office, U.S. Environmental Protection Agency, National Headquarters, Washington, D.C.

[49] The quote and other meeting details were reported in a *Washington Times* article, the first of what would become a deluge of stories on the subject in the Washington, D.C., newspapers. (Dan Vukelich, "Employees Charge EPA's Own House Needs Cleaning Up," *Washington Times,* 28 April 1988.) This quote was also corroborated in interviews I conducted with Local 2050 leaders in 1996. The air monitoring was eventually written up into a final report. (R. Singhvi, R. D. Turpin, and S. M. Burchette, *A Final Summary Report on the Indoor Air Monitoring Performed at USEPA Headquarters, Washington, D.C., on March 4 and 5, 1988* [Edison, N.J., 1988].)

The irony of this expression of environmental anxiety quickly captured media and congressional attention.

Local 2050's protest tactics stood in sharp contrast to official EPA policy on indoor exposures. The EPA had, under duress, recently established an Indoor Air Division, which remained silent about the events happening in its own building. In general, the Indoor Air Division used the term "indoor pollution" as a discursive strategy to remove the problem of chemical exposures in nonindustrial workplaces from the realm of labor disputes.[50] The division had been added by Congress in response to a large-scale study of "total exposures," that is, a study of accumulated personal chemical bombardments.[51] The study, headed by EPA scientist Lance Wallace, had unexpectedly concluded that time spent indoors, not proximity to industrial sites, was most strongly correlated with high accumulated exposures. While this study was silent about the social locations of its research subjects—university students—and thus the possible effects privilege might have on the significance of indoor exposures, it did move the white noise of errant chemicals in "ordinary" spaces into the realm of perception, if only momentarily. Because of this study, Wallace acquired a reputation as the "father" of indoor air pollution.

In response to Local 2050's lobbying, Wallace was asked to head a study of Waterside Mall. Almost two years had passed since the carpet had first been laid, and the building had been aired out many times since. Any acute emissions from the carpet were long gone. Provided only with funding for stationary air monitors, Wallace expected to detect little: "more monitoring wouldn't have told us anything anyway. We were there a year after the fact and even studies where people have gotten there fairly quickly tend to fail to show anything."[52] Months after the initial complaints, air sampling measured no acute doses of a specific chemical, and therefore no acute chemical signal could be found amid the daily noise.

Predicting that no physical evidence would be found, the study also included an elaborate questionnaire, which had been given to all EPA employees in the building and had 3,955 responses. Trained in atmospheric physics, not sociology, Wallace (like the "housewives" turned toxic waste activists who practiced popular epidemiology) found himself spending his days analyzing a survey and assembling hundreds of pages of quantitative analysis, eventually published in four volumes between 1989 and 1991.[53] Ultimately, however, the survey allowed Wallace only to make some vague suggestions about dealing with the high prevalence of health symptoms at the EPA. No single cause could be extracted from the white noise of Waterside Mall. The search for a single cause generated imperceptibility in two mutually constitutive directions. First, any specific single exposure years ago was masked by the complex and accumulated bombardments, both social and physical, to which inhabitants of Waterside Mall were regularly subjected. Second, the environmental study designed as a search for a single toxic culprit obscured the social and political circumstances through which the EPA site had become so degraded.

Local 2050 doggedly persevered in its attempts to make visible the antagonistic

[50] Michelle Murphy, "Sick Buildings and Sick Bodies: The Materialization of an Occupational Illness in Late Capitalism" (Ph.D. diss., Harvard University, 1998).

[51] Lance Wallace, *The Total Exposure Assessment Methodology (TEAM) Study: Summary and Analysis,* vol. 1 (Washington, D.C., 1987).

[52] Wallace, interview by author, Washington, D.C., 30 April 1996.

[53] EPA, *Indoor Air Quality and Work Environment Study,* 4 vols. (Washington, D.C., 1989–1991).

workplace conditions under which the study was conducted. Invoking a clause in their collective bargaining agreement that guaranteed a role for EPA scientists–union members in studies of their own workplaces, they added a supplement to the published findings. Agency internal memos, newspaper articles, earlier monitoring attempts, and independent research on carpet filled the appendix. By publishing these documents, EPA scientists aspired to reframe the study in terms of the political conditions of its production. The lack of conclusion about the toxicity of carpet, which otherwise would have scripted indoor chemical exposure into its typical role of imperception, instead was held up as an example of the effects of corruption on EPA science. For Local 2050, at least the administration's maneuverings within the EPA were laid bare in the appendix.

Meanwhile, six of the sickest EPA employees sued the building's owner. At first, the jury awarded them $948,000 in damages, the biggest indoor pollution ruling to date. As with many other claims of low-level indoor chemical exposures made by the relatively privileged and disadvantaged alike, the defense reframed the scientists' illnesses as the result of anxiety—a psychological rather than a physical response. In 1995, the District of Columbia Superior Court overturned the damages, ruling that the building's owner could not be held responsible for psychogenic illnesses.[54] Not even their professional authority as EPA scientists could effectively give witness to the ill-health effects of toxic exposures occurring in their own bodies. Their claims were struck down with the same dismissal of hysteria usually saved for women and soldiers. The ground of professional authority and privilege shifted beneath their feet.

Local 2050's scientists-turned-activists had fashioned themselves as objective producers of knowledge because they believed their laboratory and technical procedures held the potential of disassociation from the power relations swirling through questions of chemical toxicity. Forming a union in defense of scientific ethics in the face of Reaganomics was an exceptional and even radical act for a group of scientists who saw themselves as defenders of objectivity. Their outspoken criticism of corporate influence on state science was exceptional in an era of intensified boosterism for economic versus environmental calculations of benefit and risk. Union leaders and the whistleblowers they defended persistently criticized official EPA findings before numerous congressional hearings on such issues as fluoride, asbestos, and aerosol propellants. Yet the scientists did not question the technical terms of their scientific practice, just its context. As a result, what became imperceptible were the ways chronic and low-level chemical exposures, as well as their uneven distribution, were consistently rendered invisible by the narrow criteria of toxicological proof that their discipline and the courts had developed. This was true even when scientists worked to the best of their abilities, in good faith.

EPILOGUE: SEEING RACE AT THE EPA

While Jack Gladney, the privileged protagonist of *White Noise,* had a hard time envisioning himself fleeing from a toxic airborne event, other people had less trouble recognizing their own vulnerability to toxic exposures. With the rise of the environmental

---

[54] *Bahura v. S. E.W Investors et al.,* No. 90CA10594 (D.C. Super. Ct. 1995). The jury came to its conclusion with the help of expert witness Abba Terr, who testifies around the country that multiple chemical sensitivity is a form of psychosomatic illness.

justice movement, reports by government scientists were regularly pitted against claims made by laypeople whose views of the world were shaped by analysis of their disenfranchised positions.

Much changed at the EPA in the years following formation of the union. During President Bill Clinton's administration, the number of "minorities" working in Grades 13 and higher at the agency more than doubled nationally, from 1,086 in 1993 to 2,348 in 2000.[55] The EPA headquarters' staff moved out of Waterside Mall in 1998 and, ironically, into the new Ronald Reagan Building. The departure of almost 5,000 EPA employees left Waterside Mall empty and in danger of dereliction. Local congresswoman Eleanor Holmes Norton compared the negative economic impact of EPA's departure on the southwest neighborhood to that suffered by communities after a "military base closure."[56] After considerable effort by environmental justice activists, Clinton signed an executive order in 1994 mandating that the agency develop "environmental justice strategies," including use of Title VI of the Civil Rights Act of 1964, which prohibited federal agencies from discriminating on the basis of race, color, or national origin. Before long, a group of African American EPA scientists and staff charged that endemic and virulent racism existed *within* the agency's headquarters. Scientists were finally naming the work of race that had been inside all along.

Marsha Coleman-Adebayo, an African American senior scientist and an expert in African studies, sued the EPA on grounds of racial and gender discrimination, winning a $600,000 settlement.[57] The Republican-chaired House Committee on Science was quick to hold a hearing titled "Intolerance at EPA: Harming People, Harming Science?" At the hearing, the NAACP came forward with the disturbing case of another agency employee, a midlevel administrator and African American woman, who had been ordered by her manager to clean a toilet in preparation for a visit from Carol Browner, the EPA's head administrator under Clinton.[58] In her testimony, Coleman-Adebayo compared the agency to a "21st century plantation."[59]

Coleman-Adebayo and a handful of others established a new activist organization for EPA employees, the EPA Victims Against Racial Discrimination (EPAVRD). Describing itself as "walking the last mile to freedom," EPAVRD voiced its activism through discourse and strategies well known from the early phase of the civil rights movement and its Social Gospel Christianity. The group even drew Reverend Al Sharpton to its protests and referred to Coleman-Adebayo as the "Rosa Parks of the EPA."[60] In contrast to Local 2050, EPAVRD strongly linked hostile workplace conditions to longstanding prejudice against African Americans and women, not to the administration's deregulation ideology or corporate influence. Moreover, they charged that systemic racism enacted by some of their peers not only discriminated against individuals but also distorted the agency's work. How could the EPA hope to address

[55] Carol Browner, Administrator of the USEPA, Statement before the Committee on Science, House of Representatives, 4 Oct. 2000, http://www.house.gov/science/106_hearing.htm.

[56] Senate Banking Committee, "Senator Gramm's Letter to Acting SEC Chairman Laura Unger Regarding SEC Plan to Move Headquarters," press release, 19 July 2001.

[57] *Coleman-Adebayo v Browner, U.S. Environmental Protection Agency,* No. 1.98cv1939 (D. D.C., 5 Aug. 1998).

[58] Leroy W. Warren Jr., chairman of the NAACP Federal Sector Task Force, Statement before the Committee on Science, House of Representatives, 4 Oct. 2000, http://www.house.gov/science/106_hearing.htm.

[59] Jack White, "How the EPA Was Made to Clean Up Its Own Stain—Racism," *Time,* 23 Feb. 2001.

[60] Header of the EPAVRD Web site: http://www.epavard.org (accessed 16 April 2001).

environmental racism when the agency itself was racist? For EPAVRD, discrimination against African American scientists was a manifestation of the same discrimination that produced racialized distributions of hazard.

EPAVRD saw the government as complicit in the historically continuous performance of discrimination. Instead of playing out its opposition in the details of science as Local 2050 had, EPAVRD lobbied for a legislative change that would constrain the government itself. They dubbed the legislation "the first civil rights law of the 21st century." The NO FEAR bill (Notification of Federal Employees Anti-Discrimination and Retaliation Act) was a legislative strategy to protect minority employees within government agencies from discrimination. Though the lobbying in favor of this bill overwhelmingly focused on racial discrimination, tucked in the legislation was the muted inclusion of protection for whistleblowers.

The NO FEAR bill enjoyed bipartisan support in a way no criticism of corporate influence could. It was premised on deterring discrimination by making agencies pay the money for settlements out of their budgets, whereas previously the money had come out of a common federal fund. In the preamble, preventing discrimination and protecting whistleblowers was defended on the basis that "good science requires a tolerance of opposing viewpoints."[61] The argument implicitly made was that subjugated standpoints provided valuable "viewpoints" that differed from those produced by insisting on a single "neutral" standpoint. Opposing viewpoints were necessary for the EPA to analyze environmental racism without prejudice. For EPAVRD, diverse subject positions were necessarily constitutive, rather than corrupting, of knowledge production. For EPAVRD, however, discrimination at the unsettled turn of the millennium was not an act reserved only for those who benefited from white privilege—senior black administrators were also the perpetrators of racism against their black colleagues. One's "race" and the way one exercised racialized arrangements were anything but straightforward equations. The bill was signed into law by President George W. Bush on 15 May 2002.

In this paper I have tried to show that racialized privilege shaped science in the 1980s, even when "race" was not explicit to the scientists' self-fashioning or their subjects of study. The insecurity of white privilege and the authority of a "view from nowhere" within the late twentieth century have been, I would argue, intimately connected during a historical moment when *what* one could see was increasingly linked to *where* one was seen to stand. The regimes of imperceptibility I have tried to link— the unmarked exercise of racialized privilege and the indetectability of chemical exposures in 1980s environmental science—did not just coexist but also touched and sustained one another. The oppositional stances of Local 2050 and EPAVRD enacted the work of race, one by drawing on the unspoken terms of racialized privilege and the other by marking racialized disadvantage. The scientists of Local 2050 fought for the restoration of their right to a neutral workplace without undue corporate or administration influence; the members of EPAVRD fought for civil rights protection against racial discrimination within the government, thereby fostering "opposing viewpoints."

---

[61] *Notification and Federal Employee Anti-Discrimination and Retaliation Act of 2001,* HR 169, Introduction to the House, 106th Cong., 2d sess., *Congressional Record,* 3 Jan. 2001. (The same language had been used when the legislation was introduced as HR 5516 on 19 Oct. 2000.) This language was subsequently deleted in the version passed.

Whether "race" is named or not, I have tried to understand it here not as the property of individuals but as a product of uneven and changing distributions of privilege and disadvantage at the end of the twentieth century. The work of race, seen and unseen, permeated the day-to-day arrangements of science. The EPA was not exceptional or worse in this regard. The work of racialized privilege at the agency is discussed here as an example of the pervasive force of race in all "normal" science. Race, science, and chemical exposure were being made through each other amid sedimented and yet shifting distributions of power that shaped not just who was authorized to make knowledge but also the very distributions of hazard being studied. These included the landscapes of chemical exposure that differed between a bourgeois mall and a maquiladora, or between an office building and a hazardous waste dump. While "race" may have been a subject rarely spoken of in the labs and offices of the EPA in the 1980s, race wove the physical fabric of neighborhoods and buildings, exposures and protections, questions and methods that scientists lived with daily.

From Bhopal to the Informating of Environmentalism:

Risk Communication in Historical Perspective

By Kim Fortun*

ABSTRACT

This essay describes the development of information technology and culture in the environmental field since the 1980s and how this has led to new understandings of risk communication. The essay also describes how environmental information systems operate as instruments of power, in the way they configure and provide access to knowledge, in the way they manage uncertainty, and in the way they build in and project particular modes of subjectivity. The goal is to provide a brief yet compelling glimpse into the "informating of environmentalism."

INTRODUCTION

In August 2001, the U.S. Environmental Protection Agency (EPA) hosted a conference that brought together people involved in the design of environmental information systems for use at the local level.[1] Many of the participants worked in state environmental agencies that had partnered with community groups to obtain EPA grants to develop real-time systems to monitor environmental conditions important to local residents—the water quality of local beaches, contamination in drinking water, ozone levels, lead in the yards of homes. The grant program's goal was to foster the development of innovative technological solutions to the need for timely, locally relevant information collection, processing, and distribution. The conference was intended to foster this innovation by encouraging collaboration between communities and between different kinds of environmental experts.

Overall, the conference offered a remarkable display of a new form of environmental expertise, centering on informatics and computer-mediated visualization. It was not, however, without contradictions, as illustrated by the plenary presentation

* Department of Science and Technology Studies, 5408 Sage Laboratory, Rensselaer Polytechnic Institute, 110 8th Street, Troy, NY 12180; fortuk@rpi.edu.

Special thanks to Gregg Mitman, Michelle Murphy, and Chris Sellers for providing thematic orientation and commentary and for hosting the workshop for which this paper was originally written.

[1] The conference, which I attended as an ethnographer, was the 2001 EMPACT (Environmental Monitoring for Public Access and Community Tracking) conference, "Bringing It Home," Philadelphia, 7–9 Aug. EMPACT was established in 1998 as a "nonregulatory, community-driven program" to help people "be *informed* and *proactive* about protecting their health and the environment." The EMPACT program lost its funding shortly before the August 2001 conference. At that time, almost 100 communities had received EMPACT grants.

given in an opening session. The speaker, identified as a risk communication specialist, was to provide insight on how users of risk information think and act, thereby helping designers understand how their systems would actually be used. Many environmental information system designers voice frustration about their disconnection from
users, aware that data on the number of hits on a Web site reveals little about how a
system actually works. They sat in the audience listening carefully.

The speaker told the designers that their goal should be to control, or at least mitigate,
panic through "one-click" access to reassuring information. A user's need to "drill-
down" on a Web site, for example, to customize the information retrieved, should be
avoided. While new technologies have created new opportunities for risk communication, its basic principles have not changed: information designers need to provide
simple, clear, key messages to deter users from over-reading or distorting the facts.
Lots of background information should be provided to make the strange familiar because people are afraid of what they do not understand and do not respond well to facts
when emotionally upset.

The speaker also said that the provision of competing information claims should be
avoided, insisting that a basic lesson of Business 101 is that an organization should
not go public until its members can present a unified front. She encouraged designers
to withhold information from their systems until they could present one authoritative
statement describing the problem and offering a remedy.

That disagreement among experts is confusing and should be avoided was one of
many arguments this speaker made that echoed the "seven cardinal rules of risk communication" articulated in a 1988 EPA report.[2] The report highlighted the need for
a risk communicator to coordinate credible sources and present a stable, consensus-
backed picture of a problem as well as the need for listening and dialogue. The purpose of the recommended dialogue is not to be open-ended and thereby productive of
new questions, however. It is to allow risk communicators to wrap the voices of those
they dialogue with around the picture of risk the communicators want to convey. The
purpose of dialogue is assimilation.[3]

The *Seven Cardinal Rules of Risk Communication* was published in 1988, the
same year a new flow of environmental information became available for public
scrutiny as a result of the U.S. government's passage of the Emergency Planning and
Community Right-to-Know Act. For the first time, companies in key polluting sectors had to report the quantity of their emissions in a given year. Management theo-
rist Bruce Piasecki says the effect was like opening the windows of a house and invit-
ing people to peer inside.[4] The owners—plant managers—needed a strategy for
explaining what people saw. The *Seven Cardinal Rules of Risk Communication* was
intended to help.

The need for professional risk communication intensified in the late 1980s as
new information about pollution became available and was circulated by journalists, environmental organizations, community groups, and labor unions. The tac-

[2] V. T. Covello and F. Allen, *Seven Cardinal Rules of Risk Communication* (Washington, D.C.,
1988).

[3] I learned what these rules of risk communication looked like in practice through ethnographic
fieldwork in high-risk communities in the United States in the mid-1990s. See Kim Fortun, *Advocacy
After Bhopal: Environmentalism, Disaster, New Global Orders* (Chicago, 2001).

[4] Bruce Piasecki, *Corporate Environmental Strategy: The Avalanche of Change Since Bhopal* (New
York, 1995), 11.

tics of professional risk communication, however, had taken shape much earlier, in years of work to institutionalize a scientific approach to environmental risk assessment and management. During the 1970s, the EPA undertook the challenge of understanding and regulating for risk—potential problems rather than those immediately observable. By the end of the 1970s, the agency was under serious attack by industry feeling the cost of regulation to minimize risk and important scientific bodies such as the National Research Council.[5] In response, the EPA spent the early 1980s formalizing an approach to risk assessment and management that could be called "scientific," which was formally endorsed by EPA administrator William Ruckelshaus in 1984.[6] With this approach, the agency aimed to produce certifiable conclusions that could resist challenge and deconstruction. The results are supposed to be stable and solid and are intended to be circulated intact, even if dialogue is required to generate a similar interpretation by all interested parties. Official approaches to risk assessment and management thus implied a certain mode of risk communication.

It was this mode of risk communication the speaker at the August 2001 EPA conference advocated to environmental information system designers. The effect proved powerful, if confusing. Most designers in the audience took lots of notes and walked out talking about how their sites were too complicated, loaded with too much information, and demanded too many choices and too much navigation. Although the speech discounted the informational and technical realities they worked with every day, the designers were still open to its message. The single metric for judging an environmental information system, they were told, is simplicity, and it should be the gold standard against which anything of worth must be judged.

Most of the environmental information systems designers who attended that conference specialized in the provision of real-time monitoring data in accessible formats. Their challenge was to make streaming data on air or water quality available in real time, integrating as much as possible with other data sets and streams. Time taken to convene expert opinion on the information's validity would undermine the goal of making real-time data available. Since a unified front was therefore impossible, these designers had to build warning signs into their systems, reminding users that the data could have a range of problems. Monitoring equipment could have lost its calibration. Data transfer could have mangled results. Competing opinions on the types of data needed to understand and respond to a given problem are likely to exist. The information systems the designers created could not, by definition, be simple. The purpose of such a system is not to provide certifiable results but to provide decision makers with workable (even if inchoate) information. Though open to the conference speaker's delineation of what good risk communication should consist of, the designers did believe in this mission. For most of them, information is itself a good, not a hazard.

What became visible at this EPA conference were two different ways of conceptualizing the subject of environmental risk. The risk communication expert conceived of the subject of environmental risk as needing to *know* conclusively, in order

    [5] Sheila Jasanoff, "Science, Politics, and the Renegotiation of Expertise at EPA," *Osiris* 7 (1992): 195–217.
    [6] Richard Andrews, "Risk-Based Decision Making," in *Environmental Policy: New Directions for the Twenty-First Century,* 5th ed., ed. Norman Vig and Michael Kraft (Washington D.C., 2003).

to act, in order to divert panic. The environmental information system designers were building tools to help subjects of environmental risk visualize, spacialize, and prioritize risk. The goal was to draw people into complex data sets, enabling as many connections as possible, providing access to working data, often riddled by uncertainties.

The risk communication expert discussed here is a resolutely modernist figure, programmed to understand science as a guarantor of truth and political consensus. Uncertainty is a liability. Information in excess is a hazard. Public opinion is something to be controlled. The function of the expert is containment. The ideal world is conceived as a closed system in which all parameters are known and regulated through technocracy.[7] Environmental information system designers, however, tend to be different. Through daily work with network technology, they tend to equate information flow and function. A closed system is one that does not leverage its own possibility. The task of the information technologist is to enable people to track environmental problems across spatial scales, at the intersection of natural, political, and technological systems. Environmental information designers exhibit technological optimism, not in the possibility of full understanding and control of systems, but in the possibility that information systems can support working knowledge and tentative solutions within inevitably open systems.

The disjuncture between established risk communication experts and environmental information system designers is one index of what I think of as the "informating of environmentalism," brought about through the development of information technology and culture in the environmental field, particularly since the 1980s. In this historical shift it is possible to see how environmental ideas and practices have differed across time and within a shared temporal space because of different conditions of production. What becomes evident is that it is not only nature, or the environment, per se that makes a difference but also the information systems through which the environment becomes accessible to understanding and governance. The materiality of nature may not be produced by these systems themselves, but these systems do determine—quite literally—what counts and what does not.[8]

In this essay, I describe the development of information technology and culture in the environmental field and how this has led to new understandings of risk communication. I also describe how environmental information systems operate as instruments of power, in the way they configure and provide access to knowledge, in the way they manage uncertainty, and in the way they build in and project particular modes of subjectivity. My goal is to offer a brief yet compelling glimpse into the informating of environmentalism. Let me turn now to the 1980s, and to the Bhopal disaster, where the story can be said to have begun.

[7] Note that computers are implicated in idealization of closed systems as they are (albeit in different ways) in idealization of open systems. For a general account, see Paul Edwards, *The Closed World: Computers and the Politics of Discourse in Cold War America* (Cambridge, Mass., 1996).

[8] For a sampling of important recent work on environmental information systems, see G. C. Bowker, "Biodiversity Datadiversity," *Social Studies of Science* 30 (2001): 643–84; Paul Edwards, "Global Climate Science, Uncertainty, and Politics: Data-Laden Models, Model-Filtered Data," *Science as Culture* 8 (1999): 437–72; R. E. Sieber, *Computers in the Grassroots: Environmentalists, GIS, and Public Policy* (Ph.D. diss., Rutgers Univ., 1997); D. Sarewitz, R. Pielke Jr., and R. Byerly Jr., eds., *Prediction: Science, Decision-Making, and the Future of Nature* (Washington, D.C., 2000).

BHOPAL AND NEW GLOBAL ORDERS

During the 1980s, information processing and sharing capabilities grew dramatically, as did information culture, understood here to revolve around the belief that more information circulated among more actors will stimulate solutions to complex social problems. Exhibiting a now infamous "nonlinear success," for example, both Microsoft and Federal Express reached tipping points that resulted in explosive growth and significantly increased communication capabilities for the general population of the United States. At the same time, the price of telecommunications was falling, pushed by the U.S. judiciary in a 1982 decision to break up AT&T. In 1983, cell phone networks were established in the United States, and Time magazine featured the computer as "Man of the Year." In 1984, Apple released the Macintosh computer with advertisements showing how information-rich individuals could smash all specters of authoritarianism. Meanwhile economists theorized about how economic growth is driven by ideas, which are "nonrival" and thus can be shared without loss of value. Unlimited prosperity, some said, was on the horizon. Faith in democracy could be restored.

Nineteen eighty-four was also the year a U.S. plant leaked toxic gas into the city of Bhopal. Poor circulation of information exacerbated the tragedy in many ways. Circulation of information between Union Carbide's headquarters in the United States and its Indian subsidiary was problematic, as was circulation of information between workers and managers in the plant. Many information systems within the plant were nonfunctional the night of the gas leak, due in part to lack of maintenance resulting from plans to dismantle the plant and move it to another country—one component of a major corporate restructuring plan intended to prepare Union Carbide for a globalizing economy. Indeed, on the very day Carbide had to hold a press conference announcing the Bhopal disaster, a press conference explaining how Union Carbide had restructured for a global era had already been scheduled.[9]

No alarm announced the release of forty tons of toxic gas over the sleeping city of Bhopal in the early hours of December 3, 1984. People woke to a smell like that of burning chilies, which quickly affected their breathing and sight. Because there was no evacuation plan, many people ran from their homes into the wind. As many as 600,000 people may have been exposed; as many as 10,000 of them may have died within the first few days. Union Carbide representatives told doctors in Bhopal that the effects of methyl isocyanate, the bulk chemical released, could be treated by giving the victims antacids and washing their eyes with water. When Representative Steven Solarz (D-N.Y.) visited Bhopal shortly after the gas leak, he was shocked to discover that the city's mayor had no idea of the potential dangers the plant had posed.[10]

On the first day of the Ninety-ninth Congress (January 3, 1985), Senator Frank Lautenberg (D-N.J.) outlined questions raised by the Bhopal disaster: What percentage of the U.S. public lives in close proximity to facilities that produce or use hazardous chemicals? Is it known what these materials are, and what hazards they present, to adjacent communities? How adequate are the emergency procedures

[9] Wil Lepkowski, "The Restructuring of Union Carbide," in *Learning from Disaster: Risk Management After Bhopal,* ed. Sheila Jasanoff (Philadelphia, 1994).

[10] Janice Long and David Hanson, "Bhopal Triggers Massive Response from Congress, the Administration," *Chemical and Engineering News,* 11 Feb. 1985, 59.

established by the federal and state government to respond to environmental disasters? The overarching question was basic: Can "it" happen here, in the United States? Union Carbide said it could not, emphasizing the low probability of simultaneous multiple systems failure as occurred in Bhopal. But the cat was out of the bag. Clearly, the chemical industry was operating open systems and would face demands for open accounting for what leaked out and what was being done about it.[11]

Representative Henry Waxman (D-Calif.) emphasized this shift in perspective in congressional hearings begun ten days after the Bhopal gas leak: "We're being told on the one hand that it's a sealed system. But on the other, all these chemicals are leaking into the air on a routine basis. I find that troubling. The federal government doesn't know anything about it and that's outrageous enough. The state government hasn't the ability to regulate. We rely on you to regulate yourself. And if you are regulating yourself, it doesn't seem to me that your own people know why these chemicals are going into the air and what effect they're having on the public."[12]

Representative Waxman argued that it was a discredit to the EPA that it did not know what was going on, pointing out that "EPA didn't mention the fact that there are no standards because EPA hasn't set any. After 14 years it has regulated only eight toxic pollutants. Methyl isocyanate is not considered a hazardous pollutant. . . . Aren't 2,500 deaths enough to convince EPA that methyl isocyanate is hazardous?" Waxman also described the contradictory process by which the agency determined whether hazards "counted": "EPA doesn't call something a hazard until it's ready to regulate it and it doesn't regulate something until it calls it a hazard. EPA has been chasing its tail for far too long."[13]

Waxman's response to the Bhopal disaster draws out many problems for which solutions are still being worked out: lack of information about risks both within the companies that produce them and within government agencies; the dependence of government agencies on companies for knowledge about risks; and uncertainty about what should be officially listed as hazardous. The most basic, serious problem of all, however, is that the complex technologies dotting and interconnecting the contemporary landscape are open, rather than sealed, systems. In other words, these systems routinely leak and occasionally blow up. "Information strategies" have been our response.

Information strategies—efforts, often led by governments, to increase the availability of information on particular phenomena—are being instituted around the world as a way of dealing with complex problems within democratic frameworks. This is particularly the case in the environmental domain. Information strategies are now being relied upon to address pollution, loss of biodiversity, climate change, and a range of other issues involving entangled social, technical, and natural systems. Conceptual support for information strategies can be found in Kantian constructions of the subject who knows and therein becomes both capable and responsible; in John Stuart Mill's arguments in *On Liberty* about the need for informed decisions and subsequent need for freedom of the press; in arguments that supported passage of the U.S. Freedom of Information Act and the growth of the consumer rights movement in the

[11] Ibid., 53.
[12] Wil Lepkowski, "Bhopal Disaster Spotlights Chemical Hazard Issues," *Chemical and Engineering News,* 24 Dec. 1984, 20.
[13] Long and Hanson, "Bhopal Triggers Massive Response" (cit. n. 10), 56.

1970s; and in arguments that supported the democratization campaigns of the late 1980s.[14]

Information strategies are structured by ideas about the effects of information circulation and about the (ethical) good of such effects. In short, more information in more hands is assumed to be a good thing. This can imply a rational actor model of behavior and democracy: information strategies increase the knowledge base from which judgments and decisions are derived, resulting in rational actors and rational societies. Other logics, based on quite different constructs of what is real and possible, also exist. Information strategies can, for example, be perceived as an imperfect but best possible way to respond to high levels of uncertainty about both the present and the future: circulate lots of information to lots of people, hoping that dumb parts become a smart network.

Information strategies were not new in the 1980s, even within the environmental domain. The 1970 National Environmental Policy Act, for example, led to the publication of annual reports on the environment for the president and Congress and mandated that all federal agencies publish Environmental Impact Statements before starting new projects. Belief in such strategies accelerated in the 1980s as the information era became a public phenomenon. Simultaneously, protection of human health became the explicit goal of environmental legislation.[15] Efforts to protect environmental health thus became entangled with the beliefs and technologies of informationalism, and the "right-to-know" became the dominant legislative strategy for protecting human health.[16]

INFORMATING ENVIRONMENTAL POLICY

"Information strategies" for dealing with environmental risk became the explicit focus of law in the United States in 1986, through passage of the Emergency Planning

---

[14] According to Tom Tietenberg (an economist at Colby College, Waterville, Maine) and David Wheeler (lead economist for the Development Research Group at the World Bank), "Disclosure strategies form the basis of what some have called the third phase in pollution control policy—after legal regulation [emissions standards], and market-based instruments [tradable permits, emissions charges]. While these strategies have become commonplace in natural resource settings—forest certification programs, for example—they are less familiar in a pollution control context." Tietenberg and Wheeler argue that first phase approaches (legal regulations) were excessively costly or incapable of achieving stipulated goals—especially in developing countries, where legal and regulatory institutions are weak. The second phase (market-based) approaches are said to have done better, but—even in industrial countries—have not been able to handle the sheer number of substances to be controlled. Tietenberg and Wheeler explain that third-phase pollution control policy (meant to counter problems faced by the first two phases) "involves investment in the provision of information as a vehicle for making the community an active participant in the regulatory process. . . . The timing seems to emanate from a perceived need for more regulatory tools in the regulatory community, from a demand for environmental information from communities and markets and from falling costs of information collection, aggregation and dissemination." See Tietenberg and Wheeler, "Empowering the Community: Information Strategies for Pollution Control" (paper presented at "Frontiers of Environmental Economics Conference," Airlie House, Virginia, 23–25 Oct. 1998), 1.

[15] In particular through passage, in 1976, of the U.S. Resource Conservation and Recovery Act (RCRA). This legislation required "cradle-to-grave" tracking of hazardous wastes and controls on hazards waste facilities. RCRA was amended in 1984, partly in response to the problems at Love Canal, which gained media attention in 1978.

[16] The right-to-know has also been part of the post–World War II development of rights discourse encompassing human and civil rights, as well as patients' rights, animal rights, and the right to a clean environment. See Carl Wellman, *The Proliferation of Rights: Moral Progress or Empty Rhetoric?* (Boulder, Colo., 1998).

and Community Right-to-Know Act, Title III of the Superfund Amendments and Reauthorization Act (SARA). Widely regarded as the United States' primary legislative response to the Bhopal disaster, the act mandated a range of initiatives to support emergency planning and public access to information.[17] High-risk facilities, for example, had to provide the information needed by local rescue personnel to plan emergency evacuations. By the time amendments to the Clean Air Act were passed in 1990, this had evolved into a mandate for worst-case scenarios for 66,000 high-risk facilities around the United States.

Another key component of the 1986 Right-to-Know Act was the Toxic Release Inventory (TRI), the first federal database Congress said must be released to the public in a computer-readable format.[18] The goal was to allow the EPA as well as citizens to track and evaluate routine emissions. A key effect has been recognition that information itself can be a hazard—to the public image of chemical companies, in particular. In response, corporations have "gone green," and control over hazardous information has become as important as control over hazardous production.

The effects of distributing TRI data in the United States have been enormous, sparking environmental initiatives within corporations, in the communities affected by pollution, and by national and international environmental groups.[19] The first round of TRI data was submitted in July 1988. The president of Monsanto was so taken aback by the figures disclosed that he pledged to reduce emissions by 90 percent over the next five years. The following year, the Chemical Manufacturers Association initiated their Responsible Care program. "Responsible Care" is a "public commitment" to run safe plants voluntarily—beyond compliance with the law. The National Wildlife Federation responded to Responsible Care by denouncing purported progress on emissions reduction as "phantom reductions" attributable to new accounting measures and creative information manipulation.[20] Environmentalism became a struggle over how things would be categorized, counted, and represented—graphically as well as politically.

Initiatives similar to those mobilized in the United States by right-to-know legislation are now being developed around the world, as recommended in Agenda 21, the guidelines for sustainable development agreed to at the 1992 Earth Summit. Informational strategies have become a major focus at the World Bank and within UN programs, leading to initiatives to provide greater access to environmental information in many developing countries, including Mexico and Indonesia.[21] In Europe, the right-

---

[17] Susan Hadden, "Citizen Participation in Environmental Policy Making," in Jasanoff, *Learning from Disaster* (cit. n. 9).

[18] John Young, "Using Computers for the Environment," in *State of the World 1994,* ed. L. Brown (New York, 1994).

[19] On the emergence of corporate environmentalism, particularly in relation to the TRI, see Fortun, *Advocacy After Bhopal* (cit. n. 3); J. Fillo and C. Keyworth, "SARA Title III—a New Era of Corporate Responsibility and Accountability," *Journal of Hazardous Materials* 31 (1992): 219–31; D. Grant, "Allowing Citizen Participation in Environmental Regulation: An Empirical Analysis of the Effects of Right-to-Sue and Right-to-Know on Industry's Toxic Emissions," *Social Science Quarterly* 78 (1997): 859–73; S. Santos, V. T. Covello, and D. McCallum, "Industry Response to SARA Title III: Pollution Prevention, Risk Reduction, and Risk Communication," *Risk Analysis* 16 (1996): 57–66; J. C. Terry and B. Yandle, "EPA's Toxic Release Inventory: Stimulus and Response," *Managerial and Decision Economics* 18 (1997): 433–43.

[20] Chris Bedford, *Out of Control: The Story of Corporate Recklessness in the Petrochemical Industry* (Boulder, Colo., 1988), video.

[21] See S. Afsah, B. Laplante, and David Wheeler, "Controlling Industrial Pollution: A New Paradigm," Working Paper 1672, Policy Research Dept., World Bank, May 1996.

to-know is the focus of the Aarhus Convention—the UN Economic Commission for Europe Convention on Access to Information, Public Participation in Decision-Making and Access to Justice in Environmental Matters. Originally signed in Aarhus, Denmark, in the summer of 1998, the convention establishes legally binding instruments guiding the creation of national Pollutant Release and Transfer Registers (PRTRs) in the UN/EEC region, as recommended by Chapter 19 of Agenda 21. PRTRs are databases containing information about pollution from industrial facilities, similar to the U.S. TRI.[22] Environmental organizations such as the WorldWatch Institute considered PRTRs to be a key goal of the Earth Summit held in Johannesburg, South Africa, in August 2002. WorldWatch reports that there has been serious opposition to PRTRs by manufacturers since the Earth Summit in 1992 and that only twenty countries have set up PRTRs as a result. WorldWatch considers the registers a priority because they "pinpoint the most affected communities, and the most polluting industries, thereby identifying targets for action."[23]

Around the world, in different ways, right-to-know initiatives are structuring the ways power operates and is challenged.[24] This raises difficult questions: What information must be provided to fulfill the right-to-know about the environment? How must information be provided? Must information be accessible through the Internet? Has access been realized if information is not organized for efficient use and correlated with other information that reveals its significance? Is the right-to-know, in effect, the right to computer models and interactive, Web-based maps built with Geographic Information System (GIS) software?[25]

What should environmental information system designers assume about the system users? Should designers aim to provide simple, clear messages, or should they provide complex, open-ended data sets users can manipulate, download, and incorporate into their own designs?

Must scientific knowledge be stable and uncontested to be useful? How much science can "ordinary citizens" take? How does the communication of complex information actually happen, and what are the likely effects?

These questions raise difficult practical, conceptual, and ethical issues. They are regularly discussed and debated—at conferences sponsored by government agencies, at community meetings, and on email listservs that interconnect diverse stakeholders. They are also addressed through creative technological applications.

[22] See E. Petkova, with P. Veit, "Environmental Accountability Beyond the Nation-State: The Implications of the Aarhus Convention," *Environmental Governance Note* (Washington, D.C., April 2000), http://pubs.wri.org/pubs_description.cfm?PubID=3013.

[23] A. P. McGinn, "From Rio to Johannesburg: Reducing the Use of Toxic Chemicals Advances Health and Sustainable Development," *World Summit Policy Briefs* (WorldWatch Institute), 25 June 2002 (emailed edition), 3.

[24] It has become commonplace to recognize how information access will affect global order generally. Manuel Castells, for example, argues that the "multimedia world will be populated by two essentially distinct populations: the interacting and the interacted, meaning those who are able to select their multidirectional circuits of communication, and those who are provided with a restricted number of prepackage choices. And who is what will be largely determined by class, race, gender and country." (Castells, *The Rise of Network Society* [Malden, Mass., 2000], 371.) One of the goals of the broad project partially developed here is to draw out how global ordering through informationalism will affect the environmental field in particular.

[25] Note, e.g., Alberto Melucci's argument that "[i]n the contemporary context, we can define domination as a form of dependent participation in the information flow, as the deprivation of control over the construction of meaning." Melucci, *Challenging Codes: Collective Action in the Information Age* (New York, 1996), 182.

ENVIRONMENTAL DEFENSE'S SCORECARD WEB SITE

One response to the recognition that people have a right to know about environmental problems is www.scorecard.org. When the Environmental Defense Fund (EDF) launched the site in April 1998, *Chemical Week* described it as the "Internet Bomb" because of its potential effect on the reputation of chemical companies.[26] Greenpeace refers to Scorecard as the "gold standard" of environmental information systems and decided to follow EDF's lead in using the open source version of ArsDigita Community Systems (OpenACS) software for the new Greenpeace Planet Web site, launched in June 2002. Greenpeace applauds Scorecard because it "bridges the gap between setting up passive information and creating a collaborative environment for action."[27]

Scorecard.org is an interactive Web site supported by a relational database containing profiles of more than 6,800 chemicals. The site integrates local pollution information for the United States with information on health risks and information on relevant environmental regulations. It allows users to produce customized reports and encourages communication with the EPA or with a polluting company.

The goal of Scorecard is to provide the information base for a sustained effort to reduce pollution risks. Putting pressure on polluting facilities through disclosure of their emissions is a key strategy. EDF also wants it to be commonplace for people to use local environmental information when making decisions about what city or neighborhood to live in or what products to buy. A critical side effect will be greater recognition of the uneven distribution of pollution risk among social groups. Fred Krupps, president of EDF, wrote in an introductory letter posted on the Web site that the organization's goal was "to make the local environment as easy to check on as the local weather."[28]

Pollution maps form the centerpieces of the Scorecard site. Based on zip codes, these maps display the manufacturing facilities in a particular area that report their emissions to the EPA as part of the Toxic Release Inventory. It is carefully noted that the maps do not cover all pollution sources and—even for those it does cover—only accounts for the approximately 650 chemicals reported under the TRI. The information listed, however, is sufficient to provide a glimpse into pollution and health hazard realities—while also reminding users that important information gaps and uncertainties remain.

More than a billion pages can potentially be produced in Scorecard. Each user can access up-to-date, customized information, from more than 7 gigabytes of data, distilled from more than 100 gigabytes of contributing government and scientific databases.[29] Information from these databases is in different units of analysis and structured for a variety of uses. The data must be extensively massaged to be compatible

[26] P. R. Fairley and A. Foster, "Scorecard Hits Home: Web Site Confirms Internet's Reach," *Chemical Week,* 3 June 1998, 24–6

[27] See "The Story Behind Greenpeace Planet," 24 June 24 2002, http://www.greenpeace.org/features/details?features%5fid=14977 (accessed 1 July 2002).

[28] Fred Krupp, "A Letter from EDF's Executive Director," April 1999, http://www.scorecard.org/about/about-why.tcl (accessed 5 July 2002).

[29] Scorecard runs on a combination of open-source and very high-end proprietary software. At the outset, Oracle (the second largest software company in the world, after Microsoft) donated its relational database manager (an industry standard) to EDF to support Scorecard's dynamic publication system.

with Scorecard's data model. In turn, this model allows Scorecard users to see phenomena otherwise inaccessible because of disjunctive data.

In addition to providing extraordinary integration of data sets, the site provides rankings of health risks from pollution. The EDF developed the ranking system, which *Environmental Science and Technology* peer reviewed. Viewers are told how many pounds of toxins were released in a given year by a given facility. They also are told about probable risk—body system by body system—based on a hazard-ranking system that relates all chemicals to the risk of benzene, a known carcinogen (to indicate "cancer potential"), or to toluene, a developmental toxin (to indicate "noncancer risk"). This ranking system provides users with relatively stable reference points for thinking about an otherwise confusing array of health risks. It is a purposeful simplification. At the same time, users are also encouraged to learn "what we don't know about chemical safety and harm" within the Scorecard site itself. Usable information is cast as something between "the truth of the matter" and that which doesn't matter— because tied to the real in uncertain terms.

Scorecard's health-ranking system offers a useful way of configuring data widely acknowledged to be resistant to interpretation. It responds to the enrichment of data to the point of incomprehensibility by providing sense-making heuristics. It also offers a way to think around data gaps. It provides working knowledge, even if it is not entirely certain knowledge.

The goal is not to reassure the user but to connect her or him with different types of information, with the regulatory process, with people in similar and different locales, and with ways of visualizing and spatializing phenomena usually represented in abstract, impersonal terms. Getting "straight to the point" is not the goal. Instead, users are encouraged to wander through different kinds of information, much of it flagged as uncertain, visualizing comparisons, piecing together a picture of reality with which to work.[30] High levels of information literacy are required and cultivated.

The Scorecard Web site encourages users to take on many of the most recalcitrant problems within environmental politics: the need to deal with too little, as well as too much, information; the need to deal with contested scientific findings and intractable uncertainty about long-term effects; and the need to think locally as well as comparatively and globally. It also embodies a critique of conventional risk communication and the modernist subject that risk communication specialists so often presume.

---

[30] The argument that discursion—a meandering through material—produces a differently-valued outcome than linear movement toward a stable conclusion is now often made with regard to the value of hypertext, as in G. P. Landow's argument that there has been a "convergence" between critical theory and technology. The argument, however, is not new. In his introduction to a collection of personal essays, for example, Philip Lopate describes how the essay's "unmethodological method" has been utilized across time, from Montaigne and Bacon through the Frankfurt School, and in different cultures. Theodor Adorno, for example, is said to have seen rich, subversive possibilities in the "anti-systemic" properties of the essay—in the way an essay wanders through information and thought rather than working within dominant frameworks of thinking straight through to a conclusion. For Adorno, the essay was a technology for thinking outside the grand philosophical systems of his time. Environmental information systems, in my view, have a critically similar potential. See Landow, *Hypertext: The Convergence of Contemporary Critical Theory and Technology* (Baltimore, 1992); Lopate, *The Art of the Personal Essay: An Anthology from the Classical Era to the Present* (New York, 1995); and Adorno, "The Essay as Form," *Notes to Literature,* vol. 1 (New York, 1991).

DISCLOSURE, FORECLOSURE, AND THE POLITICS OF PERCEPTIBILITY

Scorecard exemplifies the way information technology and culture have developed in the environmental field, shifting conceptualization and enactment of science, of communication, and of political action. Scorecard also mobilizes a particular mode of subjectivity, which can be differentiated from the modernist subject assumed by conventional risk communication. The subject mobilized by Scorecard is engaged with rather than protected from uncertainty, and encouraged to understand science as a contested, iterative resource that rarely offers straightforward conclusions. Scorecard feeds desire for information as well as skepticism about information. It directs a user's gaze by providing heuristics for making sense of data so enriched that it has become incomprehensible. Scorecard encourages users to act, at many possible scales, without full knowledge as a guide. Basic questions, such as what counts as "accurate" or what counts as "safe," have to be continually negotiated.

Scorecard produces a flood of information that continually threatens to overwhelm the user while also providing heuristics for making judgments and establishing hierarchies of significance. This double move is a characteristic gesture in the informating of environmentalism. Such moves make the environment available for understanding and governance while establishing what Michelle Murphy calls "regimes of imperceptibility."[31] They disclose and foreclose, highlight and efface.

The U.S. Toxic Release Inventory, for example, provides quantitative data (in tons) of polluting emissions. Prior to 1989, when the TRI first became available, much of this pollution was invisible since air quality standards established in the 1970s had already greatly reduced the most visible pollution. What became visible through the TRI, however, was limited. The TRI only lists about 650 of thousands of chemicals manufactured today, emitted only from large plants in the most obvious sectors. Furthermore, what is reported to the TRI is clearly subject to creative accounting, which could explain the apparent reduction in TRI-reported emissions since 1989. Critics refer to these as "phantom emissions" and point out that TRI data is rarely audited and there is no liability when companies provide inaccurate information. Some companies do not even submit correct coordinates for the latitude and longitude of their plants.

Different forms of auditing have been proposed and have been opposed by industry on grounds that they would require disclosure of trade secrets. Meanwhile, there is continual industry activism to delist substances from the TRI, which can result in immediate and quite dramatic reductions in the total tonnage of pollution reported.

Though critically aware of these problems with TRI data, Scorecard's designers chose to build a complex information system to make it more accessible. Their logic is also complex: By using industry-reported data, they sidestep industry claims that the information is politically motivated. When they map realities industry would rather not have publicized, they can say that industry itself provided the basic data. Scorecard simply pulls it together.[32] The designers of Scorecard also work with an in-

---

[31] See Michelle Murphy, "Uncertain Exposures and the Privilege of Imperception: Activist Scientists and Race at the U.S. Environmental Protection Agency" (this volume).

[32] This description is drawn from interviews and materials published by Bill Pease, who was the project director for Scorecard at the Environmental Defense Fund. See Erich Schienke, "Bill Pease/An Original Developer of Scorecard.org," *CECS Working Interviews* 5 (Troy, N.Y., 2001); and Michael Stein, "Environmental Defense, from Brochureware to Actionware: An Interview with Bill Pease, April 11, 2001," http://www.benton.org (accessed 1 July 2001).

teresting understanding of information itself. Rather than depending on the validity of single data points, they build up an information assemblage that functions as an interactive system, to which single data points make a contribution, not unlike the way pixels contribute to the overall effect of a digital image. The image is clearest when every pixel is precise but is still readable when they are not.

Scorecard makes information readable even when information quality cannot be controlled and glaring data gaps exist. Scorecard itself, however, also produces imperceptibles. To make health risks readable, for example, Scorecard provides a ranking system that reductively equates all cancer risks to the risk of one iconic chemical—benzene—and equates all developmental risks to toulene. Health risks become meaningful by sweeping their particularities aside. Information on particular chemicals *is* a crucial part of the analysis. Indeed, the integration of different information sources on particular chemicals is a significant accomplishment of Scorecard. Accessibility, however, depends not just on the information being there but also on the prosthetics available for making the information sensible. Phillip Greenspun, an MIT computer scientist who designed the software for Scorecard, puts it simply: disclosure is good. Disclosure plus interpretation is better.[33]

Interpretation, however, requires foreclosures. This fact is not new, but it has taken on a new intensity with the development of technologically driven information systems. Such systems work by managing the tension between data overload and data sensibility. Criticizing them because they foreclose and efface as they disclose and highlight therefore misses the critical point. What many environmental information system designers emphasize is the need to evaluate different modes of disclosure, noting how a given mode accomplishes the reduction necessary for sensibility, while also disclosing how the interpretive prosthetics provided are themselves designed. It is for this double move that Scorecard is so well regarded.

Scorecard illustrates how the environment emerges in a high-tech landscape. The materiality of the environment blinks through the screen on which one observes it. For anything at all to be visible, other things are not. The "truth" is always technologically mediated. Environmentalism, a fundamentally materialist project, also has to be recognized as a virtual project.

INFORMATING ENVIRONMENTALISM

The Bhopal disaster provoked new awareness of industrial risk at a time when the growth of information technology and culture was also dramatic. This simultaneity is noteworthy because it provides an opportunity to observe how available discourses and technologies shape the definition and construction of solutions to problems. Bhopal did, in fact, demonstrate the need for better circulation of information about environmental risk. The magnitude of investment in "information strategies" since Bhopal can only be understood, however, in the context of technological and cultural change. It is this context that sets the stage for what I call the "informating of environmentalism."

The informating of environmentalism is evident in the massive investment in information technology projects by nongovernmental organizations of all sizes, by government agencies, and in various fields of the environmental sciences—from ecology

---

[33] Phillip Greenspun, "Better Living through Chemistry," *Phillip and Alex's Guide to Web Publishing* (Cambridge, Mass., 1999), chap. 16, 4, http://phillip.greenspun.com/panda.

to toxicology. The informating of environmentalism can also be seen in the pervasiveness throughout the environmental field of an information culture revolving around belief that increasing the production and circulation of information will help solve a range of complex problems. Though not new, information culture has gained particular force in the contemporary period.

Information culture is not a simple phenomenon. Many different logics can undergird belief in information strategies; a range of effects cascade from the development of information culture in particular organizations and fields; the way that information technology and culture drive each other makes it very difficult to determine what caused what. Examining the combined effects of information technology and culture is nonetheless fruitful. In the environmental field, it allows one to understand how basic, politically-charged ideas—about communication, science, nature, citizenship, and so on—are being challenged. It also draws out the powerful ways that both technology and culture determine what is perceptible and what is not, and how the gray matter between certain truth and the unknowable is made a part of our thinking and decision-making. Examining the combined effects of information technology and culture is thus a way to map and evaluate the historical production of significance, marginality, and legitimacy.

In this essay, I have examined the combined effects of information technology and culture in the environmental field to draw out different ways that risk communication can be constructed. The ways environmental information systems designers understand risk communication is clearly shaped by their daily work with information technology and by an amorphous, but powerful, information culture. These influences no doubt create particular blindnesses and biases. Yet they also open up a way of thinking about and operationalizing risk communication that engages, rather than denies, the extraordinary complexity—material and informational—of the environment. This is an important and encouraging dimension of the informating of environmentalism.

Notes on Contributors

Warwick Anderson is Robert Turell Professor of Medical History and Population Health and Chair of the Department of Medical History and Bioethics at the University of Wisconsin–Madison. He is the author of *The Cultivation of Whiteness: Science, Health, and Racial Destiny in Australia* (New York, 2003). When he finishes his book on American colonial medicine in the Philippines, he plans to write more on the history of disease ecology.

Giovanna Di Chiro teaches Environmental Studies and Women's Studies at Mount Holyoke College. She has published widely on the topics of community-based knowledge production and environmental justice. Her publications include contributions to the edited collections *Uncommon Ground* (New York, 1996), *Reclaiming the Environmental Debate* (Cambridge, Mass., 2000), and *The Environmental Justice Reader* (Tucson, Ariz., 2002). She is a coeditor of the forthcoming collection *Appropriating Technology: Vernacular Science and Social Power* (Minneapolis, Minn.).

Kim Fortun is Associate Professor in the Department of Science and Technology Studies at Rensselaer Polytechnic Institute. Her 2001 book, *Advocacy After Bhopal: Disaster, Environmentalism, New Global Orders,* was published by the University of Chicago Press. She is currently writing a book about the ways information technology, theory, and culture have shaped the environmental field since the 1980s.

Ronnie Johnston is a Lecturer in History at Glasgow Caledonian University, Scotland. He has published extensively on the history of occupational health in Scotland and is the author of *Clydeside Capital, 1870–1920* (East Linton, Scotland, 2000) and (with Arthur McIvor) of *Lethal Work* (East Linton, Scotland, 2000). He is currently working with Arthur McIvor on a new book, *Miners' Lung: A History of Dust Disease in British Coal Mining* (London, [2004/5]).

Susan D. Jones is Assistant Professor of History at the University of Colorado–Boulder, where she teaches the history of science, medicine, and technology. She is also a practicing veterinarian and the author of *Valuing Animals: Veterinarians and Their Patients in Modern America* (Baltimore, 2003). Her current projects include a transnational history of bovine tuberculosis and a study of anthrax in biomedical science, livestock production, and public health.

Nicholas B. King is a Robert Wood Johnson Health and Society Scholar at the University of Michigan. He received his master's degree in medical anthropology and his doctorate in history of science from Harvard University. His research examines emerging diseases, biological terrorism, surveillance and information technology, and the commodification of medicine. He is currently preparing a manuscript based on his doctoral dissertation, *Infectious Disease in a World of Goods* (Harvard University, 2001).

Scott Kirsch is Assistant Professor in the Department of Geography, University of North Carolina at Chapel Hill. His research explores the spatiality of nineteenth- and twentieth-century American science and technology, the interrelations of science and government, and the place of science in environmental politics. Some of his work has been published in *Society and Space, Antipode, Social Text,* and the *Annals of the Association of American Geographers.*

Arthur McIvor is a Reader in History at the University of Strathclyde, Glasgow, Scotland, and Joint Director of the Scottish Oral History Centre. He has published extensively on the history of occupational health in the United Kingdom and is the author of *Organised Capital* (Cambridge, 1996), *A History of Work in Britain, 1880–1950* (New York, 2001), and (with Ronald Johnston) *Lethal Work: A History of the Asbestos Tragedy in Scotland* (East Linton, Scotland, 2000).

Gregg Mitman is Professor of History of Science, Medical History, and Science & Technology Studies at the University of Wisconsin–Madison. He has written widely on the history of ecology and nature in twentieth-century America and is currently completing a book on the ecological history of allergy in the United States that brings together his interests in environmental history, history of science, and medical history.

Michelle Murphy is Assistant Professor of History and Women's Studies at the University of Toronto and is editor of the RaceSci Web site. She is currently finishing a book titled *Toxic Privilege: Sick Buildings, Gender, and the Politics of Imperceptibility in Postwar America.*

Linda Nash is Assistant Professor of History at the University of Washington, where she specializes in the environmental and cultural history of the United States in the twentieth century. This article is drawn from her manuscript *The Nature of Health: Environment and Illness in a Western American Landscape.*

Adriana Petryna is Assistant Professor of Anthropology at the Graduate Faculty of Political and Social Science at the New School. She is author of *Life Exposed: Biological Citizens after Chernobyl* (Princeton, N.J., 2002) and coeditor of a forthcoming volume, *Global Pharmaceuticals: Ethics, Markets, Practices.*

Harold L. Platt is Professor of History at Loyola University Chicago. In 1992, *The Electric City: Energy and the Growth of the Chicago Area, 1880–1930,* was published by the University of Chicago Press. It will also publish his book *Shock Cities: The Environmental Transformation and Reform of Manchester and Chicago.*

Christopher Sellers is Associate Professor of History at Stony Brook University. In 1992, he received a Ph.D. in American Studies from Yale University and an M.D. from the Medical School at the University of North Carolina at Chapel Hill. He is the author of *Hazards of the Job: From Industrial Disease to Environmental Health Science* (Chapel Hill, N.C., 1997) as well as articles on environmental, medical, and general American history. He is currently completing an "eco-cultural" history of post–World War II suburbanization in the United States.

Helen Tilley is Assistant Professor in the History Department at Princeton University, with affiliations to the History of Science Program and African Studies. She specializes in the history of science in colonial Africa, placing particular emphasis on environmental, medical, and anthropological sciences. Her manuscript for a forthcoming book, based on her doctoral dissertation, has the provisional title *Africa as a Living Laboratory* and covers the period between 1860 and 1960.

Conevery Bolton Valenčius is Assistant Professor of History, American Culture Studies, and Environmental Studies at Washington University in St. Louis. Prof. Valenčius wrote *The Health of the Country: How American Settlers Understood Themselves and Their Land* (New York, 2002), which won the 2003 George Perkins Marsh Prize from the American Society for Environmental History, and is at work on *The River Ran Backward*, a history of the great New Madrid earthquakes of 1811–12.

Luise White is Professor of History at the University of Florida. She is the author of *The Comforts of Home: Prostitution in Colonial Nairobi* (Chicago, 1990), *Speaking with Vampires: Rumor and History in Colonial Africa* (Berkeley, Calif., 2000), and *The Assassination of Herbert Chitepo: Texts and Politics in Zimbabwe* (Bloomington, Ind., 2003).

Index

SUGGESTIONS FOR CONTRIBUTORS TO OSIRIS

OSIRIS is devoted to thematic issues, conceived and compiled by guest editors who submit volume proposals for review by the OSIRIS Editorial Board in advance of the annual meeting of the History of Science Society in November. For information on proposal submission, please write to the Editor at Osiris@georgetown.edu.

1. Manuscripts should be submitted electronically in Rich Text Format using Times New Roman font, 12 point, and double-spaced throughout, including quotations and notes. Notes should be in the form of footnotes, also in 12 point and double-spaced. The manuscript style should follow *The Chicago Manual of Style*, 14th ed.

2. Bibliographic information should be given in the footnotes (not parenthetically in the text), numbered using Arabic numerals. The footnote number should appear as superscript. "Pp." and "p." are not used for page references.

   a. References to books should include the author's full name; complete title of book in *italics*; place of publication; date of publication, including the original date when a reprint is being cited. *Example*:

   [1] Mary Lindemann, *Medicine and Society in Early Modern Europe* (Cambridge, 1999), 119.

   b. References to articles in periodicals or edited volumes should include the author's name, title of article in quotes, title of periodical or volume in *italics*; volume number in Arabic numerals; year in parentheses; page numbers of article; and, if required, number of the particular page cited. Journal titles are spelled out in full on the first citation and then abbreviated subsequently according to the journal abbreviations listed in *Isis Current Bibliography*. *Example*:

   [2] Lynn K. Nyhart, "Civic and Economic Zoology in Nineteenth-Century Germany: The 'Living Communities' of Karl Möbius," *Isis* 89 (1999): 605–30, on 611.

   c. Journal articles are given in full in the first reference. For succeeding citations, use an abbreviated version of the title with the author's last name. *Example*:

   [3] Nyhart, "Civic and Economic Zoology" (cit. n. 2), 612.

3. Special characters and mathematical and scientific symbols should be entered electronically.

4. A small number of illustrations, including graphs and tables, may be used in each volume. Hard copies should accompany electronic images. Images must meet the specifications of The University of Chicago Press "Artwork General Guidelines" available from the Editor.

5. Manuscripts are submitted to OSIRIS with the understanding that upon publication copyright will be transferred to the History of Science Society. That understanding precludes consideration of material that has been previously published or submitted or accepted for publication elsewhere, in whole or in part. OSIRIS is a journal of first publication.

OSIRIS (SSN 0369-7827) is published once a year.

Subscriptions are $50.50 (hardcover) and $33.00 (paperback).

Address subscriptions, single issue orders, claims for missing issues, and advertising inquiries to *Osiris*, The University of Chicago Press, Journals Division, P.O. Box 37005, Chicago, Illinois 60637.

Postmaster: Send address changes to *Osiris*, The University of Chicago Press, Journals Division, P.O. Box 37005, Chicago, Illinois 60637.

OSIRIS is indexed in major scientific and historical indexing services, including *Biological Abstracts*, *Current Contexts*, *Historical Abstracts*, and *America: History and Life*.

Hardcover edition, ISBN 0-226-53249-6
Paperback edition, ISBN 0-226-53251-8

Osiris

**A RESEARCH JOURNAL DEVOTED
TO THE HISTORY OF SCIENCE
AND ITS CULTURAL INFLUENCES**

**A PUBLICATION OF THE
HISTORY OF SCIENCE SOCIETY**

EDITOR
KATHRYN OLESKO

MANUSCRIPT EDITOR
JARELLE S. STEIN

PROOFREADER
JENNIFER PAXTON

EDITORIAL OFFICE
BMW CENTER FOR GERMAN & EUROPEAN STUDIES
SUITE 501 ICC
GEORGETOWN UNIVERSITY
WASHINGTON, D.C. 20057-1022 USA
osiris@georgetown.edu